Catchment and Water Resources Management

Catchment and Water Resources Management

Edited by Gloria Shaw

Larsen & Keller
www.larsen-keller.com

Catchment and Water Resources Management
Edited by Gloria Shaw
ISBN: 979-8-88836-079-8 (Hardback)

Larsen & Keller
Published by Larsen and Keller Education,
5 Penn Plaza,
19th Floor,
New York, NY 10001, USA

Cataloging-in-Publication Data

Catchment and water resources management / edited by Gloria Shaw.
p. cm.
Includes bibliographical references and index.
ISBN 979-8-88836-079-8
1. Watershed management. 2. Water-supply--Management.
3. Water resources development. I. Shaw, Gloria.
TC413 .C38 2023
333.91--dc23

For more information regarding Larsen and Keller Education and its products, please visit the publisher's website www.larsen-keller.com

Table of Contents

Preface

The world is advancing at a fast pace like never before. Therefore, the need is to keep up with the latest developments. This book was an idea that came to fruition when the specialists in the area realized the need to coordinate together and document essential themes in the subject. That's when I was requested to be the editor. Editing this book has been an honour as it brings together diverse authors researching on different streams of the field. The book collates essential materials contributed by veterans in the area which can be utilized by students and researchers alike.

A river catchment is an area of land where the rainwater collects. Catchment management requires managing all the components of the hydrological cycle that operate within a catchment. Reservoirs are artificial or natural lakes that are used as a source of water supply. River catchments and reservoirs play a critical role in ensuring water security, community wellbeing, and socio-economic prosperity. In many parts of the world, the reservoirs are under increasing pressure to meet the challenges from population growth, economic activities, and changing climates. Water resource management refers to the activity of developing, planning, distributing and managing all such water resources efficiently. This book is a valuable compilation of topics, ranging from the basic to the most complex advancements in catchment and water resources management. It is appropriate for students seeking detailed information in this area as well as for experts.

Each chapter is a sole-standing publication that reflects each author´s interpretation. Thus, the book displays a multi-facetted picture of our current understanding of application, resources and aspects of the field. I would like to thank the contributors of this book and my family for their endless support.

Editor

1

Integrated Water Resources Management Approaches to Improve Water Resources Governance

Juliet Katusiime * and Brigitta Schütt

Department of Earth Sciences, Freie Universität Berlin, Physical Geography, Malteserstr. 74-100, Haus H, 12249 Berlin, Germany; Brigitta.Schuett@fu-berlin.de

* Correspondence: juliet.katusiime@fu-berlin.de.

Abstract: The water crisis can alternatively be called a governance crisis. Thus, the demand for good water governance to ensure effective water resources management and to attain specific water goals is growing. Many countries subscribe to the Integrated Water Resources Management (IWRM) approach to achieve this goal. The Integrated Water Resources Management approach aims to ensure a process that promotes the coordinated development and management of water, land, and related resources in a drainage basin to maximise economic and social welfare equitably without compromising the sustainability of vital ecosystems. The design of the Integrated Water Resources Management approach, including its pillars and principles, aspires to good water governance and effective resource management. However, empirical studies examining this hypothesis and analysing the impact of the Integrated Water Resources Management approach on water resources governance are limited, especially in developing countries. Therefore, we characterised and compared the water resources governance aspects of two catchments in Uganda's Lake Albert basin. One of the catchments was exposed to integrated water resources management projects, while the other had no exposure to integrated water resources management projects. Some of the factors that supported the comparability of the two sites included spatial proximity linking into a related hydrological and social-economic setup, common water needs and belonging to the same water administration zone. Comparing both areas led us to analyse whether there was a difference in water resources governance actions, as well as in the quality of water resources governance, under the same overall water management and administrative zone. The data were based on field surveys using questionnaires and information guides in both catchments. The results show that the performance of water resources governance is markedly better in the catchment with Integrated Water Resources Management practices than the base catchment unaffected by these practices. Key themes examined include water resources governance styles, water resources governance systems presence, functionality, the performance of good governance principles, and water resources management effectiveness. The findings contribute to the aspirations for the promotion of integrated water management approaches for improved water resources governance, and the concept that the effectiveness of water resources management measures depends on governance effectiveness. Water governance is significant, as it spells out the power, rights, decisions, and priorities relating to given water resources and communities.

Keywords: IWRM; good water governance; catchment management; effectiveness

1. Introduction

Water forms and also plays a vital role in both the environment and human life [1]. With only 2.5% as freshwater and the rest saline, little water is readily available for the many demands of humankind, testing the illusion of inexhaustibility [2]. Consequently, the global water crisis, which is characterised by increased water demand, limited access to clean water, ineffective water resources management and uncertainties, can alternatively be called a governance crisis [3].

Water possesses ecological, social-cultural, economic, political, spiritual interests and potential uses. The advancement of integrated resources management and or catchment-based resource management approaches aims at bringing all watershed components, water resource users, managers and respective interests together for holistic consideration [1,4–6]. One such approach is the Integrated Water Resources Management (IWRM) approach, which follows the recommendations of the landmark United Nations Conference on Environment and Development Agenda 21 of 1992, Section 2. These recommendations resulted in the establishment of the Global Water Partnership (GWP) in 1996 to foster an integrated approach to water resources management while advancing governance and management of water resources for sustainable and equitable development [4–6]. Integrated water resources management is defined as a process that promotes the coordinated development and management of water, land and related resources to maximise economic and social welfare equitably without compromising the sustainability of vital ecosystems [7]. The implementation of an integrated water resources management approach requires the establishment of an enabling environment, including appropriate policies, strategies and legislation, institutional framework and management instruments [8], while applying the four water principles of [8,9]:

- *Freshwater is a finite and vulnerable resource, essential to sustain life, development and the environment;*
- *Water development and management should be based on a participatory approach, involving users, planners and policymakers at all levels;*
- *Women play a central part in the provision, management and safeguarding of water; and*
- *Water has an economic value in all its competing uses and should be recognized as an economic good as well as a social good.*

Although integrated water resources management and water governance concepts are related, they are not mutually exclusive [10–13]. Indeed, the concept of integrated water resources management and or basin approaches result from the desire to transform water governance [14]. The interest in improving water governance, including while implementing integrated water resources management approaches is also indicated in the United Nations Sustainable Development Goal 6 and target reports [15–17]. Possible reasons for focusing on water governance include the documented limitations and challenges of implementing the integrated water resources management concept around the world [9,10]. To others, addressing the symptoms of inadequate provision of water services and dwindling water resources while neglecting the root causes of unequal power balances, unfair patterns between and within countries, and deficits in democratization is questionable [3]. The deficits extend to exclusion of informal and customary systems of water management and governance in developing countries reforms [18,19]. Therefore, water management challenges cannot be exclusively solved through infrastructural means, but also through addressing water resources governance. The integrated water resources management approach deals with water governance and water management, which differ conceptually but are interrelated. Water management refers to the primary mechanism through which actions are implemented to achieve set goals [20] and involves the application of structural interventions like soil erosion, flood control infrastructures as well as non-infrastructural interventions like behavioural change, education, water resources assessment, allocation, pollution monitoring and control, financial management, information management, and planning for human and environmental purposes [12]. Water governance refers to the mechanisms through which rules that guide the water actions and plans are established and enforced [21].

Several countries in Eastern Africa, Mediterranean, Caribbean, Central and Eastern Europe, South Asian among other regions practice or subscribe to the integrated water resources management approach through the Global Water Partnership [22]. In this contribution, we focus on cases from Uganda in Eastern Africa, where integrated water resources management pilot projects were set up in selected catchments to examine the contribution of the approach to improving water resources governance. The study (a) evaluates whether there was a significant difference in water resources governance in catchments where integrated water resources management practices were implemented, and (b) assesses whether catchments experienced good water governance, and thus, the possibility of water resources management effectiveness. We hope the findings contribute empirical evidence regarding the potential of integrated water resources management approaches in alleviating the water governance crisis for the purpose of scale-up.

1.1. The Concept and Context of Water Resources Governance

Water resources governance is believed to influence water resources management by spelling out power, ownership, boundaries, decision making and course of actions. As such, when governance is ineffective, management is likely to be ineffective, as evidenced by resource quality and quantity deterioration, limited access to operational resources, high costs of service delivery and implementation inefficiency [23]. Water resources governance roles specifically include: (a) supporting the formulation and implementation of water resources related institutions, legislations, and policies; (b) offering clarification on the roles and responsibilities of government, civil society and the private sector water resources and services, enabling inter-sectoral dialogue and co-ordination, stakeholder participation and conflict management; and (c) defining water rights and regulation [24]. Water use systems are, therefore, supported by management and governance components [23,24].

Governance concepts and definitions including water governance vary widely [14,16,25]. The concepts of governance may include varied systems of power and decision making, whether developed and enforced by markets, hierarchies, or networks. Governance may also encompass activities of social, political, and administrative actors seen as purposeful efforts to guide, steer, control, or manage the pursuance of public goods [26,27]. At a higher hierarchical level, governance includes the exercise of economic, political, and administrative authority to manage the country's affairs at all levels and viewed as a composition of the mechanisms, processes and institutions through which citizens and civic groups articulate their interests, exercise their legal rights, meet their obligations, and mediate their differences—a view that adds an element of the functionality of prevailing systems and structures [28]. Water resources governance is also defined variably, as documented in [16], though we particularly relate to the view of water governance as encompassing a range of political, social, economic and administrative systems that are in place to develop and manage water resources, and the delivery of water services in society [29]. The diverse views about water resources governance are reported to often inform divergent policy strategies and decisions. Some of the views mimic political processes, characterised by the confrontation of rival political theories grounded on different values and principles or synonymous to the government [14,25]. Recently, many view water resources governance as a process of pragmatic "pluralism". The pluralism process envisages different types of interaction resulting from (a) the articulation of the classic forms of authority embodied in the state (hierarchical organisations), (b) the private sector (driven by market competition), and (c) the voluntary sector or "civil society" (characterised by citizens' voluntary action, reciprocity, and solidarity) by water actors. These thus encompass the notions of "public-private partnership" and "tri-partite partnership" [12,14,25]. Other arguments include appreciating water resources governance as a subset of a country's general governance system and consistent with other resources sectors' governance for effective management of water resources [30].

Good governance, a term used about a system that adheres to certain principles in water resources governance, demands representation of various interests in water-related decision-making and the recognition of the role of power and politics as important components [12]. Examples of water governance principles include transparency, accountability, responsiveness, equity and inclusion, stakeholder participation, rule of law, integrity, effectiveness and efficiency in service delivery [31]. Several institutions and stakeholders highlight varying governance principles of interest but underscore accountability, participation that seeks to involve every stakeholder in the decision-making process, and transparency where all relevant information is shared in a timely fashion [11,13,21,29,31,32].

Consequently, good governance is associated with effective water resources governance and leading to effective water resources management. Similarly, poor governance is associated with failing a range of technical solutions for water problems [23].

However, water resources governance status monitoring and studies find measuring descriptive and relative terms such as 'good', 'weak', 'bad' and the compound 'good governance principles' challenging [29] when it comes to generalisable consensus. Other considerations in the discussion on governance include the scale and level of water-related actions. Thus, the need to design a governance system fit for the spatial and jurisdictional scale of the resource [26,31,33,34]. Some of the approaches to address scale issues in water governance resulted into (a) the integrated water resources management approach to focus at the different functional, operational, organizational and constitutional levels in water resources management which interlink as decision-making levels [29]; (b) the idea of multilevel governance to facilitate administrative and ecological scales at supranational, national, regional, and local level, including the threefold displacement of state power and control upwards to international actors and organizations, downwards to regions, cities, and communities, and outwards to civil society and non-state actors [26,35]; and (c) adopting flexible governance styles like adaptive governance, network governance and earth system governance to solve scale, flexibility and certainty issues in governance [36–38]. The above-mentioned innovations ensure the adaptation of governance responses to territorial specificities, context-informed design and according to needed solutions [38,39].

The scale and level of action issue in natural resources management possibly influenced some of the conceptual shifts from the monocentric forms of governance [26]. Monocentric governance mainly features the state as the centre of power, authority and control over society, economy, and resources compared to multi-stakeholder approaches in which state authority appreciates mutual interdependences. Accordingly, monocentric governance involves setting the agenda of societal problems, deciding upon policy goals and means, and the top-down implementation of policies unlike polycentric, networking, multilevel, earth system, adaptive governance systems, collaborative governance systems [24,26]. In Table 1, we highlight the water resources governance systems focused on further in the study.

Table 1. A summary description of select resources systems of governance.

Water Resources System of Governance	Operational Meaning
Conventional	Also known as monocentric, "the government perspective", "hierarchical governance", "command and control systems of governance", or the "classical modernist approach of governance", where government centralises most powers at the top and commands from top-down while governing resources [24,40].
Collaborative	A governance arrangement where one or more public agencies directly engages non-state stakeholders in a collective decision-making process that is formal, consensus-oriented, and deliberative to make or implement public policy or manage public programs or assets. Collaborative governance involves criteria, actors, and decisions by consensus. The focus of the collaboration is on public policy or public management [41].
Polycentrism	Polycentric governance is characterised by an organisational structure where multiple independent actors mutually order their relationships with one another under a general system of rules. Polycentric governance comprises of multiple decision-making centres each with substantive autonomy and also located at varying levels [42–44]. Emphasis is put on deconcentrating power from political actors and instead keeping it dispersed among organised actors [24,42,45].
New Public Management (NPM)	A reformed public management administration theory or concept bred by the need to bring economics and markets to supplement governance. Easily understood as a Public-Private-Partnership [46–48].
Traditional	In some settings, indigenous, cultural or traditional form a core part of the power and, decision making regarding natural resources and water governance system [18,19].

1.2. *Case of Water Resources Governance in Uganda: Trends and Status*

Uganda's water resources include an estimated mean annual rainfall of about 1200 mm, the River Nile's annual flow, which exceeds 25 km^3, and water storage in the county's broad lake system supplied by various rivers and stream systems, with lakes Victoria, Albert, Edward, and Kyoga as major lakes. However, potential evaporation amounts to up to 75% of the annual rainfall. The predicted average increase in water use ranges between 2.8 to 14.1% in 2030 and groundwater withdraw is estimated to increase up to 15% by 2030—developments that make water resources management an important activity in Uganda [49]. The legal definition of water resources given by the Uganda Water Act (1997) includes 'water flowing or situated upon the surface of any land or contained in any river, stream, watercourse or other natural courses for water like lakes, pans, swamps, marshes or springs, whether or not it has been altered or artificially improved; groundwater; and such other water as the Minister may from time to time declare to be water' [50]. The legislation in the Uganda Water Act (1997) additionally designates the government as the overseer of water resources rights in Section 5. The Uganda Water Act of 1997 is operationalised through the Uganda Water Policy of 1999, recognising the need for the Integrated Water Resources Management (IWRM) approach. The policy actions included reforming the water resources management approach; thus, the Catchment-based Water Resources Management (CbWRM) model, composed of four water management and administration zones, was delineated in 2010 to ably implement integrated water resources management [5,50–54]. The water and environment sector policies reform process included relevant stakeholders experimenting with the integrated water resources management approach through projects since 2006 in select catchments like rivers Mpanga and Rwizi [54]. The process resulted in the gradual establishment of Catchment Management Organisations (CMO) to facilitate stakeholder-driven integrated water resources management and development. Each Catchment Management Organisation (CMO) is composed of the Catchment Stakeholder Forum (CSF), Catchment Management Committee (CMC), Catchment Technical Committee (CTC) and the Catchment Secretariat (CS). The operations of the organisations are guided by the Catchment Management Operations manuals, plans, guidelines, strategies, and the broader water resources governance framework consisting of the Constitution of Uganda and other relevant legislation, mainly the National Wetland Policy (1995), National Environment Management Policy (1994), National Environment Management Act (1995), and Land Act (1998). The 1995 Uganda Constitution specifically commits to taking all practical measures to promote a good water management system, protect the environment, ensure accountability and rules for public officeholding and foreign policy objectives beneficial to transboundary issues. It additionally stipulates the duties of citizens under the national objectives and directive principles of state policy.

Uganda still faces water resources management effectiveness challenges, regardless of sector reforms and subscription to the integrated water resources management approaches. Water management effectiveness challenges relate to institutional financing and capacity, sectors coordination, management approaches, policy implementation, enforcement of legislation. Other challenges include resource conflicts, population growth and related land water demands, conflicting political decisions, climate change and the biophysical limitations from complex and transboundary hydrological systems [13,49,51,54–56]. Therefore, solutions through research are needed.

Similarly, the water crisis progressively documented around the world is characterised by water resources management ineffectiveness due to governance. Consequently, the promotion of integrated water resources management approaches as a solution [4–6,57]. While both integrated water resources management and governance have coverage in literature, no explicit study explains whether integrated water resources management approaches improve the water resources governance. We therefore compare the water resources governance situation of two catchments of the rivers Mpanga and Semliki. The Mpanga river catchment was exposed to substantial integrated water resources management projects while the Semliki river catchment is unexposed. Exposure to integrated water resources management is expected to enhance good water governance possibilities following the conceptual pillars and principles [10,11,58]. To prove this hypothesis, we examined the water resources governance situation, comparing two catchments, one affected and one unaffected by integrated water resources management measures in the form of projects.

The prevailing water resources governance was evaluated based on the definitions provided by the Global Water Partnership [29] and the United Nations Development Program (1997) view of governance [27], the water governance principles [11,13,21,29,31,32]. Study conceptualization guidance and reference tools used include the user's manual on water governance assessment of the United Nations Development Programme [12].

1.3. Study Area

The study area is in the Lake Albert basin, administered as the Albert Water Management Zone in Uganda. The sampled catchments include that of river Mpanga as a catchment affected by integrated water resource management practices pouring into Lake George and the catchment of river Semliki as the control site unaffected by integrated water resource management practices, pouring into Lake Albert.

The major socio-economic activities in the areas include agriculture, pastoralism, tourism, fisheries and the developing oil and gas industry [56]. Geologically, the area is characterised by the pre-Cambrian Tooro-Buganda rock systems, mainly composed of intrusive rocks. Tectonically, the Albert Water Management Zone belongs to the Cenozoic rift basin system, developed along the Precambrian Mozambique orogenic belt [59,60]. The rifting was initiated during the Late Oligocene or Early Miocene; sandstones, siltstones, clay stones and shales characterize the Cenozoic basin infill [61]. Regional climate corresponds to a tropical savanna climate (as tropical wet and dry or savanna climate after Köppen-Geiger) and tropical monsoon climate (Am) around the Rwenzori Mountain ranges [62]. A spatially sharp variation in rainfall amounts occurs, with the Rift Valley and low-lying landscape of Semliki catchment receiving annual average rainfall amounts of 875 mm, while the elevated mountain ranges of Mt. Rwenzori and Mpanga catchment receive an annual average rainfall of 2500 mm [56]. The average monthly temperature varies between 27 °C and 31 °C, average monthly humidity between 60–80%; due to high evaporation, rates locally negative water balance appears [61]. The land cover consists of well-stocked and low-stocked areas, including forests, bushlands, open waters, aquatic, afro-alpine vegetation, and grasslands and woodlands [61,63,64]. The major land uses include the protected areas (national parks, wildlife reserves and forest reserves), agriculture (crops and livestock) and human settlements [61]. The Albert basin faces several water resources management challenges, especially the increasing water and land demand due to population growth increasing water resources encroachment and land-use changes [65].

The study population was drawn from primary beneficiaries and participants in the integrated water resources management projects areas of Bukuku-Karangura (upstream), Fort-portal urban area (midstream) and Nyabani-Ntara sub-counties (downstream) in river Mpanga Sub catchment. The river Mpanga (Figure 1) delineated to an area of 5203.2 km^2 sub-catchment flows from Mt. Rwenzori ranges and discharges into Lake George. The integrated water resources management projects were implemented in the sub-catchment starting in 2006 by the Directorate of Water Resources Management of the Ministry of Water and Environment and stakeholders. A range of catchment interventions implemented included research, advocacy, service delivery, information and capacity development [66–70].

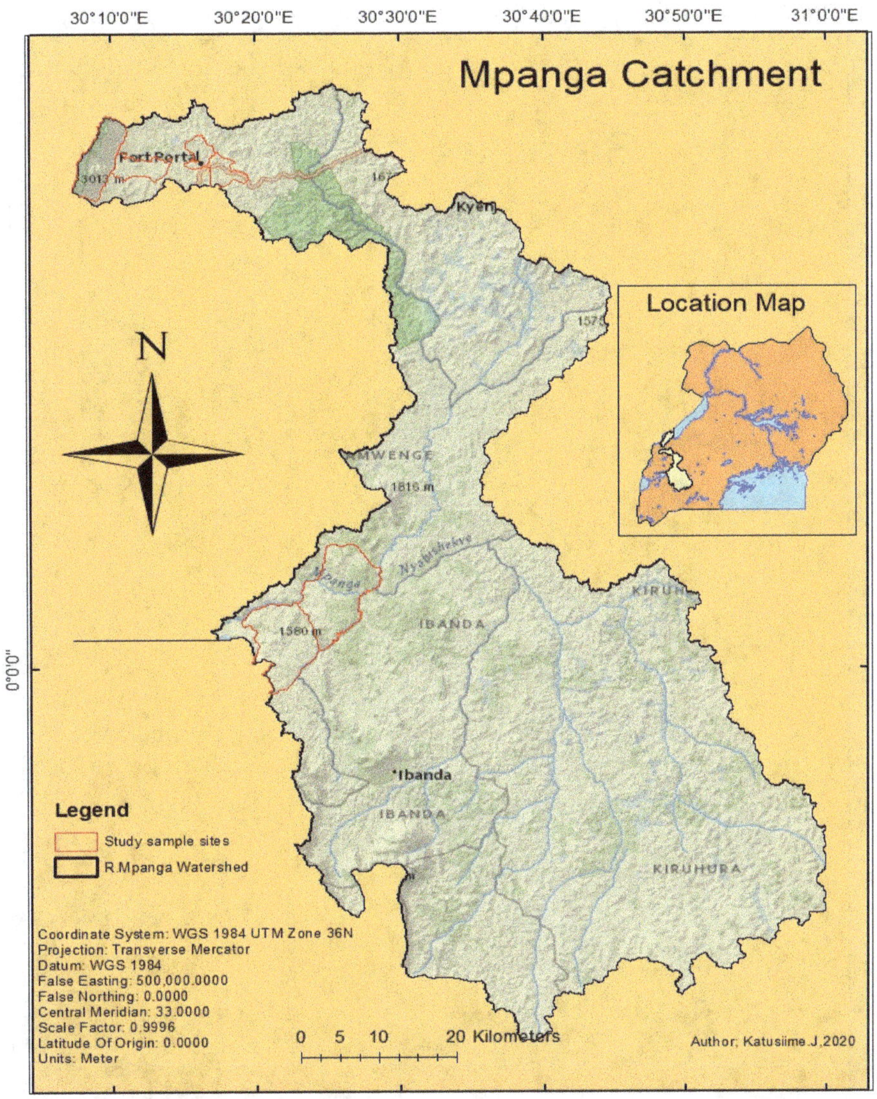

Figure 1. The study catchment Mpanga, highlighting the sample study sites. The catchment is exposed to integrated water resources management projects and measures.

The study control site of the river Semliki (Figure 2) is a transboundary catchment located between Uganda and the Democratic Republic of Congo. The river Semliki flows from Lake Edward through the rift valley floor into the Democratic Republic of Congo west of Mt. Rwenzori, having the character of a border river along the international boundary between Uganda and the Democratic Republic of Congo before pouring into Lake Albert. The estimated size of the entire catchment is 8213 km^2 and it is transboundary. However, for purposes of this study, the part more exclusive to Uganda is delineated and estimated at 833.59 square kilometres, where the study population was drawn from the areas of Rwebisengo, Kanara and Bweramule towards Lake Albert in Ntoroko District. The river Semliki sub-catchment is predominantly known for protected areas like the Tooro-Semliki game reserve, a relatively flat landscape that is rich in biodiversity. Although facilitating agriculture, agro-pastoralism, fishing and small-scale border trading, the Semliki sub-catchment is threatened by deforestation, overgrazing, flooding and deteriorating quality and quantity of water [49,61,64–66]. Watersheds are known to be hydrologically unique elements, so the Mpanga and Semliki relate basing on the interconnected hydrological system, common socio-economic practices that include agriculture, pastoralism, fisheries, tourism, peri-urbanism, cultural system and thus, shared water interests.

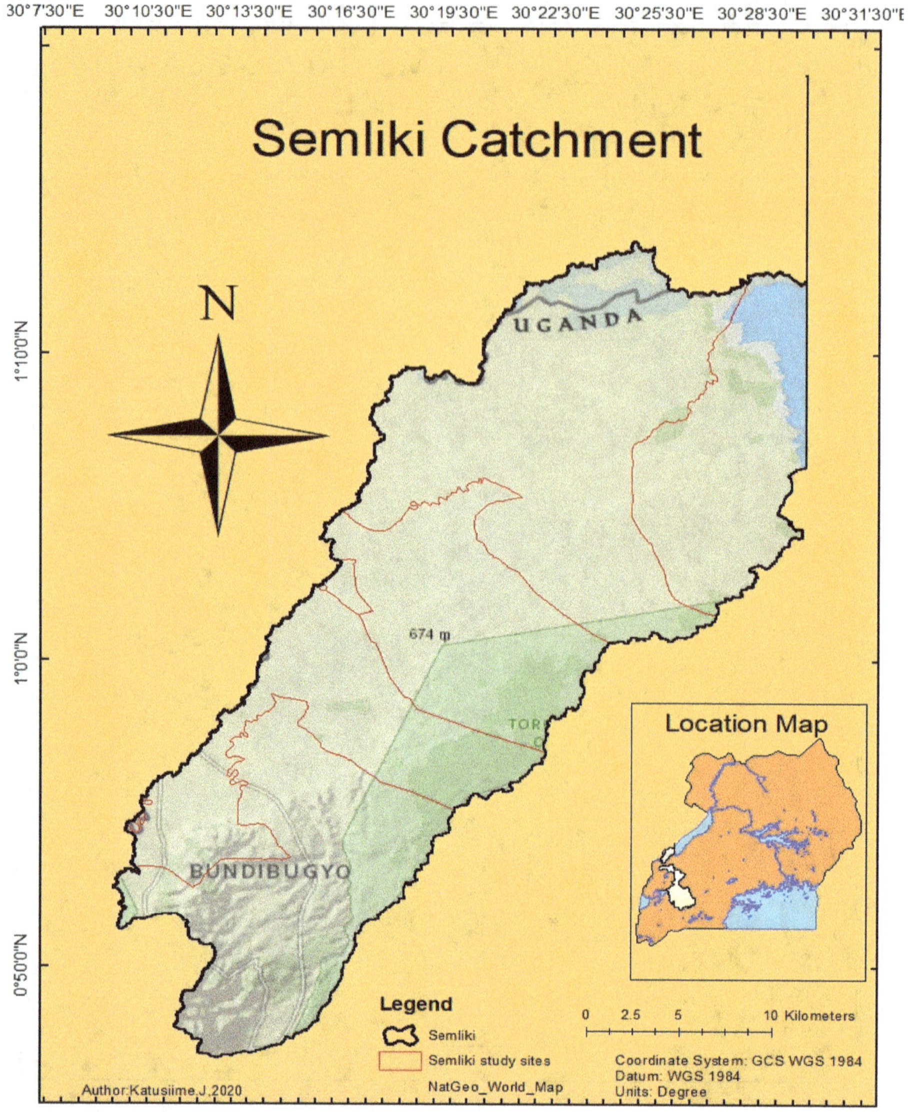

Figure 2. A section of the river Semliki catchment indicating the sample study site's relief topography and a portion of the Uganda-Congo (DRC) Boarder.

2. Materials and Methods

Survey research was conducted to compare the water governance situation in the two purposively selected study catchments. Aware of the limitations when comparing less homogeneous catchments, the Semliki catchment was preferred as the control or base site, given its proximity to the Mpanga catchment, which had been exposed to integrated water resources management projects. The catchment also shares a related hydrological, social-cultural, economic, water administration system, leading to anticipation of a close water resources governance performance. To consistently compare the two catchments, the indicators and characters measured focused on respondent knowledge and capacity, water resources governance styles, the presence and functionality of water resources governance systems, performance of good governance principles, and water resources management effectiveness [29,71]. The assessment was further guided by the water governance definitions and the user's guide for assessing water governance of the United Nations Development Program (2013), which emphasizes actors and institutions, governance principles, and performance as a basic framework for assessing water governance [12]. The estimated sample size of 383 refers to the procedure suggested by Krejcie and Morgan [72], though only 342 questionnaires of the 383 received were fully completed, with 156 questionnaires responding from the Mpanga river catchment and

186 questionnaires responding from the Semliki river catchment. The sample size was drawn from the population data on a sub-county level provided by the government in the study areas, estimated at 139,583 (Mpanga 117,774 residents, Semliki 21,809 residents) [73], and was proportionately distributed between the two study catchments. A simple random sampling of households using target village records as primary respondents to the research questionnaires was performed. Sampling using village-level data increased data collection feasibility and reduced the possibility of bias. The population sampling and selection of target villages targeted the project intervention sites which themselves mainly targeted communities near the main water resources at the upper (hilly or undulating landscape), middle (mostly peri-urban), and lower (relatively flat landscape and some fishing communities) segments of the catchments in Mpanga. A similar arrangement was assumed in Semliki catchment. The survey questionnaire with both open and close-ended questions was used as the main data collection tool, targeting households. The focus group discussions (FDG), key informants guide, and transect walk checklist were used as complimentary data. The data collection exercise took place in July 2015 involving field household interviews, physical and telephone key informant interviews with 15 sector stakeholders (political leaders, cultural leaders, technical resources managers, policymakers, Water User Committee leader, and Civil Society Organisation members). Additionally, six focus group discussions (FGDs), each composed of 10–25 gender-equitable participants, were organised considering upstream, midstream, and downstream catchment zones as much as possible. During transect walks in the catchment, we rapidly evaluated the physical characteristics and validated some of the respondents' information. The focus group discussions and key informant interviews provided information considering issues at community and large spatial scale. The approach involving "multiple-levels and respondents checked for consistency as learned from validation methods relating to environmental governance [74]. Secondary data and information were derived from government reports, plans, strategic papers, and policies. The methodology provided both qualitative and quantitative data from various sources, enabled in-depth exploration of issues while allowing credibility testing of the research findings [75–79].

Qualitative data were numerically coded to allow tabulation and computation of descriptive statistics using the STATA statistical package (StataCorp LLC: College Station, TX, USA). During analysis, the neutral responses were controlled (out) as per variable attributes scale of measurements to generate the site-specific means and t-statistics. Data from Semliki river as the base study catchment was coded zero (0), while data from the Mpanga river catchment affected by integrated water resources management measures was coded one (1). A two paired t-test on the mean proportions (%) from the descriptive statistics was carried out, the interpretation of which made it possible to establish the water resources governance status and whether a significant difference in water resources governance existed between Mpanga and Semliki river catchments. Qualitative data from observation checklists, interviews and focus group discussions notes were analysed, applying content analysis techniques [80].

The catchments maps (Figures 1 and 2) were delineated based on a Digital Elevation Model (DEM) based on Shutter Rudder Topography Mission (STRM) with 30 × 30 m resolution (NASA Earth Data accessed: https://earthdata.nasa.gov/). DEM data were processed applying QGIS 3.12.2 and Arc Map 10.7 (ESRI, Redlands, CA, USA). Additional spatial datasets used were acquired from arc GIS base maps by National Geographic and Esri, World Resources Institute (Accessed: https://datasets.wri.org/dataset/waterbodies-in-uganda) and the United Nations OCHA Humanitarian Data Exchange database (Accessed: https://data.humdata.org/dataset/uganda-administrative-boundaries-admin-1-admin-3).

We present the study results in an order that reveals the respondents' knowledge and capacity regarding the research theme, the respective evaluation of the governance situation and resource management effectiveness outcome. We later discuss the findings, offering more information on the studied variables and local context.

3. Results

3.1. Respondents' Water Resources Governance Knowledge and Capacity

Catchment stakeholder's knowledge and capacity are important aspects of integrated water resources management. Therefore, based on the possible implementation of the capacity development strategy of the water and environment sector [81], we evaluated respondents' knowledge and capacity. Variables examined included knowledge of basic water sector issues like the relevant actors and institutions, water resources governance elements like water rights and legislations, individuals' community capacity to contribute to improved water resources management and governance in the community and access to capacity development opportunities. The results show a statistically significant difference in knowledge and capacity between the respondents of the Mpanga river catchment, which was influenced by integrated water resources management projects, and the respondents of the Semliki catchment, which was without any influence of integrated water resources management projects. For instance, the Mpanga catchment respondents' knowledge of the relevant institutions and actors was significantly different ($\alpha < 0.05$), and the same applies to water resources governance issues ($\alpha < 0.01$) and the individuals' community resource management capacity ($\alpha < 0.01$) (Table 3).

3.2. Water Resources Governance in the Catchments

Water sector reforms and the adoption of the integrated water resources management approach promise to transform the water resources systems to those coherent with good water management and governance principles. The water resources systems of governance, also known as styles, are characteristically defined by decision-making procedures and approaches at the local level. The characteristic features were categorised as collaborative, polycentric, new public management (NPM), traditional or customary, and conventional systems. The results indicate the prevalence of the conventional governance system, following the percentage of observations in both catchments, in comparison with other systems of governance (Table 2). Efforts to engage stakeholders through collaborative means like signed Memoranda of Understanding (MOU) while implementing integrated water resources management were observed in the Mpanga catchment.

Table 2. Water resource systems of governance.

System of Governance	Semliki Catchment % of the Total Observations	Mpanga Catchment % of the Total Observations
Conventional	30	28
NPM	24	17
Collaborative	22	23
Polycentrism	7	18
Traditional	17	14

Water governance, defined as a range of political, social, economic, and administrative systems present for managing water resources and subsequent services, is explored. Guided by contextual examples, the results indicate the Mpanga river catchment area, which has been affected by integrated water resources management projects, has markedly more socio-economic, political, and administrative structures than the base catchment of Semliki river. Descriptive statistics of the occurrence of water governance components comparing the two catchments show that the presence of political systems supporting water resource systems occurs significantly more frequent in Mpanga river catchment than in Semliki river catchment ($\alpha < 0.01$). Correspondingly, economic systems present in both catchment areas strongly differ, and are more advance in the Mpanga river catchment than in the Semliki river catchment ($\alpha < 0.01$). In contrast, the locally established administrative systems, as well as traditional resource management systems, do not significantly differ ($\alpha > 0.05$) (Table 3). Although the political systems and structures were widely reported in both catchments, the traditional systems were less engaged, the administrative systems exhibited enforcement challenges, and multi-stakeholder financing was the prevalent economic system.

Table 3. Water resources governance characters in Mpanga (n = 156) and Semliki catchments (n = 186).

Variable	Semliki Catchment		Mpanga Catchment		Both Catchments		t-Statistics
	Mean	Std	Mean	Std	Mean	Std	
Knowledge and capacity							
Knowledge of water resources stakeholders, institutions	1.432	0.497	1.577	0.496	1.498	0.501	−2.679 *
Capacity building opportunities accessed	2.211	0.409	2.380	0.487	2.287	0.453	−3.423 *
Individual knowledge	1.092	0.339	1.368	0.484	1.227	0.438	−5.279 *
Community resource user' capacity	−0.398	2.091	0.234	2.263	−0.114	2.189	−2.531 **
Water resources governance							
Political systems presence	2.135	0.343	2.618	0.487	2.374	0.485	−10.052 *
Traditional systems presence	2.068	0.254	2.164	0.372	2.110	0.314	−2.665 **
Economic systems presence	2.000	0.000	2.191	0.395	2.085	0.279	−6.435 *
Administrative systems presence	2.989	0.756	3.205	0.649	3.088	0.717	−2.801 **
Functionality of systems							
Systems to report concerns and handle disputes	2.531	0.500	2.600	0.491	2.563	0.496	−1.242
Systems enabling water legal rights	2.150	0.358	2.322	0.468	2.228	0.421	−3.702 *
Systems enabling obligations	2.068	0.253	2.285	0.453	2.175	0.380	−5.286 *
Systems for resource management (Water Committees)	2.367	0.483	2.454	0.501	2.396	0.490	−1.3683
Water resources management							
Water resources management effectiveness	1.622	1.221	2.473	1.506	2.008	1.419	−4.9038 *

Note: ** = Significant at 5%, * = Significant at 1%: Neutral responses were controlled during analysis. The negative means realised at with bi-polar Likert scales. The means of 2.0 in Semliki were a result of the same response in negation by all respondents, thus no variation in the mean, after controlling out the 'neutral' responses.

The functionality of the water governance systems is examined from the resource user perspective as a mechanism or institution through which citizens articulate interests, exercise legal rights, meet obligations, and mediate their differences. The resource governance structures at the lower local government level mostly multi-function with the main objective guided by the establishing institution. Therefore, context-specific system functionality was analysed with respect to practical aspects of handling concerns, mediating conflicts, invoking social obligations, fulfilling water legal rights and local resources management. The functionality of systems differed statistically ($\alpha < 0.01$) between the two catchments, with the realisation of water rights and obligations being more supported in the Mpanga catchment, influenced by integrated water resources management projects. However, systems for handling water-related concerns were rarely present in either catchment, and the occurrence of its components like users' associations was not statistically different ($\alpha < 0.05$). Similarly, the local resources management committees, which included the water user committees, beach management units and environmental committees' functionality, was not significantly different between Mpanga and Semliki catchments ($\alpha < 0.05$) (Table 3).

3.3. *Water Resources Governance Principles*

The good water governance principles offered governance quality assessment indicators. Analysis of selected good water resources governance principles indicated a statistically significant difference between integrated water resources management projects catchment and the base catchment ($\alpha < 0.01$), including the principles of participation, responsiveness, equity and inclusion, effectiveness, rule of law, and transparency. However, aspects of integrity and anti-corruption did not show a significant difference ($\alpha < 0.05$; see Table 4). The governance effectiveness principle also showed a significant difference ($\alpha < 0.01$). Integrated water resource management projects established multi-stakeholder forums for participation, local multipurpose water user committees and associations (WUC/A), organised dialogue meetings, facilitated capacity building, carried out policy dissemination, and projects also offered physical solutions to catchment challenges. In return, the differences in water resources governance effectiveness (Table 4) and management effectiveness (Table 3) were statistically significant between the two catchments analysed ($\alpha < 0.01$). The results, therefore, point to the possible relationship between water resources governance quality or status and resource management effectiveness.

Table 4. Water governance principles significance tests in Mpanga and Semliki river catchments.

	Mpanga Catchment (IWRM Pilot Projects Site)		**Semliki Catchment (No IWRM Pilot Projects Sites)**		**Both Catchments**		
Principles	mean	std.dev.	mean	std.dev.	mean	std.dev.	t-statistics
Participation	1.127	1.109	−0.051	1.295	0.472	1.349	−8.103 *
Accountability	−0.509	1.759	−1.392	1.330	−1.107	1.533	−3.502 *
Equity & Inclusion	0.789	1.309	−0.564	1.302	−0.050	1.458	−7.123 *
Transparency	0.308	1.281	−0.681	1.502	−0.268	1.492	−4.809 *
Rule of law	−0.282	1.495	−100	1.316	−0.723	1.427	−3.418 *
Integrity & anti-corruption	−0.572	1.806	−0.935	1.682	−0.813	1.728	−1.439
Responsiveness	0.910	1.538	−1.020	1.545	−0.169	1.813	−8.266 *
Effectiveness (governance)	0.207	1.507	−1.430	1.046	−0.801	1.474	−7.869 *

Note: * = Significant at 1%: Neutral responses were controlled during analysis. The negative means realised are due to bi-polar Likert scales. The means of 2.0 in Semliki were a result of the same response in negation by all respondents after controlling out the 'neutral' responses.

Overall, the t-test results comparing the water resources governance characters of Mpanga catchment and Semliki river catchments are compiled in Tables 3 and 4; statistics indicate a significant difference in water resources governance between the two study catchments for most variables at at least a 5% level ($\alpha < 0.05$). The results hence indicate a significant difference in water resources governance and management effectiveness between areas affected by integrated water resources management projects and areas without integrated water resources management measures.

4. Discussion

4.1. *Observations of Respondent's Knowledge and Capacity*

Survey responses and findings are dependent on several factors to be aware of, including respondents' knowledge, ignorance, mood, the sensitivity of the matter, opinion, surrounding circumstances, anticipated rewards, and methods [82–85]. Integrated water resources management and water governance are relatively new and complex concepts among natural resources practitioners and local resource users. Therefore, respondents' basic knowledge about water resources governance and management, relevant sector stakeholders and institutions are important hints for effective participation in water resources management and participatory assessments thereof. Notably, integrated water resources management approaches are characterised by knowledge development, sharing, learning and transfer, including through online toolboxes and sites [86,87]. Implementing partners are, thereafter, expected to apply the knowledge through policies and practice. Such policies and mechanisms include the Uganda Water and Environment Sector Capacity Development Strategy [81].

Capacity may be defined as the capability of a society or a community to identify and understand its development issues, to act and address them, learn from experience, and accumulate knowledge for the future. Knowledge, on the other hand, can be viewed as awareness, acquaintance, skill, and familiarity with the facts surrounding water resources management issue and information. However, knowledge can both be input into capacity development and a product [88,89]. Depending on interests, knowledge and capacity, measurements tend to take diverse approaches [89]. Measuring individuals' basic knowledge of aspects relating to water resources management and governance in the country, actors and institutions were of interest for this study. Respondents' knowledge and capacity regarding water resources governance were more limited in the base catchment than in the catchment influenced by integrated water resources management projects. In addition to the difference in knowledge and capacity-building opportunities, other dynamics such as stakeholder involvement, structuring and related institutional power dynamics are documented [90]. Stakeholders' knowledge and capacity compliments water and related resources management in practice, the participatory monitoring and learning in integrated approaches, as well as ensuring adequate survey evaluations [82,84,91–93]. Community water resource users may have both traditional and conventional knowledge about water resources governance and management and can access adequate information with the advancement in network technologies and multi-stakeholder collaborations to compel action. However, the flow of knowledge and information is patchy and disconnected at the local levels to effectively impact decision making and action [94,95]. As such, water governance and water management knowledge monitoring, organisation and accommodation in integrated water resources management approaches are crucial [96].

4.2. Catchments' Water Resources Governance

The study examines the water resources systems of governance defined as an agreed relational and engagement mechanism of the various stakeholders in a defined framework [27,97]. The stakeholders include government, civil society organisations, resource users and markets (private sector). The 'top-down' governance style dominated in both catchments, and some cases of collaboration and intent to allow multi-stakeholder participation. Despite the top-down mechanism, also known as the conventional governance style, resource users reported holding informal power and decisions about water resources within their community jurisdictions and land. However, the informal power dispersal is less harnessed in existing centralised governance legislation and approaches. Other system de-linkages exhibit when conventional water management systems tend to prioritise professional and scientific "expert" knowledge and views more than indigenous experiences and knowledge [23]. Remediation to informal power and knowledge exclusion includes good governance processes recognising formal and informal power decision-making actors and networks while demanding less centralised systems to accommodate 'down-up' stakeholder-driven interests and approaches [23]. Examples of Stakeholder driven governance styles designed include collaboration among multiple stakeholders and public agencies without necessarily creating independence [41]. Other models such as the public-private partnerships leverage financial resources, the polycentric systems allow issue-based independence and power clusters, while catchments with customary systems present equally use indigenous norms, practices, and knowledge. Water resources systems or styles of governance in Uganda (Table 2) mostly bear characteristics of a centralized system. However, the results pointing to multiple governance styles affirm the argument that resource governance systems need to be designed while considering interests, biophysical resource scale and function levels [26,31,33,98]. For instance, a transboundary resources governance system and integrated water resources management actions befit the Semliki catchment. Actions would include transboundary local water committees with participation from Uganda and the Democratic Republic of Congo. The inter-state transboundary catchment committee would enhance the regional Cooperative Framework Agreement (CFA) and the Nile Basin Initiative Act (2002) coordinated by the Nile Basin Initiative (NBI) among other international water instruments aimed at controlling state behaviour and promote cooperation [99]. Governance of transboundary catchment aquifers is also critical, because over-abstraction, contamination and degradation of recharge areas threaten the sustainability of aquifers worldwide according to the International Groundwater Resource Assessment Centre [100,101].

The water resources governance process includes establishing and making functional a range of socio-economic, political, and administrative systems to develop and manage water resources, and the delivery of water services [12,28,29]. Using the integrated water resources management approach pillars for system examples, the social dimensions pillar focuses on equity of access to and use of water resources, equitable distribution of water resources and services among various social and economic groups. The economic dimensions pillar highlights efficiency in water allocation and uses, while the political dimensions pillar focuses on providing stakeholders with equal rights and opportunities to take part in various decision-making processes. The environmental dimensions pillar emphasizes the sustainable use of water and related ecosystem services [10–12,29]. The conventional political systems could include a set of legal institutions that constitute either "government", "state", or a set of "processes" or interactions with other non-political sub-systems [102]. Thus, the different pillars offer a basis to define what constitutes the socio-economic, political, and administrative context of each system. A range of examples of political, administrative, and socio-economic systems in the national context were used to guide system presence identification. For example, the elective and non-elective political appointees composed the range of political systems. The relevant state agencies and instruments, like water taxes, permits and fees, policies, bylaws, Acts of Parliament, guidelines, procedures, all mainly administered by a central or local government formed the administrative system. The range of socio-economic systems identified included Payment for Ecosystems Services (PES) schemes like the REDD+ (Reducing Emissions from Deforestation and forest Degradation) program, water, or environmental funds. Other market and non-market schemes like mineral, oil and gas royalties, industrial and domestic water user fees, water extraction and effluent discharge permit fees as well as related project grants were also included. The traditional systems of water resources governance identified with social norms, practices, and cultural leadership involvement. Systems examples were mostly aligned according to the main objective for the establishment to avoid double reporting. The occurrence of water governance systems was more pronounced in the Mpanga catchment, influenced by integrated water resources management projects. However, political systems and structures prevailed more than the administrative, economic, and traditional systems in both study catchments, in addition to multi-purposing especially at the lower administration level. The involvement of political systems in natural resources development and management in developing countries, including Uganda, in particular, is, however, to be regarded with caution, given the possibility of political sabotage from interests and power surpassing scientific knowledge and effective resources management [103–105]. The identified socio-economic-political and administrative water systems need to be capable of serving different purposes. While there are possible variations in the measurement of functionality, the study related functionality to the presence of governance systems as compositions of mechanisms, processes and institutions through which citizens and civic groups articulate their interests, exercise their legal rights, meet their obligations, and mediate their differences [12,20,28,29]. We regard the functionality of systems significantly different with respect to the realisation of water resources use rights and the facilitation of user obligations and exclusive conflict management. Possible reasons observed for limited functionality include the prevalence of a 'top-down' style of governance, the limited reach and scale of projects, and the absence of competent local resources committees in most areas. The findings are related to previously observed gaps in governance systems and challenges during the implementation of policies and enforcement of legislation in the water and environment sector in Uganda [51,54,104–107].

Water governance principles are key measurement indicators of water governance quality [12,29,31,39]. The quality of water resources governance, as indicated by the water governance principles, was significantly different between the two study sites, as was the anti-corruption principle (Table 4). Stakeholders' participation was enhanced, according to the results, while the perceived deficiencies in integrity and fighting graft concurred with earlier studies on transparency, integrity and accountability in the water and sanitation sector in Uganda [13]. The results indicate that the integrated water resources management approach has a high potential to ensure good water governance; this finding is consistent with chances of success documented elsewhere, like the case of Zambia [108,109].

Water challenges persist to varying degrees in both study catchments, independent from already practised integrated water management approaches and the national policy direction. The mixture of successes and challenges observed is to some extent consistent with the successful evidence and approach limitations documented elsewhere [5,11,54,58,108], including the view of Integrated Water Resources Management as a fixed concept rather than understanding it as a flexible and adaptive concept with the capacity to accommodate alternative resource management approaches [110]. The results also point to the possible mutuality in ensuring successful implementation of the integrated water resources management approaches and improving water resources governance. Thus, there is an absence of a very defined linear order of 'cart and horse' [10]. Ineffective water resources governance stifles water resources management effectiveness, as observed in the Semliki catchment, which was unaffected by integrated water resources management projects [23,28,57]. The shortfalls observed in water resources governance in this study equate to policy failure, a possibility also observed in managing interconnected and complex natural systems using approaches incompatible with the broader approaches to governance and public management [109]. It is argued that policy failure (fully or to some degree) is a normal element of natural resources governance, including water resource governance; however, public policymakers fail to contemplate and manage this contingency [25,111]. Policy failure as a constituent of governance failure may be due to the insensitivity of water management institutional (instruments and agencies) and decision-making processes to stakeholder needs at all levels. In consequence, this leads to failure in administration, technical services delivery, financial and economic management and political oversight in the water and related resources sector [112]. Other documented reasons for governance failure are related to state and market failure, given the involvement of the state and the private sector in water resources management and development. The state induced failures might result from over- and or under-regulation, ill-defined rights, nonfulfillment of set resources management goals, corruption, poor leadership, inadequate policy, and legislation responses. The co-existence of formal and informal power structures characterised by political sabotage, power struggles and withholding of useful resources also affects good governance efforts [113]; strategies to overcome policy failure include appropriate instruments to improve institutional credibility and efficiency where markets and regulatory instruments require efficient legal and administrative systems trusted by the community [111]. Additional strategies could include the development of more effective water governance and related land governance regimes designed to overcome government failure, market failure and system failure or a combination of the three [29]. Importantly, water governance regimes are expected to be cognizant of the Dublin water principles and related management approach principles [9,86,87].

The implementation of the integrated water resources management approach in selected watersheds drew lessons about the costs of work at varying scales conducted by multiple stakeholders and institutions [114,115]. As study results indicate the likely effect of unmatched financing for water resources management and governance of catchments, a case in literature demonstrates how the integrated management approach comes with an increase in governance costs, but a drop in infrastructural costs (Figure 3).

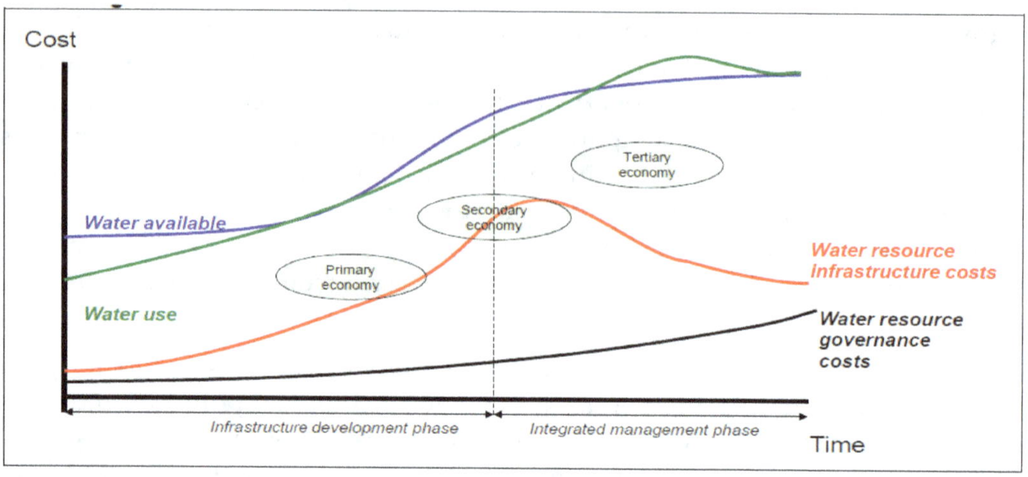

Figure 3. Costs dynamics of water resource governance (black line), water resource infrastructure (red line) and XYZ (green line) as a function of time and related to infrastructure development and implementation of integrated water resource management (graph adapted from EU Water Initiative—Finance Working Group (EUWI-FWG), Report of 2012 on Financing Water Resources Management/Experiences from Sub-Saharan Africa [116]).

5. Conclusions

The water crisis can alternatively be called a water governance crisis. Thus, integrated water resources management approaches and concepts for alleviating both crises have evolved since the 1990s. This study contributes to the increasing consensus regarding the interconnectivity of resources, systems and sustainable development, viewing the integrated water resources management approaches as a key block to successful land and water resources management and governance [15]. The case of the integrated water resources management (IWRM) approach relies on the three cornerstones of establishing an enabling environment, institutional framework, and management instruments. The existing literature indicates the possibility of success or failure of this approach in attaining the desired goals. The adoption of integrated water resources management in Uganda has included water sector reforms and experimental projects implemented in selected catchments since 2006. In this study, the water resources governance situation of two catchments as compared. Both located in the Albert Water Management Zone of the Lake Albert Basin, the river Mpanga catchment has been influenced by the implementation of integrated water resource management projects and was evaluated and compared with a base-catchment that has been unaffected by integrated water resource management projects, that is, the river Semliki catchment (focusing on the Ugandan side). The results show the water resources governance situation to be significantly different between the two catchments under the same water administration. Recognising integrated water resources management as a nationally agreed water management strategy in the water policy, the study findings further demonstrate the effects of differentiated policy translation. Statistical differences in the characteristics of water resources governance and the resulting management effectiveness were highlighted, clearly showing the impact of integrated water resources management on improving water resources governance in the Mpanga river catchment, while water resources governance was largely poor in the Semliki river catchment. Thus, integrated water resources management improves water resources governance and management effectiveness. We observe the integrated water resources management concept to embody both water management and governance tenets; thus, it is double-pronged, ensuring mutual effectiveness. The comparison between both catchments additionally indicates the necessity of including various aspects such as spatial scale, level of action, policy, and institutionalisation for the successful implementation of the integrated water resources management approach.

Persistent challenges observed in both study catchments include the increasing demand for arable land, leading to riverbank encroachment, catchment deforestation and degradation, in conflict with natural resource rights and tenure. These challenges, coupled with limited enforcement of legislation, threaten resource sustainability and related management objectives in both Semliki and Mpanga catchments. We thus recommend an appreciation of the integrated water resource management approach that is highly cognizant of the local context at the time, to improve water resources governance. The aspirations of integrated water resources management approach include holistic management of land and water resources, which implies future research to consider the examination of water and land resources governance for a holistic view.

Author Contributions: Conceptualization, Methodology, Software, Validation, Formal Analysis Investigation, Resources, Data Curation, Visualization, Original Draft Preparation, J.K.; Writing, J.K., B.S., Review and Editing, B.S. All authors have read and agreed to the published version of the manuscript.

Acknowledgments: We thank the German Academic Exchange Service (DAAD) for providing the fellowship during which this study was done. We also appreciate the guidance and supervision of Moses M.Tenywa and Gerd Förch during the field study period, anonymous colleagues for review and comments. We appreciate the various institutions that implemented IWRM, engaged in water and environment management for the information provided and support during the field survey, the respective funders, and the community members who positively responded.

References

1. Heathcote, I.W. *Integrated Watershed Management: Principles and Practices*, 2nd ed.; Wiley: Hoboken, NJ, USA, 2009; ISBN 978-0-470-37625-6.
2. Shiklomanov, I.A. Appraisal and Assessment of World Water Resources. *Water Int.* **2000**, *25*, 11–32. [CrossRef]
3. UNESCO, World Water Assessment Programme (United Nations) (Ed.) *Water: A Shared Responsibility*; United Nations world water development report; United Nations Educational, Scientific and Cultural Organization (UNESCO): Paris, France; Berghahn Books: New York, NY, USA, 2006; ISBN 978-92-3-104006-1.
4. United Nations Sustainable Development. *United Nations Conference on Environment & Development Rio de Janerio, Brazil, 3 to 14 June 1992 AGENDA21*; United Nations Department of Economic and Social Affairs (DESA): New York, NY, USA, 1992; p. 351.
5. United Nations; Environment Programme; United Nations; Development Programme. *Status Report on the Application of Integrated Approaches To Water Resources Management*; United Nations Environment Programme: Nairobi, Kenya, 2012; ISBN 978-92-807-3264-1.
6. United Nations. *Report of the United Nations Water Conference: Mar. del Plata, 14–25 March 1977*; United Nations: New York, NY, USA, 1977; p. 188.
7. Brachet, C.; Valensuela, D. *The Handbook for Integrated Water Resources Management in Transboundary Basins of Rivers, Lakes and Aquifers*; International Network of Basin Organizations (INBO) and the Global Water Partnership (GWP): Paris, France, 2012; ISBN 978-91-85321-85-8.
8. Agarwal, A.; Global Water Partnership. *Integrated Water Resources Management*; Global Water Partnership: Stockholm, Sweden, 2000; ISBN 978-91-630-9229-9.
9. Solanes, M.; Gonzalez-Villarreal, F. *The Dublin Principles for Water as Reflected in a Comparative Assessment of Institutional and Legal Arrangements for Integrated Water Resources Management*; Global Water Partnership Technical Advisory Committee: Stockholm, Sweden, 1999; ISBN 978-91-586-7668-8.
10. Lautze, J.; de Silva, S.; Giordano, M.; Sanford, L. Putting the cart before the horse: Water governance and IWRM: Putting the cart before the horse: Water governance and IWRM. *Nat. Resour. Forum* **2011**, *35*, 1–8. [CrossRef]

11. Allan, A.; Rieu-Clarke, A. Good governance and IWRM—A legal perspective. *Irrig. Drain. Syst.* **2010**, *24*, 239–248. [CrossRef]
12. Jacobson, M.; Meyer, F.; Ingvild, O.; Paavani, R.; Tropp, H. *User's Guide on Assessing Water Governance*; United Nations Development Programme: Stockholm, Sweden, 2013.
13. Jacobson, M.; Mutono, S.; Nielsen, E.; O'Leary, D.; Rop, R. *Promoting Transparency, Integrity and Accountability in the Water and Sanitation Sector in Uganda*; Water Integrity Network: Berlin, Germany, 2010; p. 24.
14. Tropp, H. Water governance: Trends and needs for new capacity development. *Water Policy* **2007**, *9* (Suppl. 2), 19–30. [CrossRef]
15. UN Water. *Sustainable Development Goal 6: Synthesis Report 2018 on Water and Sanitation*; United Nations Publications; United Nations: New York, NY, USA, 2018; ISBN 978-92-1-101370-2.
16. Jiménez, A.; Saikia, P.; Giné, R.; Avello, P.; Leten, J.; Liss Lymer, B.; Schneider, K.; Ward, R. Unpacking Water Governance: A Framework for Practitioners. *Water* **2020**, *12*, 827. [CrossRef]
17. Rosa, W. (Ed.) Transforming Our World: The 2030 Agenda for Sustainable Development. In *A New Era in Global Health*; Springer Publishing Company: New York, NY, USA, 2017; ISBN 978-0-8261-9011-6.
18. Malzbender, D.; Goldin, J.; Turton, A.; Earle, A. Traditional Water Governance and South Africa's "National Water Act"—Tension or Cooperation? Presented at African Water Laws: Plural Legislative Frameworks for Rural Water Management in Africa, Johannesburg, South Africa, 26–28 January 2005; p. 13.
19. Kapfudzaruwa, F.; Sowman, M. Is there a role for traditional governance systems in South Africa's new water management regime? *Water SA* **2009**, *35*, 10. [CrossRef]
20. Grigg, N.S. *Governance and Management for Sustainable Water Systems*; IWA Pub.: New York, NY, USA; London, UK, 2011; ISBN 978-1-78040-150-8.
21. Lautze, J.; de Silva, S.; Giordano, M.; Sanford, L. Key concepts in water resource management: A review and critical evaluation. In *Water Governance*; Routledge-Earthscan: Oxon, UK, 2014; pp. 25–38.
22. Vieira, E.d.O.; Sandoval-Solis, S.; Pedrosa, V.d.A.; Ortiz-Partida, J.P. (Eds.) *Integrated Water Resource Management: Cases from Africa, Asia, Australia, Latin America and USA*; Springer International Publishing: Cham, Switzerland, 2020; ISBN 978-3-030-16564-2.
23. Bucknall, J. Good Governance for Good Water Management. *Annu. Rev.* **2006**, *FY 2005–2006*, 4.
24. Kooiman, J. (Ed.) *Modern Governance: New Government-Society Interactions*; Sage: London, UK; Newbury Park, CA, USA, 1993; ISBN 978-0-8039-8890-3.
25. Castro, J.E. Water governance in the twentieth-first century. *Ambiente Soc.* **2007**, *10*, 97–118. [CrossRef]
26. Termeer, C.J.A.M.; Dewulf, A.; van Lieshout, M. Disentangling Scale Approaches in Governance Research: Comparing Monocentric, Multilevel, and Adaptive Governance. *Ecol. Soc.* **2010**, *15*, art29. [CrossRef]
27. Pahl-Wostl, C. *Water Governance in the Face of Global Change: From Understanding to Transformation*; Springer: Berlin/Heidelberg, Germany, 2015; ISBN 978-3-319-21855-7.
28. United Nations Development Report (Ed.) *Human Development to Eradicate Poverty*; Oxford Univ. Press: New York, NY, USA, 1997; ISBN 978-0-19-511997-8.
29. Rogers, P.; Hall, A.W.; Global Water Partnership. *Effective Water Governance*; Global Water Partnership: Stockholm, Sweden, 2003; ISBN 978-91-974012-9-6.
30. Tortajada, C.; Joshi, Y.K. Water Resources Management and Governance as Part of an Overall Framework for Growth and Development. *Int. J. Water Gov.* **2013**, *1*, 285–306. [CrossRef]
31. Michael, R.; van der Valk, V.; Penelope, K. Principles of good governance at different water governance levels. Presented at the International Hydrological Programme (IHP) of UNESCO, Delft, The Netherlands, 22 March 2011.
32. Havekes, H.; Hofstra, M. *Building Blocks for Good Water Governance*; Water Governance Centre: The Hague, The Netherlands, 2016; p. 160.
33. Nunan, F. Navigating multi-level natural resource governance: An analytical guide: Fiona Nunan/Natural Resources Forum. *Nat. Resour. Forum* **2018**, *42*, 159–171. [CrossRef]
34. Tai, H.-S. Cross-Scale and Cross-Level Dynamics: Governance and Capacity for Resilience in a Social-Ecological System in Taiwan. *Sustainability* **2015**, *7*, 2045–2065. [CrossRef]
35. Peters, B.G.; Pierre, J. Developments in intergovernmental relations: Towards multi-level governance. *Policy Politics* **2001**, *29*, 131–135. [CrossRef]
36. Pahl-Wostl, C.; Sendzimir, J.; Jeffrey, P.; Aerts, J.; Berkamp, G.; Cross, K. Managing Change toward Adaptive Water Management through Social Learning. *Ecol. Soc.* **2007**, *12*, art30. [CrossRef]

37. Folke, C.; Hahn, T.; Olsson, P.; Norberg, J. Adaptive Governance of Social-Ecological Systems. *Annu. Rev. Environ. Resour.* **2005**, *30*, 441–473. [CrossRef]
38. Huitema, D.; Mostert, E.; Egas, W.; Moellenkamp, S.; Pahl-Wostl, C.; Yalcin, R. Adaptive Water Governance: Assessing the Institutional Prescriptions of Adaptive (Co-)Management from a Governance Perspective and Defining a Research Agenda. *Ecol. Soc.* **2009**, *14*, art26. [CrossRef]
39. OECD Organisation for Economic Co-operation and Development -Principles-on-Water-Governance 2015.
40. Tantoh, H.B. Complexity and uncertainty in water resource governance in Northwest Cameroon: Reconnoitring the challenges and potential of community-based water resource management. *Land Use Policy* **2018**, *75*, 237–251. [CrossRef]
41. Ansell, C.; Gash, A. Collaborative Governance in Theory and Practice. *J. Public Adm. Res. Theory* **2007**, *18*, 543–571. [CrossRef]
42. Tarko, V. Polycentric Governance: A Theoretical and Empirical Exploration. Ph.D. Dissertation, George Mason University, Fairfax, VA, USA, 2015.
43. Thiel, A. *The Polycentricity Approach and the Research Challenges Confronting Environmental Governance*; IRI THESys—Integrative Research Institute on Transformations of Human-Environment Systems Humboldt-Universität zu Berlin: Berlin, Germany, 2016; p. 30.
44. Aligica, D.P.; Tarko, V. Polycentricity: From Polanyi to Ostrom, and Beyond. *Gov. Int. J. Policy Adm. Inst.* **2012**, *25*, 26. [CrossRef]
45. Boettke, P.J.; Lemke, J.S.; Palagashvili, L. Polycentricity, Self-governance, and the Art & Science of Association. *Rev. Austrian Econ.* **2015**, *28*, 311–335. [CrossRef]
46. Gruening, G. Origin and theoretical basis of New Public Management. *Int. Public Manag. J.* **2001**, 25. [CrossRef]
47. Manning, N. The Legacy of the New Public Management in Developing Countries. *Int. Rev. Adm. Sci.* **2001**, *67*, 297–312. [CrossRef]
48. Asif, R.; Dawood, M. Does New Public Management Practices Lead to Effective Public Welfare Responses in Pakistan. *Turk. Econ. Rev.* **2017**, *5*, 150–173.
49. Ministry of Water and Environment (MWE); Directorate of Water Resources Management. *Uganda National Water Resources Assessment*; Ministry of Water and Environment (MWE): Kampala, Uganda, 2013; ISBN 978-9970-467-00-6.
50. Government of Uganda. *The Water Act*; Ministry of Water and Environment (MWE): Kampala, Uganda, 1995; p. 50.
51. Songa, P.; Rumohr, J.; Musota, R. *Policy and Institutional Framework Considerations in the Implementation of Catchment-Based Water Resources Management in Uganda: Highlights from the River Rwizi Catchment*; A Coruña: Southampton, UK, 2015; pp. 15–26.
52. Mehta, L.; Derman, B.; Manzungu, E. (Eds.) *Flows and Practices: The Politics of Integrated Water Resources Management in Eastern and Southern Africa*; Weaver Press: Harare, Zimbabwe, 2017; ISBN 978-1-77922-320-3.
53. Rubarenzya, M.H. Integrated Water Resources Management in Uganda: Past, Present, and a Vision for the Future. In Proceedings of the World Environmental and Water Resources Congress 2008; American Society of Civil Engineers: Honolulu, HI, USA, 2008; pp. 1–11.
54. Global Water Partnership (GWP). *Integrated Water Resources Management in Eastern Africa: Dealing with "Complex." Hydrology*; Technical focus paper; Elanders: Stockholm, Sweden, 2015; ISBN 978-91-87823-21-3.
55. Ruettinger, L.; Taenzler, D. *Water, Crisis and Climate Change in Uganda: A Policy Brief*; ADELPHI: Berlin, Germany, 2011; p. 18.
56. Ministry for Water and Environment Catchment based Water Resources Management (CbWRM) in Uganda. Available online: https://www.mwe.go.ug/library/catchment-based-water-resources-management-cbwrm (accessed on 26 November 2020).
57. United Nations Educational, Scientific and Cultural Organization (UNESCO). *Water a Shared Responsibility: The United Nations World Water Development Report 2*; UNESCO: Paris, France, 2006.
58. Biswas, A.K. Integrated Water Resources Management: Is It Working? *Int. J. Water Resour. Dev.* **2008**, *24*, 5–22. [CrossRef]
59. Schlüter, T.; Hampton, C. *Geology of East Africa*; Beiträge zur regionalen Geologie der Erde; Borntraeger: Berlin, Germany, 1997; ISBN 978-3-443-11027-7.
60. Schlüter, T. *Geological Atlas of Africa*; Springer: Berlin/Heidelberg, Germany, 2008; ISBN 978-3-540-76324-6.

61. National Environment Management Authority. *Environmental Sensitivity Atlas for the Albertine Graben*, 1st ed.; NEMA: Kampala, Uganda, 2009; ISBN 978-9970-881-01-7.
62. Peel, M.C.; Finlayson, B.L.; Mcmahon, T.A. Updated world map of the Köppen-Geiger climate classification. *Hydrol. Earth Syst. Sci.* **2007**, *4*, 439–473. [CrossRef]
63. Plumptre, A.J.; Davenport, T.R.B.; Behangana, M.; Kityo, R.; Eilu, G.; Ssegawa, P.; Ewango, C.; Meirte, D.; Kahindo, C.; Herremans, M.; et al. The biodiversity of the Albertine Rift. *Biol. Conserv.* **2007**, *134*, 178–194. [CrossRef]
64. Winterbottom, B.; Eilu, G. *Uganda Biodiversity and Forest Assesment*; USAID: Kampala, Uganda, 2006; p. 64.
65. National Environment Management Authority. *Environmental Sensitivity Atlas for the Albertine Graben*, 2nd ed.; NEMA: Kampala, Uganda, 2010.
66. National Environment Management Authority. *The National Environment (Wetlands, River Banks and Lake Shores Management) Regulations*; NEMA: Kampala, Uganda, 2000; p. 33.
67. National Environment Management Authority. *The National Environment (Hilly And Mountainous Area Management) Regulations*; NEMA: Kampala, Uganda, 2000; p. 20.
68. Butsel, V.J.; Donoso, N.; Gobeyn, S.; Troyer, D.N.; Echelpoel, V.W.; Lock, K.; Goethals, P.L.M.; Peeters, L.; Bwambale, G.; Muganzi, E.; et al. *Ecological Water Quality Assessment of the Mpanga Catchment, Western Uganda*; Protos: Ghent, Belgium, 2017; p. 37.
69. Reinhardt, J.; Liersch, S.; Abdeladhim, M.A.; Diallo, M.; Dickens, C.; Fournet, S.; Hattermann, F.F.; Kabaseke, C.; Muhumuza, M.; Mul, M.L.; et al. Systematic evaluation of scenario assessments supporting sustainable integrated natural resources management: Evidence from four case studies in Africa. *Ecol. Soc.* **2018**, *23*, art5. [CrossRef]
70. Taylor, T.; Markandya, A.; Mwebaze, T.; Sebbit, A.; Rautenbach, H. *Economic Assessment of the Impacts of Climate Change in Uganda: Case Study on Water and Energy Sector Impacts in the Mpanga River Catchment*; Ministry of Water and Environment (MWE): Kampala, Uganda, 2015.
71. UNDP. *Human Development Report*; UNDP: New York, NY, USA, 1997.
72. Krejcie, R.V.; Morgan, D.W. Determining Sample Size for Research Activities. *Educ. Psychol. Meas.* **1970**, *30*, 607–610. [CrossRef]
73. Uganda Bureau of Statistics. *The National Population and Housing Census 2014—Main Report*; UBOS: Kampala, Uganda, 2016; p. 108.
74. O'Neill, K.; Weinthal, E.; Marion Suiseeya, K.R.; Bernstein, S.; Cohn, A.; Stone, M.W.; Cashore, B. Methods and Global Environmental Governance. *Annu. Rev. Environ. Resour.* **2013**, *38*, 441–471. [CrossRef]
75. Kayser, G.L.; Amjad, U.; Dalcanale, F.; Bartram, J.; Bentley, M.E. Drinking water quality governance: A comparative case study of Brazil, Ecuador, and Malawi. *Environ. Sci. Policy* **2015**, *48*, 186–195. [CrossRef]
76. Pahl-Wostl, C.; Lebel, L.; Knieper, C.; Nikitina, E. From applying panaceas to mastering complexity: Toward adaptive water governance in river basins. *Environ. Sci. Policy* **2012**, *23*, 24–34. [CrossRef]
77. Gable, G.G. Integrating Case Study and Survey Research Methods: An Example In Information System. *Eur. J. Inf. Syst.* **1994**, *3*, 17. [CrossRef]
78. Stecher, B.; Borko, H. Integrating findings from surveys and case studies: Examples from a study of standards-based educational reform. *J. Educ. Policy* **2002**, *17*, 547–569. [CrossRef]
79. Barabas, J.; Jerit, J. Are Survey Experiments Externally Valid? *Am. Polit. Sci. Rev.* **2010**, *104*, 226–242. [CrossRef]
80. Onwuegbuzie, A.J.; Dickinson, W.B.; Leech, N.L.; Zoran, A.G. A Qualitative Framework for Collecting and Analyzing Data in Focus Group Research. *Int. J. Qual. Methods* **2009**, *8*, 1–21. [CrossRef]
81. Ministry of Water and Environment. *Uganda Water and Environment Sector Capacity Development Strategy 2012–2017*; MWE: Kampala, Uganda, 2012.
82. Ferber, R. The Effect of Respondent Ignorance on Survey Results. *J. Am. Stat. Assoc.* **1956**, *51*, 576–586. [CrossRef]
83. Heide, M.; Gronhaug, K. Respondents' Moods As a Biasing Factor in Surveys: An Experimental Study. *Assoc. Consum. Res.* **1991**, *18*, 566–575.
84. Helgeson, J.G.; Voss, K.E.; Terpening, W.D. Determinants of mail-survey response: Survey design factors and respondent factors. *Psychol. Mark.* **2002**, *19*, 303–328. [CrossRef]

85. Sauer, C.; Auspurg, K.; Hinz, T.; Liebig, S.; Schupp, J. Method Effects in Factorial Surveys: An Analysis of Respondents' Comments, Interviewers' Assessments, and Response Behavior. *SSRN Electron. J.* **2014**. [CrossRef]
86. Global Water Partnership (GWP Integrated Water Resource Management (IWRM) Toolbox. Available online: https://www.gwp.org/en/learn/iwrm-toolbox/About_IWRM_ToolBox (accessed on 12 April 2020).
87. Schulte, A.; Suthfeld, R.; Vogt, B. E-Learning Project IWRM—Integrated Water Resources Management. Department of Earth Sciences, Freie Universitaet Berlin. Available online: https://www.geo.fu-berlin.de/en/v/iwrm/index.html (accessed on 12 April 2020).
88. Alaerts, G.; Dickinson, N. (Eds.) Water for a Changing World—Developing Local Knowledge and Capacity. In Proceedings of the International Symposium "Water for a Changing World Developing Local Knowledge and Capacity", Delft, The Netherlands, 13–15 June 2007; CRC Press: Boca Raton, FL, USA, 2008. ISBN 978-0-429-20688-7.
89. Alaerts, G.J.; Kaspersma, J.M. Progress and Challenges in Knowledge and Capacity Development. In *Capacity Development for Improved Water Management*; Blokland, M.W., Alaerts, G.J., Kaspersma, J.M., Hare, M., Eds.; CRC Press: Boca Raton, FL, USA, 2019; pp. 3–30. ISBN 978-0-203-84930-9.
90. Yinusa, S.O.; Wehn, U. Institutional dynamics in national strategy development: A case study of the capacity development strategy of Uganda's water and environment sector. *Water Policy* **2016**, *18*, 1174–1193. [CrossRef]
91. Wagenet, L.P.; Pfeffer, M.J.; Sutphin, H.D.; Stycos, J.M. Adult Education and Watershed Knowledge In Upstate New York. *J. Am. Water Resour. Assoc.* **1999**, *35*, 609–621. [CrossRef]
92. Olsson, P.; Folke, C. Local Ecological Knowledge and Institutional Dynamics for Ecosystem Management: A Study of Lake Racken Watershed, Sweden. *Ecosystems* **2001**, *4*, 85–104. [CrossRef]
93. Vargas, V.; Carrasco, N.; Vargas, C. Local Participation in Forest Watershed Management: Design and Analysis of Experiences in Water Supply Micro-Basins with Forest Plantations in South Central Chile. *Forests* **2019**, *10*, 580. [CrossRef]
94. Schiffer, E.; McCarthy, N.; Birner, R.; Waale, D.; Asante, F. *Information Flow and Acquisition of Knowledge in Water Governance in the Upper East Region. of Ghana*; The International Food Policy Research Institute (IFPRI): Washington, DC, USA, 2008; p. 25.
95. Pedregal, B.; Cabello, V.; Hernández-Mora, N.; Limones, N. Information and Knowledge for Water Governance in the Networked Society. *Water Altern.* **2015**, *8*, 20.
96. Buuren, A. Knowledge for water governance: Trends, limits, and challenges. *Int. J. Water Gov.* **2013**, *1*, 157–175. [CrossRef]
97. Pahl-Wostl, C. An Evolutionary Perspective on Water Governance: From Understanding to Transformation. *Water Resour. Manag.* **2017**, *31*, 2917–2932. [CrossRef]
98. Mwangi, E.; Wardell, A. Multi-level governance of forest resources. *Int. J. Commons* **2012**, *6*, 79–103. [CrossRef]
99. Brels, S.; Coates, D.; Loures, F.; Secretariat of the Convention on Biological Diversity; Center for International Forestry Research. *Transboundary Water Resources Management: The Role of International Watercourse Agreements in Implementation of the CBD*; Secretariat of the Convention on Biological Diversity: Montreal, QC, Canada, 2008; ISBN 978-92-9225-092-8.
100. Conti, K.I.; Gupta, J. Global governance principles for the sustainable development of groundwater resources. *Int. Environ. Agreem. Polit. Law Econ.* **2016**, *16*, 849–871. [CrossRef]
101. Villholth, K.G.; López-Gunn, E.; Conti, K.; Garrido, A.; van der Gun, J.A.M. (Eds.) *Advances in Groundwater Governance*; CRC Press/Balkema: Leiden, The Netherlands, 2018; ISBN 978-1-315-21002-5.
102. Heslop, A.D. Political system. Available online: https://www.britannica.com/topic/political-system (accessed on 20 October 2019).
103. Collier, P. The Political Economy of Natural Resources. *Soc. Res.* **2010**, *77*, 29.
104. Saito, F. *Local Council Commons Management in Uganda: A Theoretical Reassessment*; Nagoya University: Nagoya, Japan, 2007; p. 22.
105. Muhereza, F. Decentralising Natural Resource Management and the Politics of Institutional Resource Management in Uganda's Forest Sub-Sector. *Afr. Dev.* **2006**, *37*, 67–101.
106. Akello, C.E. *Environmental Regulation in Uganda: Successes and Challenges*; Law, Environment and Development Journal: New Delhi, India, 2007; p. 20.
107. Naiga, R.; Penker, M.; Hogl, K. Challenging pathways to safe water access in rural Uganda: From supply to demand-driven water governance. *Int. J. Commons* **2015**, *9*, 237–260. [CrossRef]

108. Uhlendahl, T.; Salian, P.; Casarotto, C.; Doetsch, J. Good water governance and IWRM in Zambia: Challenges and chances. *Water Policy* **2011**, *13*, 845–862. [CrossRef]
109. Lenton, R.L.; Muller, M. (Eds.) *Integrated Water Resources Management in Practice: Better Water Management for Development*; Earthscan: London, UK; Sterling, VA, USA, 2009; ISBN 978-1-84407-649-9.
110. Giordano, M.; Shah, T. From IWRM back to integrated water resources management. *Int. J. Water Resour. Dev.* **2014**, *30*, 364–376. [CrossRef]
111. Martin, P.; Williams, J. *Water Governance: A Policy Risk Perspective*; WIT Press: Southampton, UK, 2013; pp. 73–84.
112. Bakker, K.; Kooy, M.; Shofiani, N.E.; Martijn, E.-J. Governance Failure: Rethinking the Institutional Dimensions of Urban Water Supply to Poor Households. *World Dev.* **2008**, *36*, 1891–1915. [CrossRef]
113. Pahl-Wostl, C.; Conca, K.; Kramer, A.; Maestu, J.; Schmidt, F. Missing Links in Global Water Governance: A Processes-Oriented Analysis. *Ecol. Soc.* **2013**, *18*, art33. [CrossRef]
114. MWE. *Joint Water and Environment Sector Support Programme (JWESSP, 2013–2018)*; Ministry of Water and Environment: Kampala, Uganda, 2013; p. 215.
115. Ministry for Water and Environment. *Joint Water and Environment Sector Support. Programme Phase II 2018–2023*; Ministry for Water and Environment: Kampala, Uganda, 2018; p. 158.
116. Winpenny, J.; Hall, A.; Lindgaard-Jørgensen, P. *Financing of Water Resources Management: Experiences from Sub-Saharan Africa*; European Union Water Initiative Finance Working Group (EUWI-FWG): Stockholm, Sweden, 2012.

2

Impact Assessment and Geochemical Background Analysis of Surface Water Quality of Catchments Affected by the 2017 Portugal Wildfires

Mário David Sequeira [1,*,†], Ana Maria Castilho [2], Pedro Alexandre Dinis [3] and Alexandre Oliveira Tavares [1]

1 Centre for Social Studies, Department of Earth Sciences, Faculty of Sciences and Technology, University of Coimbra, 3030-790 Coimbra, Portugal; atavares@ci.uc.pt

2 Geosciences Centre, Department of Earth Sciences, Faculty of Sciences and Technology, University of Coimbra, 3030-790 Coimbra, Portugal; amcastil@dct.uc.pt

3 MARE-Marine and Environmental Sciences Centre, Department of Earth Sciences, Faculty of Sciences and Technology, University of Coimbra, 3030-790 Coimbra, Portugal; pdinis@dct.uc.pt

* Correspondence: mario.sequeira@uc.pt

† Current Address: Departamento de Ciências da Terra, Rua Sílvio Lima, Universidade de Coimbra—Pólo II 3030-790 Coimbra, Portugal.

Abstract: In October 2017, the Mondego hydrological basin was devastated by multiple wildfires. To understand the impacts that these had on surface water quality, monthly monitoring campaigns were implemented in the affected catchments. The campaigns oversaw the water's physical characteristics and major, minor and trace constituents. Analysis of major changes was performed, as well as comparisons with geochemical backgrounds, determined using publicly available data collected prior to the fires. The backgrounds allowed us to verify which water parameters were most influenced by the fires. Wildfires were responsible for increasing turbidity on all watercourses, and electrical conductivity increased in watercourses with flatter, granitoid/sedimentary-dominated watersheds. Cl^-, NO_3^-, and SO_4^{2-} surpassed background values in most monitored basins. Changes in Al, Fe, Mn, As, Ba, and Zn concentrations were observed in all watercourses after the wildfires and decreases after intense rainfall. Background values were particularly exceeded for Al in all monitoring points and As in basins with extensive agricultural areas. Burnt plant material and organic matter partially account for the increases in Cl^-, NO_3^-, SO_4^{2-}, Al, As, Fe, and Mn content. High Al, Fe and Zn with low water conductivity suggests that these elements were associated with sediment loads after the removal of vegetation.

Keywords: surface water monitoring; physical and chemical parameters; rural fires; geochemical background; land-use change

1. Introduction

Portugal rural fires have been a major point of national concern due to their incidence and severity [1–5]. The regional distribution of forest fire ignitions and burnt areas in mainland Portugal reflect a close relationship between biophysical and human variables [6,7]. Although the meteorological conditions are crucial in the ignition and during the fire spread, there is a positive trend for increasing burnt areas related to changes in farming and land use [8]. Because of the changes in climatic patterns [9–12] and in the way land is used, especially land abandonment and introduction of pine and eucalypt [6,13–16], fire activity in the Mediterranean region has escalated in the last two decades [17], with Portugal being the European country most affected by fires [18]. The impacts of rural fires on public policies have been highlighted and extensively covered by the media [3,19,20]. In 2017, when the country was suffering a severe drought, two major fire events devastated around 5000 km^2 [21]. While the first was mostly characterized by the massive number of deaths, the second one, in October, under the influence of Hurricane Ophelia, was marked by the sheer size of the burnt area [21].

Streamflow transfer and erosion processes is affected by wildfires [22,23]. The precipitation after forest fires tends to augment the runoff and erosion, due to partial or complete destruction of vegetation and litter cover [14,24] and changes of soil properties [25,26], sometimes with enhanced water repellency [27,28].

Fires have also been linked to changes in the water's chemical and biological constituents, in particular after the beginning of rainfall as sediments, ashes, debris and products of combustion, carried by the runoff reach the watercourses [22,24]. Contributions of N and P to watercourses resulting from fire activity were reported before [29–31], as well as SO_4^{2-} [29,31], due to sulfur oxidation during combustion of the organic matter [22] and release of Cl^- from burnt vegetation [31]. Studies on the effects on trace elements and other compounds are less prominent [22]. As happened after the Cerro Grande Fire in Los Alamos (USA) [32], significant amounts of Al, As, Fe, and Mn were measured in small basins of central Portugal after the forest fires of 2017 [33]. It is, however, unclear whether the measured concentrations can be considered abnormal and how they vary with geomorphological conditions.

A particular strategy to perceive changes in water quality after forest fires is a comparison of a burnt area with an unburnt control region [30,31,34,35]. This procedure is not feasible where fire effects in regions with diversified geomorphological features are meant to be investigated. Establishing normal concentrations in watercourses is a practical alternative to detect abnormal element concentrations [36]. Many methods have been applied to identify normal concentration in a medium, namely a "threshold" calculation [37], geochemical background [38], and the natural background levels [39]. Using the median ± median absolute deviation is a procedure that does not rely on statistical assumptions and has the advantages of being particularly effective against data outliers commonly found on datasets, delivering conservative values [38], and being preferred in ecological studies [40].

In order to understand the surface water changes on the Mondego River after the forest fires of October 2017, 10 surface water sampling points, representing catchment basins with different physiographic, lithological, land use features and extension of fire-affected areas, were selected for a series of seven monthly monitoring campaigns (Table 1). This work served to test a methodology that would allow the determination of a surface water geochemical background using 30-year data. Combining the results obtained from a series of monitoring campaigns on affected areas, with background levels for the same watercourses, should provide safe information on the effects that the rural fires of 2017 had on the watercourses of the Mondego River hydrological basin.

Table 1. Monitored watercourses, their reference (Ref.), location, average slope (Av. Slp), watershed areas (Area), spring distance (Spr.), and their respective burn area percentages (B.A.).

Watercourse	Ref.	Latitude	Longitude	Av. Slp (°)	Area (km^2)	Spr. (km)	B.A. (%)
Mondego [1]	RM1	40°24′09.36″ N	7°59′09.26″ W	6.7	1421.6	124.1	44.2
Cavalos	CAV	40°22′28.74″ N	7°59′35.23″ W	4.5	86.8	13.4	83.6
Covelo	COV	40°19′40.22″ N	8°05′36.89″ W	4.1	20.9	3.5	68.3
Pomares	POM	40°17′32.14″ N	7°54′13.21″ W	18.3	44.7	10.8	99.2
Cerdeira	CRD	40°15′58.21″ N	7°59′14.21″ W	14.0	43.0	9.7	99.2
Alva [1]	RA1	40°16′12.83″ N	7°59′18.64″ W	14.9	433.3	50.0	52.0
Alva [2]	RA2	40°17′42.47″ N	8°14′43.94″ W	12.2	705.9	107.6	49.8
Ceira [1]	RC1	40°08′19.72″ N	7°59′54.69″ W	17.7	159.4	39.1	82.7
Ceira [2]	RC2	40°09′42.88″ N	8°16′40.08″ W	14.6	412.6	85.9	48.6
Mondego [2]	RM2	40°12′03.38″ N	8°25′39.83″ W	7.6	4905.6	188.5	39.0

[1] Upstream; [2] Downstream.

2. Materials and Methods

2.1. Area Description

The 2017 fires were responsible for a death toll of 64 citizens in July [41], with 51 in October [42], along with half of the 4400 km^2 of area that burnt that year [43]. The study region, the Mondego River drainage basin in Portugal's central region, was harshly affected by the 2017 wildfires (30% of burnt area) (Figure 1a). During the next months, it was still possible to observe ashes and other combustion products in the margins of the watercourses (Figure 2).

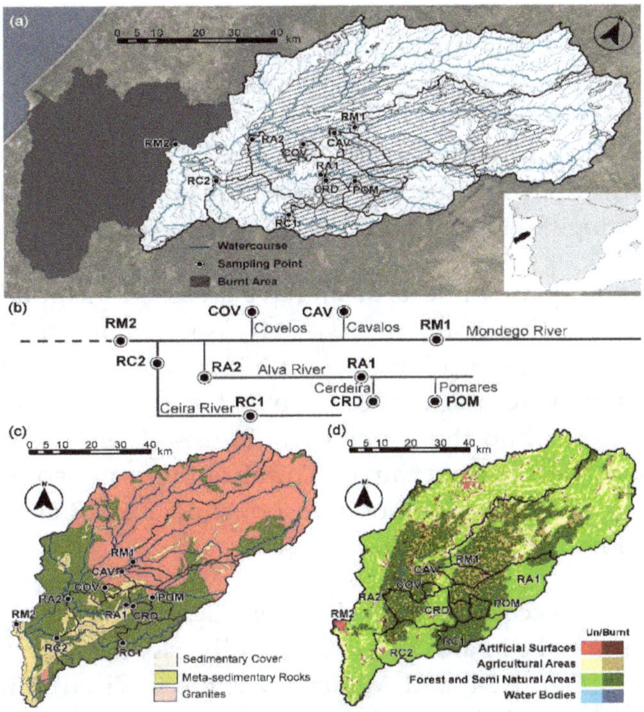

Figure 1. Location of the Mondego River hydrological basin in the Iberian Peninsula, with location of (**a**) the monitored basins and burnt area extent, (**b**) the watercourses hierarchy, (**c**) geologic context with sampling points overlay, and (**d**) land-use before the rural fires, with overlay of affected areas.

Figure 2. Presence of the remains of the fires with ashes by the Mondego River margins.

The Mondego River drainage basin is a highly contrasted area, from a geological and geomorphological standpoint. Geologically, it includes mainly metasedimentary units associated with Precambrian flysch sequences, in the southern areas, and to two-mica granitoids prevailing in the northeast (Figure 1c). Occasionally, Ordovician–Silurian metasedimentary units that include thick quartzite beds occur in the region. Some streams also drain Cretaceous–Cenozoic sedimentary units. The southeast mountain ridge (the Portuguese Central Chain), with embedded valleys and altitudes ranging from 200 m to 1200 m, diverges from the northeast, is flatter upstream (200–500 m), and is almost flat downstream (0–50 m). The study area is mainly occupied by forests and semi-natural areas (≥70%), comprising mainly maritime pines and eucalyptus. Upstream catchments also had considerable agricultural areas (Figure 1c). After the 2017 fires, some catchments (RC1, CRD, and POM) lost the majority of their forest cover (80% to 88%), while others (RM1, CAV, and COV) lost significant agricultural areas (13.7% to 40.1%) (Figure 1d).

The Mondego River flows from northeast to southwest, for 258 km, into the Atlantic Ocean, crossing the Caldeirão, Aguieira, Raiva, and Coimbra Dams. Of the main tributaries, it can be highlighted by the Dão, Corvo, Alva, and Ceira Rivers, the last two being part of the monitored area. The Alva River flows from East to west, for 110 km, crossing the Fronhas Dam, before joining the Mondego River. The Ceira River crosses the Alto do Ceira Dam, just 10 km from its spring, flows from east to west, for 100 km, joining the Mondego River near Coimbra.

Previous studies on water quality in this basin have assessed the good quality in in terms of pH, alkalinity and nitrogen species in the upstream areas [44]. Water shows higher mineralization in flatter granitic areas, contrasting with the steeper metamorphic areas [33]. Post-fire analysis of suspended sediments on the same basins have pointed to enhanced erosion, which in some sites even led to a complete removal of soil cover and the exposure of bare soil, ending up limiting local sediment supply [45]. Downstream, in the alluvial plain of Coimbra, there are some important ground water catchments, namely the Boavista wells capitation, responsible in 2017 for the distribution of water to 13 municipalities and more than 387,000 inhabitants [46].

The climate in the region is, according to the Köppen–Geiger climate classification, Mediterranean cool dry-summer [47]. The normal climate (1971–2000) for the closest stations shows the lowest rainfall in July (10.9 mm in Coimbra) (Figure 3a) and August (17.9 mm in Viseu) (Figure 3b) and the highest in December (126.2 mm to 203.4 mm) [47] (Figure 3). September of 2017 was the driest in 87 years [48]. The first five months of the hydrological year 2017/2018 were dryer than normal, with ~60% of normal rainfall, while March rainfall was 4.4–5.7 times higher than normal.

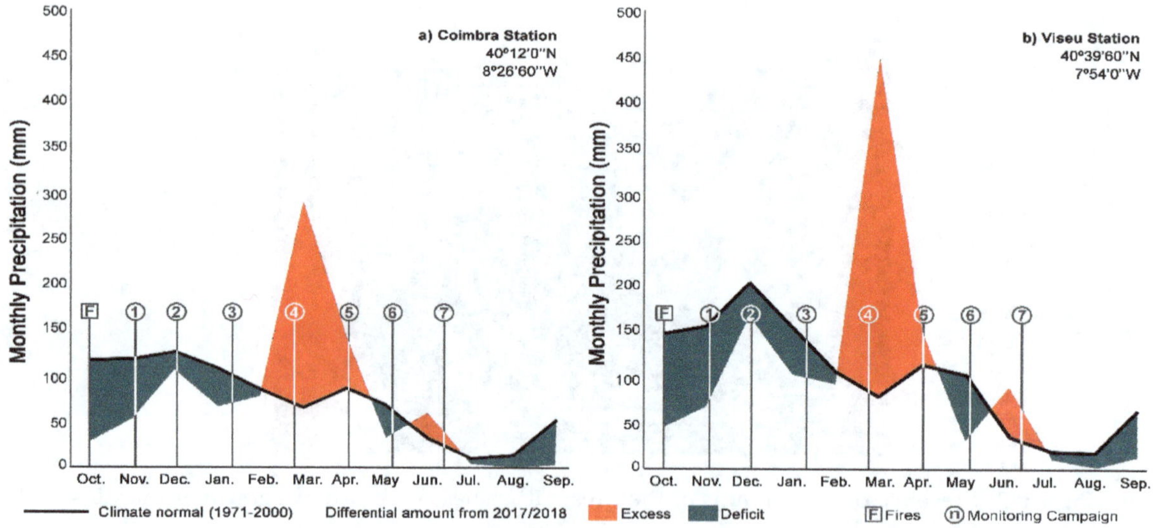

Figure 3. Normal precipitation from 1981–2010 and differential precipitation of the 2017/2018 hydrological year in the meteorological stations of: (**a**) Coimbra (40°12′0′′ N; 8°26′60′′ W) and (**b**) Viseu (40°39′60′′ N; 7°54′0′′ W). The date and number of the campaign is denoted by the circles.

2.2. *Methodology*

A monitoring strategy was designed in accordance with the ISO 5667-1 International Standard [49]. From the selected three rivers (the Mondego, Alva, and Ceira), an upstream and a downstream sampling point was chosen, while four smaller streams (the Cavalos, Cerdeira, Covelo, and Pomares) were monitored near their mouths (Figure 1a). The monitoring campaigns were initiated one month after the fires, in the beginning of the precipitation period, and ended in June of 2018, in the end of the raining season. The monitoring campaigns were conducted monthly (29 to 44 day intervals). Water sampling was done in accordance with international standards [49,50]. Full details on sampling procedures are presented in [33].

2.2.1. Field Parameters

Electrical conductivity, pH, water temperature (Tw), and dissolved oxygen (DO) were measured in situ using a multiparameter probe WTW multi 340i. Alkalinity was measured with a digital titrator, following an international standard [51]. The water's turbidity (Turb) was determined using the HANNA HI 93102 Turbidity Portable Meter (HI-93102, Hanna Instruments, Woonsocket, RI, USA).

2.2.2. Laboratory Analysis and Quality Assurance

The laboratory analysis featured some major and trace constituents. Ion chromatography and inductively coupled plasma optical emission spectrometry were used to analyze the concentrations of Cl^-, NO_3^-, SO_4^{2-}, Ca^{2+}, Mg^{2+}, Na^{2+}, and K^+ [52]. Trace constituents were determined by inductively coupled plasma optical emission spectrometry, for Al, Ba, Cu, Fe, Mn, and Zn, and by inductively coupled plasma mass spectrometry for As, Hg, and Pb. The precisions for all analyzed elements and compounds are presented as Supplementary Material (Table S1).

2.2.3. Determination of Geochemical Backgrounds

Due to the immense burnt area it was unmanageable to create control points with comparable geological/geomorphological characteristics to all sampled basins. Therefore, the collected data was instead compared with 30 years of datasets (1980–2017) from these same catchments, obtained from network stations from the Portuguese Environmental Agency (APA). To establish a comparison range, we used an approach similar to the geochemical background, which is calculated through the median and the median absolute deviation (Mdn ± 2MAD) [38] (Supplementary Material, Table S2). The median absolute deviation is calculated through the median of the absolute deviation from the dataset's median.

To compensate for the dilution–concentration effect, inherent to the relation between the precipitation and surface water, the selected historical data corresponded only to a high precipitation period in which the sampling campaigns occurred (November to April). In order to achieve statistical reliability, each calculated parameter had to have at least 50 (N) observations. Three scales were used, from the catchment size (Figure 4a), to the river (Figure 4b) and regional scale (Figure 4c). As smaller catchments have in general less observations, when this condition was not met for a parameter in a sampling site, the same methodology was applied at a higher scale. Due to the lack of previous records, it was not possible to establish backgrounds for K^+, Na^{2+}, and SO_4^{2-}.

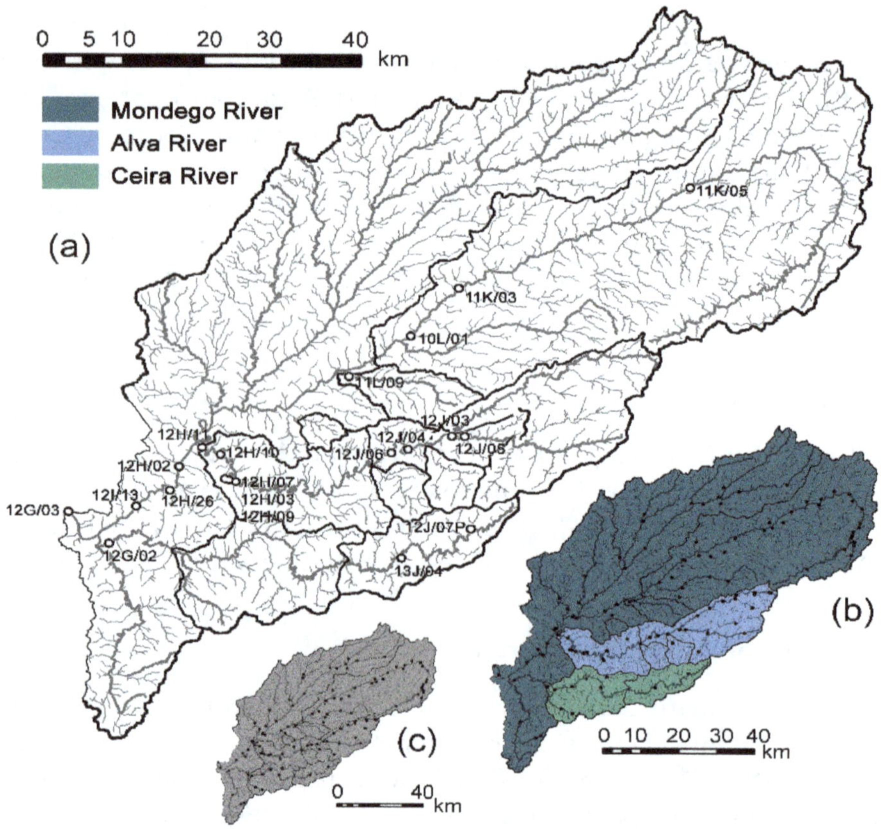

Figure 4. Location of the Portuguese Environmental Agency (APA) stations used to calculate the geochemical backgrounds, (**a**) at catchment scale, (**b**) river scale, and (**c**) regional scale.

3. Results

3.1. Field Parameters

River flowing in the granitic-dominated areas, which tend to be flatter (RM1, CAV, and RM2), has a higher electrical conductivity than in the steeper metamorphic-dominated catchments (POM, CRD, RA1, RA2, RC1, and RC2). This differentiation is valid for both the historic records and the monitoring campaigns performed for the present investigation. Most of the earlier campaigns also displayed higher values of electrical conductivity, and the intense precipitation of March and April seems to play a major role in reducing the dissolved substances in the surface water. The majority of the electrical conductivity for the sampled waters was above background between November and January (Figure 5).

The pH of the water remained similar in all sampled areas, with the month of March registering the lowest pH, above the unpolluted rain's pH (~5.2–5.5; [53]), in almost all studied watercourses (Figure 5). During the highest precipitation months, March and April, the majority of the watercourses had lower pH than determined backgrounds, reaching an average of 6.5.

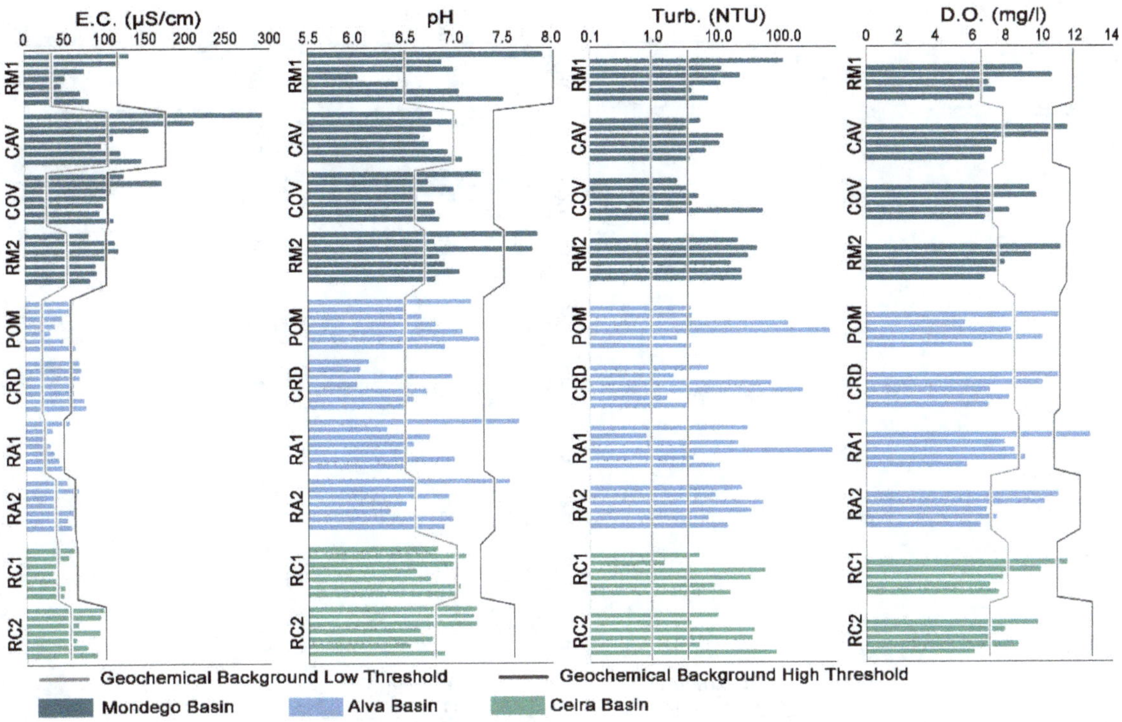

Figure 5. Temporal variations of electrical conductivity (E.C.) (µS/cm), pH, turbidity (NTU), and dissolved oxygen (D.O.) (mg/L) juxtaposed with the calculated backgrounds. The lines represent the campaigns, from earlier on top to the latest on the bottom.

In the Mondego River, a major increase in turbidity was observed in December and January, while in its smaller tributaries this increase was much later, with the intense precipitation of March and April (Figure 5). Despite this, all the monitored points presented turbidity levels above the regional turbidity background, and occasionally above historical records. Dissolved oxygen concentration does not differ much between watercourses (Figure 5). In upstream sites, it was frequently above the background during winter time; by the end of the monitoring period, it was close or below the background in most locations.

3.2. Major Ions

The major ions analysis revealed more mineralized waters in the granitic-dominated basins, which presented the majority of the values above the background. In contrast, the metamorphic-dominated basins were generally less mineralized. Over the course of the monitoring program, reductions were observed in the concentrations of Ca^{2+}, Cl^{-}, K^{+}, Mg^{2+}, Na^{2+} and SO_4^{2-} in almost all sites (Figure 6). Nevertheless, in the downstream site of Mondego River, Ca^{2+}, Mg^{2+} and SO_4^{2-} content actually increased until January 2018. Despite it was not possible to establish geochemical backgrounds for Na^{+} and K^{+}, it is observable in one the most affected basins, the Covelo watercourse, an increase in the first precipitation period, of December and January. After that, most of the concentrations decreased over the course of the monitoring campaigns.

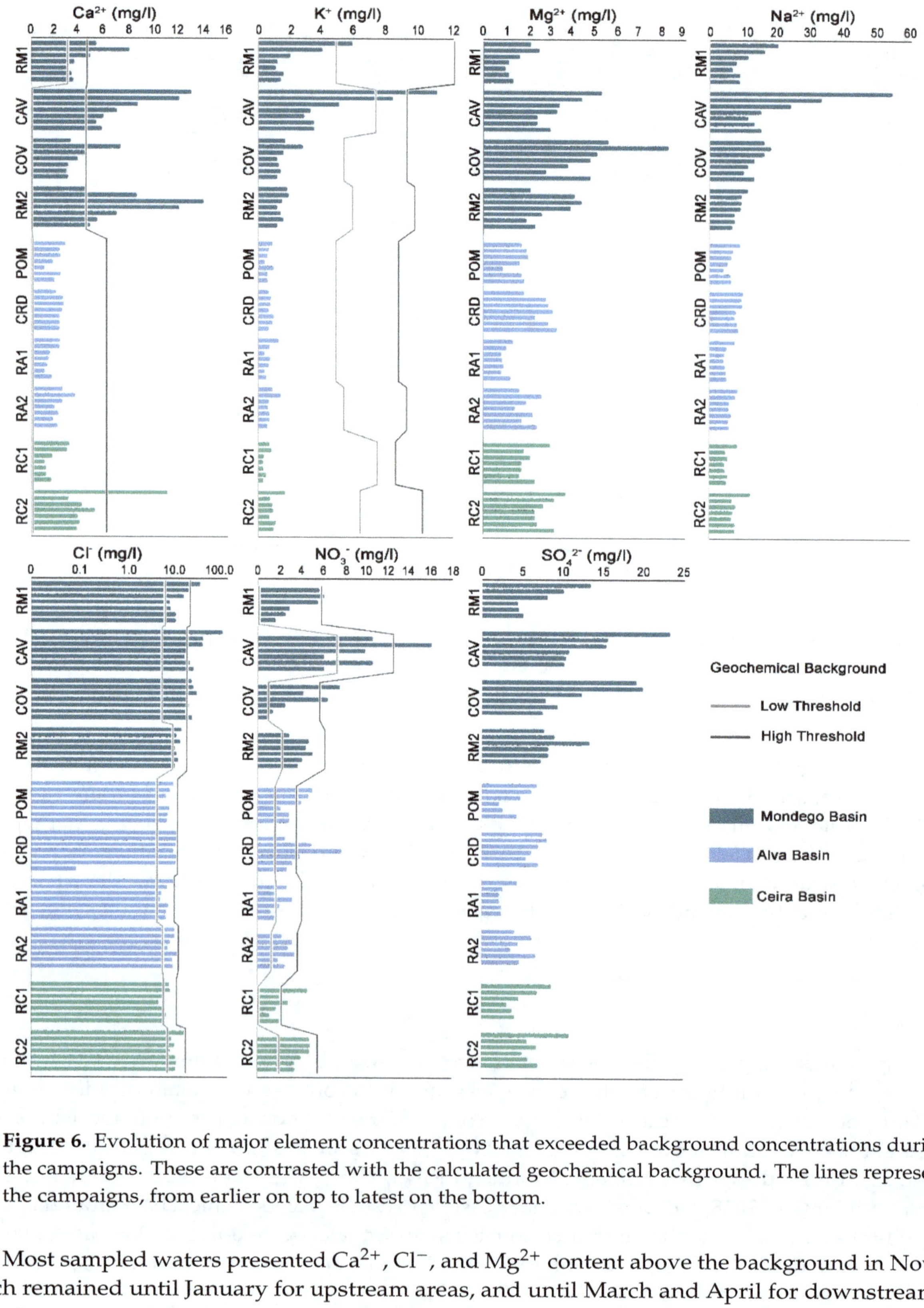

Figure 6. Evolution of major element concentrations that exceeded background concentrations during the campaigns. These are contrasted with the calculated geochemical background. The lines represent the campaigns, from earlier on top to latest on the bottom.

Most sampled waters presented Ca^{2+}, Cl^{-}, and Mg^{2+} content above the background in November, which remained until January for upstream areas, and until March and April for downstream areas. After intense precipitation occurred, there were increases in NO_3^- in all sampled watercourses, which may exceed the determined backgrounds in smaller watercourses and in upstream locations of Mondego and Ceira Rivers most clearly in January and March. After these precipitation periods, the NO_3^- concentrations return to normal concentrations.

3.3. Trace Elements

Regarding minor and trace constituents, the bigger catchments tended to yield highest concentrations of Fe, Al, and Mn, particularly immediately after the intense rainfall episodes of December. In general, a considerable change of Al content was observed from months with high precipitation to months with less precipitation. In the Mondego River, Al contents are frequently above background levels, being highest immediately after the first events of intense rainfall in December. In smaller metamorphic-dominated areas the peak values were found in March and April, when they frequently exceed background levels (Figure 7). As a general rule, higher Fe and Mn than calculated backgrounds were observed preferentially in November, December, and January.

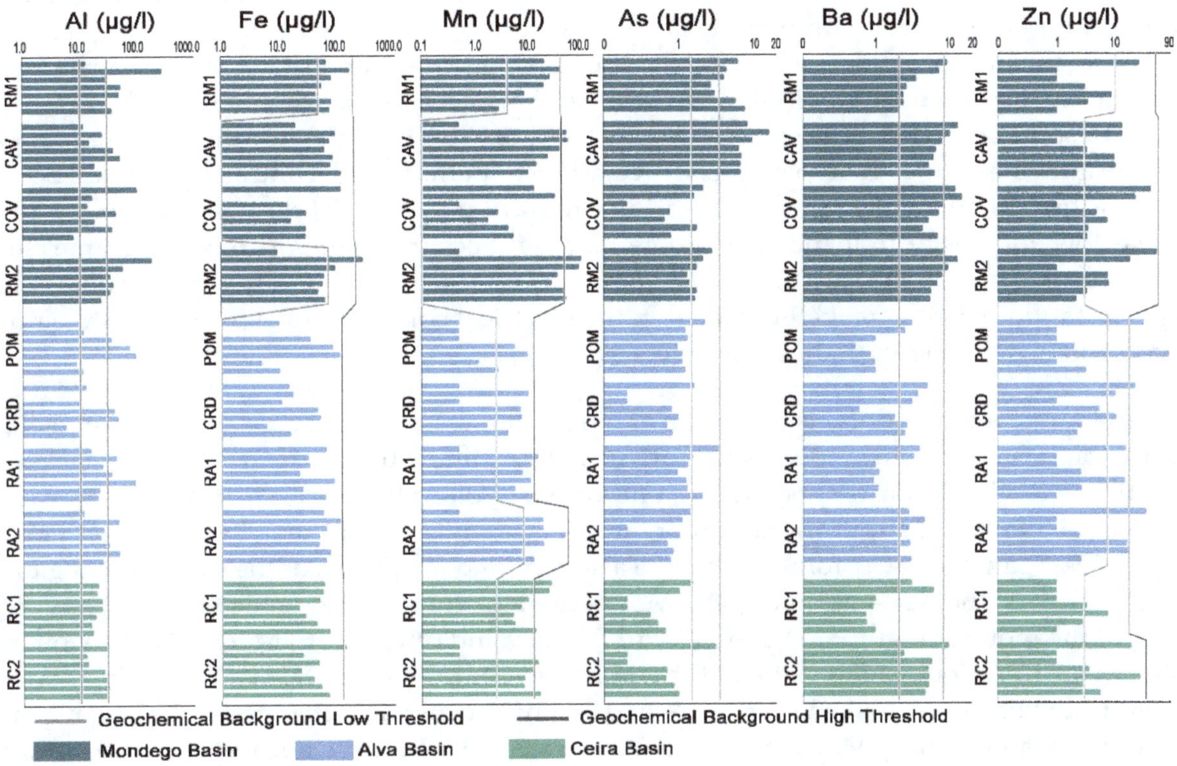

Figure 7. Evolution of trace element concentrations that exceeded background concentrations during the campaigns. These are contrasted with the calculated geochemical background. The lines represent the campaigns, from earlier on top to the latest on the bottom.

All monitored basins showed As in November, but its concentrations tended to decline through the monitoring period (Figure 6). An exception was visible in granitic-dominated catchment with intense agricultural occupation (Cavalos), where it increased in December. Actually, As only exceed background levels in catchments with extensive agricultural areas. Barium tended to be more abundant in granitic-dominated catchments, and may exceeded the background short after the wildfires, from November to January, but it is barely found on metamorphic-dominated catchments (Figure 7). Despite occasional April increases in some small rivers, earlier campaigns usually showed higher concentrations of Zn (Figure 7). Contrariwise, Pb was mostly found by the end of the monitoring period, in June, while Cu and Hg were always bellow quantifiable concentrations.

4. Discussion

In the studied area, water chemistry strongly depended on the geomorphology of the source areas. Both the campaign data as well as the historical datasets show that electrical conductivity and major and minor concentrations tend to be higher in the granite-dominated areas with low slopes

(4° to 8°), contrasting clearly with the metamorphic-dominated areas of higher slopes (14° to 18.5°). This differentiation is attributed to the higher water-soil interactions due to longer residence time of the water in soil environments in flatter areas. The concentrations of some elements should also be influenced by the forms of land use in the catchment areas. For example, river water derived from areas where agricultural occupation is more common (e.g., in Cavalos River and the upstream site of the Mondego River) yields relatively high contents of NO_3^- and As. Although increasing As concentrations have been reported after wildfires elsewhere [22], this metalloid is also commonly found in phosphate fertilizers [54]. Hence, it is reasonable to assume that the high NO_3^- and As in these locations were originally sourced by fertilizers.

Wildfires are responsible for changes in water features that should also depend on geomorphic processes. The partial or full elimination of the vegetation cover frequently promotes enhanced erosion and sediment production [24,27,55], with the transport of ash and sediments into watercourses leading to an increase in turbidity [56]. The combustion processes during the fire also increase the availability of some chemical species on the soil surface [22,57]. The shifts in water features, however, appear to be influenced by rainfall patterns. Three different periods with contrasting rainfall occurred after the wildfires of 2017 during the monitoring program (Figure 3). From November to January, rainfall was low and heavy, influenced by a severe draught occurring in Portugal. It became close to normal in February and abnormally high during March and April, returning to normal values in May. These precipitation periods are well reflected in the electrical conductivity and Ba, with the rainfall highs synchronized with lows of these parameters.

The low precipitation period is marked by an overall increase in the concentrations of elements that tend to be hosted in ashes, such as Cl^-, SO_4^{2-}, NO_3^-, and Mn. High Cl^- content was already linked with the leaching of ash deposits [22] and/or burnt plant litter [22,57], while SO_4^{2-} can also be formed by oxidation of sulfur in the soil organic matter by combustion [22]. In central Portugal, increases in SO_4^{2-} and Cl^- after forest fires were also reported in springs of the Caramulo region [35] and in tributaries of the Águeda River [58]. Most sampled streams also revealed high NO_3^- at the beginning of the monitoring program, some until January and others until March. Nitrate ion might be formed by nitrification of NH_4^+ due to the combustion of organic matter [57]. Finally, changes in Mn concentration have been linked with ashes and/or physicochemical changes of the organic matter, which provide water soluble Mn for soils in burnt areas [59]. Because eucalyptus bioaccumulates Mn and is abundant in forests of Central Portugal affected by the fires of 2017, it must have a major influence on the concentration of this element.

Despite Zn having been previously found in water in association with ashes in the suspended sediments [60], it had a minimal presence through the campaigns. Zn was observable in November in all the sampling points, achieving the highest concentration in March and April. In the next few months, it returned to unquantifiable concentrations in most of the basins, which can be related to a co-precipitation with Fe and/or Mn [54].

Elements associated with erosion and transport of soil particles, such as Al and Fe, appear to be particularly abundant during November–December. High concentrations of these elements short after the Cerro Grande Fire of 2000 were attributed to enhanced suspended loads carried by the runoff [32]. However, these findings can also be attributed to soil organic matter interaction with mineral components of the soil, such as Fe and Al oxides [61], which were then released after the combustion into the watercourses. The March–April period of extreme precipitation showed renewed increases of Al, Fe, and Mn concentrations in some rivers. At this time, the pH measured in most watercourses were similar to uncontaminated rain and the lowest values of electrical conductivity were obtained. This displays how the rain was acting as a dilution agent, reducing the concentration of dissolved elements. Nevertheless, during March and April, turbidity reached highest values for most of the watercourses. It is noteworthy that, although less parameters exceeded calculated backgrounds (e.g., turbidity, Cl^-, NO_3^-, and Al) during this timeframe, this was when some reached more than five times those reference values.

5. Conclusions

The geochemical background approach used in the present investigation allowed an assessment of fire effects on water quality, which is particularly useful where a comparison with control areas is not possible. This methodology is heavily influenced by the available data. In relatively big rivers, which tend to be better monitored by environmental agencies, the methodology provides close backgrounds for comparison, but where streams have received low attention backgrounds have to rely on data from monitored rivers in close proximity, giving less accurate estimations. Despite the used methodology is mostly utilized in economic geology, it can be used in environmental studies to detect anomalies related to contaminations. Also, although the geochemical background is mostly calculated for soils and sediments, as it was observed through the present study, it can also be used in surface waters, as long as the volume of the precipitation is taken into account.

The monthly monitoring periodicity allows the recognition of short term, direct and indirect, impacts of forest fires. The rural fires of 2017 did affect the quality of the surface water of the Mondego hydrological basin. An increase in turbidity was observed, likely due to the arrival of ashes, sediments and color-altering elements in the watercourses. Shifts in the concentrations of several elements were detected during the monitoring period. In an earlier stage shortly after the wildfires, they can be attributed to the presence of material derived from burnt organic matter (e.g., Cl^-, NO_3^-, SO_4^{2-}, Al, As, Fe, and Mn). The increment of Cl^- and SO_4^{2-} was most evident in the downstream areas. The rise in the concentrations of Mn can be associated with the extensive burnt forest and natural areas that are occupied by eucalyptus, which bioaccumulates Mn in its leaves. High NO_3^- and As were found in basins with extensive agricultural areas, suggesting that these two contaminants are partially associated to agricultural activities. High rainfall events coincided with increases in turbidity and in the concentration of elements preferentially hosted in very fine-grained sediments (e.g., Al and Fe), along with lower electrical conductivity and pH due to dilution effects by rain water.

Author Contributions: Conceptualization, M.D.S., P.A.D. and A.O.T.; Methodology, M.D.S. and A.M.C.; Formal analysis, M.D.S. and A.M.C.; Investigation, M.D.S. and A.M.C.; Resources, A.M.C. and P.A.D.; Data curation, M.D.S.; Writing—original draft preparation, M.D.S. and A.O.T.; Writing—review and editing, M.D.S., P.A.D. and A.O.T.; Visualization, M.D.S. and A.M.C.; Supervision, A.O.T.; Project administration, A.O.T.; Funding acquisition, A.O.T. All authors have read and agreed to the published version of the manuscript.

Acknowledgments: This study is part of the RiskAquaSoil project, co-financed by the European Regional Development Fund (FEDER) through the Cooperation Program INTERREG Atlantic Area, with reference EAPA_272/2016.

References

1. Carvalho, J.; Lopes, J. *Classificação de Incêndios Florestais. Manual do Utilizador*; Direcção-Geral das Florestas: Lisboa, Portugal, 2001. (In Portuguese)
2. Nunes, M.C.; Vasconcelos, M.J.; Pereira, J.M.; Dasgupta, N.; Alldredge, R.J.; Rego, F.C. Land cover type and fire in Portugal: Do fires burn land cover selectively? *Landsc. Ecology* **2005**, *20*, 661–673. [CrossRef]
3. Mateus, P.; Fernandes, P.M. Forest fires in Portugal: Dynamics, causes and policies. In *Forest Context and Policies in Portugal: Present and Future Challenges*; Reboredo, F., Ed.; Springer: Cham, Germany, 2014; pp. 97–115.
4. Tedim, F.; Remelgado, R.; Carvalho, S.; Martins, J. The largest forest fires in Portugal: The constraints of burned area size on the comprehension of fire severity. *J. Environ. Biol.* **2015**, *36*, 133–143. [PubMed]
5. Nunes, A.N.; Lourenço, L.; Meira, A.C. Exploring spatial patterns and drivers of forest fires in Portugal (1980–2014). *Sci. Total Environ.* **2016**, *573*, 1190–1202. [CrossRef] [PubMed]

6. Nunes, A.N. Regional variability and driving forces behind forest fires in Portugal an overview of the last three decades (1980–2009). *Appl. Geogr.* **2012**, *34*, 576–586. [CrossRef]
7. Ferreira-Leite, F.; Bento-Gonçalves, A.; Vieira, A.; Nunes, A.; Lourenço, L. Incidence and recurrence of large forest fires in mainland Portugal. *Nat. Hazards* **2016**, *84*, 1035–1053. [CrossRef]
8. Pereira, M.G.; Trigo, R.M.; da Camara, C.C.; Pereira, J.M.; Leite, S.M. Synoptic patterns associated with large summer forest fires in Portugal. *Agric. For. Meteorol.* **2005**, *129*, 11–25. [CrossRef]
9. Harding, A.; Palutikof, J.; Holt, T. The climate system. In *The Physical Geography of the Mediterranean*; Woodward, J., Ed.; Oxford University Press: Oxford, UK, 2009; pp. 69–88.
10. Moriondo, M.; Good, P.; Durao, R.; Bindi, M.; Giannakopoulos, C.; Corte-Real, J. Potential impact of climate change on fire risk in the Mediterranean area. *Clim. Res.* **2006**, *31*, 85–95. [CrossRef]
11. Turco, M.; Jerez, S.; Augusto, S.; Tarín-Carrasco, P.; Ratola, N.; Jiménez-Guerrero, P.; Trigo, R.M. Climate drivers of the 2017 devastating fires in Portugal. *Sci. Rep.* **2019**, *9*, 1–8. [CrossRef]
12. Nunes, L.J.R.; Meireles, C.I.R.; Gomes, C.J.P.; de Almeida Ribeiro, N.M.C. The Evolution of Climate Change in Portugal. In *Climate Change Impact on Environmental Variability in the Forest*; Nunes, L.J.R., Meireles, C.I.R., Gomes, C.J.P., de Almeida Ribeiro, N.M.C., Eds.; Springer: Cham, Germany, 2020; pp. 19–67.
13. Moreira, F.; Vaz, P.; Catry, F.; Silva, J.S. Regional variations in wildfire susceptibility of land-cover types in Portugal: Implications for landscape management to minimize fire hazard. *Int. J. Wildland Fire* **2009**, *18*, 563–574. [CrossRef]
14. Shakesby, R.A. Post-wildfire soil erosion in the Mediterranean: Review and future research directions. *Earth-Sci. Rev.* **2011**, *105*, 71–100. [CrossRef]
15. Nunes, A.N.; De Almeida, A.C.; Coelho, C.O. Impacts of land use and cover type on runoff and soil erosion in a marginal area of Portugal. *Appl. Geogr.* **2011**, *31*, 687–699. [CrossRef]
16. Pereira, P.; Úbeda, X.; Martin, D.; Mataix-Solera, J.; Cerdà, A.; Burguet, M. Wildfire effects on extractable elements in ash from a Pinus pinaster forest in Portugal. *Hydrol. Process.* **2014**, *28*, 3681–3690. [CrossRef]
17. Pausas, J.G. Changes in fire and climate in the eastern Iberian Peninsula (Mediterranean basin). *Clim. Chang.* **2004**, *63*, 337–350. [CrossRef]
18. Rego, F.; Louro, G.; Constantino, L. The impact of changing wildfire regimes on wood availability from Portuguese forests. *For. Policy Econ.* **2013**, *29*, 56–61. [CrossRef]
19. Ribeiro, C.; Valente, S.; Coelho, C.; Figueiredo, E. A look at forest fires in Portugal: Technical, institutional, and social perceptions. *Scand. J. For. Res.* **2015**, *30*, 317–325. [CrossRef]
20. Areia, N.; Intrigliolo, D.; Tavares, A.O.; Mendes, J.M.; Sequeira, M.D. The role of media between expert and lay knowledge: A study of Iberian media coverage on climate change. *Sci. Total Environ.* **2019**, *682*, 291–300. [CrossRef]
21. Viegas, D.X. Wildfires in Portugal. *Fire Res.* **2018**, *2*, 52. [CrossRef]
22. Smith, H.G.; Sheridan, G.J.; Lane, P.N.; Nyman, P.; Haydon, S. Wildfire effects on water quality in forest catchments: A review with implications for water supply. *J. Hydrol.* **2011**, *396*, 170–192. [CrossRef]
23. Rhoades, C.C.; Nunes, J.P.; Silins, U.; Doerr, S.H. The influence of wildfire on water quality and watershed processes: New insights and remaining challenges. *Int. J. Wildland Fire* **2019**, *28*, 721–725. [CrossRef]
24. Shakesby, R.A.; Doerr, S.H. Wildfire as a hydrological and geomorphological agent. *Earth Sci. Rev.* **2006**, *74*, 269–307. [CrossRef]
25. Varela, M.E.; Benito, E.; Keizer, J.J. Effects of wildfire and laboratory heating on soil aggregate stability of pine forest in Galicia: The role of lithology, soil organic matter content and water repellency. *Catena* **2010**, *83*, 127–134. [CrossRef]
26. Campos, I.; Abrantes, N.; Pereira, P.; Micaelo, A.C.; Vale, C.; Keizer, J.J. Forest fires as potential triggers for production and mobilization of polycyclic aromatic hydrocarbons to the terrestrial ecosystem. *Land Degrad. Dev.* **2019**, *30*, 2360–2370. [CrossRef]
27. Scott, D.F.; Versfeld, D.B.; Lesch, W. Erosion and sediment yield in relation to afforestation and fire in the mountains of the Western Cape Province, South Africa. *S. Afr. Geogr. J.* **1998**, *80*, 52–59. [CrossRef]
28. Keizer, J.J.; Doerr, S.H.; Malvar, M.C.; Prats, S.A.; Ferreira, R.S.V.; Oñate, M.G.; Coelho, C.O.A.; Ferreira, A.J.D. Temporal variation in topsoil water repellency in two recently burnt eucalypt stands in north-central Portugal. *Catena* **2008**, *74*, 192–204. [CrossRef]
29. Earl, S.R.; Blinn, D.W. Effects of wildfire ash on water chemistry and biotain South-Western USA stream. *Freshw. Biol.* **2003**, *48*, 1015–1030. [CrossRef]

30. Lane, P.N.J.; Sheridan, G.J.; Noske, P.J.; Sherwin, C.B. Phosphorus and nitrogen exports from SE Australian forests following wildfire. *J. Hydrol.* **2008**, *361*, 186–198. [CrossRef]
31. Mast, M.A.; Clow, D.W. Effects of 2003 wildfires on stream chemistry in Glacier National Park. *Montan. Hydrol. Process.* **2008**, *22*, 5013–5023. [CrossRef]
32. Gallaher, B.; Koch, R. Quality of Storm Water Runoff at Los Alamos National Laboratory in 2000 with Emphasis on the Impact of the Cerro Grande Fire. Los Alamos National Laboratory LA-13926. Available online: https://pdfs.semanticscholar.org/45db/26759a2b60debfb6ce5ee1d0fc487ca7889d.pdf?_ga=2.131141005.348100239.1582302245-895007879.1582302245 (accessed on 8 January 2020).
33. Sequeira, M.D.; Castilho, A.M.; Tavares, A.O.; Dinis, P. Assessment of superficial water quality of small catchment basins affected by Portuguese rural fires of 2017. *Ecol. Indic.* **2020**, *111*, 105961. [CrossRef]
34. Costa, M.R.; Calvão, A.R.; Aranha, J. Linking wildfire effects on soil and water chemistry of the Marão River watershed, Portugal, and biomass changes detected from Landsat imagery. *Appl. Geochem.* **2014**, *44*, 93–102. [CrossRef]
35. Mansilha, C.; Duarte, C.G.; Melo, A.; Ribeiro, J.; Flores, D.; Marques, J.E. Impact of wildfire on water quality in Caramulo Mountain ridge (Central Portugal). *Sustain. Water Resour. Manag.* **2019**, *5*, 319–331. [CrossRef]
36. Gałuszka, A.; Migaszewski, Z. Geochemical background-an environmental perspective. *Mineralogia* **2011**, *42*, 7–17. [CrossRef]
37. Hawkes, H.E.; Webb, J.S. Geochemistry in mineral exploration. *Soil Sci.* **1963**, *95*, 283. [CrossRef]
38. Reimann, C.; Filzmoser, P.; Garrett, R.G. Background and threshold: Critical comparison of methods of determination. *Sci. Total Environ.* **2005**, *346*, 1–16. [CrossRef] [PubMed]
39. Parrone, D.; Ghergo, S.; Preziosi, E. A multi-method approach for the assessment of natural background levels in groundwater. *Sci. Total Environ.* **2019**, *659*, 884–894. [CrossRef] [PubMed]
40. Rothwell, K.A.; Cooke, M.P. A comparison of methods used to calculate normal background concentrations of potentially toxic elements for urban soil. *Sci. Total Environ.* **2015**, *532*, 625–634. [CrossRef] [PubMed]
41. Viegas, D.X.; Figueiredo Almeida, M.; Ribeiro, L.M.; Raposo, J.; Viegas, M.T.; Oliveira, R.; Alves, D.; Pinto, C.; Jorge, H.; Rodrigues, A.; et al. O complexo de incêndios de Pedrógão Grande e concelhos limítrofes, iniciado a 17 de junho de 2017. Centro de Estudos sobre Incêndios Florestais da Faculdade de Ciências e Tecnologia da Universidade de Coimbra. Available online: https://www.portugal.gov.pt/download-ficheiros/ficheiro.aspx?v=3bb9773b-59fb-4099-9de5-a22fdcad1e3b (accessed on 6 January 2020).
42. Viegas, D.X.; Almeida, M.F.; Ribeiro, L.M. Análise dos incêndios florestais ocorridos a 15 de outubro de 2017. Análise dos incêndios florestais ocorridos a 15 de outubro de 2017. Centro de Estudos sobre Incêndios Florestais da Faculdade de Ciências e Tecnologia da Universidade de Coimbra. Available online: https://www.portugal.gov.pt/pt/gc21/comunicacao/documento?i=analise-dos-incendios-florestais-ocorridos-a-15-de-outubro-de-2017 (accessed on 7 January 2020).
43. DGAPPF. Relatório Provisório de Incêndios Florestais: 2017. Instituto da Conservação da Natureza e das Florestas. Available online: www.icnf.pt/portal/florestas/dfci/Resource/doc/rel/2017/10-rel-prov-1jan-31out-2017.pdf (accessed on 4 January 2020).
44. Ferreira, V.; Graça, M.A.; Feio, M.J.; Mieiro, C. Water quality in the Mondego river basin: Pollution and habitat heterogeneity. *Limnetica* **2004**, *23*, 295–306. [CrossRef]
45. Dinis, P.A.; Sequeira, M.; Tavares, A.O.; Carvalho, J.; Castilho, A.; Pinto, M.C. Post-wildfire denudation assessed from compositional features of river sediments (Central Portugal). *Appl. Clay Sci.* **2020**, *193*, 105675. [CrossRef]
46. AdCL. Relatório e Contas. Águas do Centro Litoral Grupo Águas de Portugal. Available online: http://www.aguasdocentrolitoral.pt/pt/menu-de-topo/comunicacao/publicacoes/downloads/pub_pdf23_pt.pdf (accessed on 27 February 2020).
47. IPMA Clima Normais. Available online: https://www.ipma.pt (accessed on 5 January 2020).
48. IPMA Boletim Climatológico Setembro 2017 Mensal e Ano Hidrológico Portugal Continental. Instituto Português do Mar e da Atmosfera, I.P., Divisão de Clima e Alterações Climáticas. Available online: https://www.ipma.pt (accessed on 6 January 2020).
49. ISO Standard No. 5667-1. *Water Quality–Sampling–Part 1: Guidance on the Design of Sampling Programmes and Sampling Techniques*; European Committee for Standardization: Brussels, Belgium, 2006.
50. ISO Standard No. 5667-3. *Water quality–Sampling–Part 3: Guidance on the Preservation and Handling of Water Samples*; European Committee for Standardization: Brussels, Belgium, 2012.

51. ISO Standard No. 9963-1. *Water quality–Determination of Alkalinity-Part 1: Determination of Total and Composite Alkalinity*; European Committee for Standardization: Brussels, Belgium, 1994.
52. Pedro, R.; Castilho, A.; Tavares, A. Surface water quality in a contrasted land-use river catchment. In IOP Conference Series: Earth and Environmental Science, Proceedings of the World Multidisciplinary Earth Sciences Symposium, Prague, Czech Republic, 3–7 September 2018; IOP Publishing: Bristol, UK, 2018. [CrossRef]
53. Santos, P.S.M.; Otero MSantos, E.B.H.; Duarte, A.C. Chemical composition of rainwater at a coastal town on the southwest of Europe: What changes in 20 years? *Sci. Total Environ.* **2011**, *409*, 3548–3553. [CrossRef]
54. Reimann, C.; De Caritat, P. *Chemical Elements in the Environment: Factsheets for the Geochemist and Environmental Scientist*; Springer Science & Business Media: Berlin/Heidelberg, Germany, 2012.
55. Rulli, M.C.; Rosso, R. Modeling catchment erosion after wildfires in the San Gabriel Mountains of southern California. *Geophys. Res. Lett.* **2005**, 32. [CrossRef]
56. Abraham, J.; Dowling, K.; Florentine, S. Risk of post-fire metal mobilization into surface water resources: A review. *Sci. Total Env* **2017**, *599*, 1740–1755. [CrossRef]
57. Khanna, P.K.; Raison, R.J. Effect of fire intensity on solution chemistry of surface soil under a Eucalyptus pauciflora forest. *Aust. J. Soil Res.* **1986**, *24*, 423–434. [CrossRef]
58. Ferreira, A.J.D.; Coelho, C.; Boulet, A.K.; Lopes, F.P. Temporal patterns of solute loss following wildfires in Central Portugal. *Int. J. Wildland Fire* **2005**, *14*, 401–412. [CrossRef]
59. Parra, J.G.; Rivero, V.C.; Lopez, T.I. Forms of Mn in soils affected by a forest fire. *Sci. Total Environ.* **2016**, *181*, 231–236. [CrossRef]
60. Burton, C.A.; Hoefen, T.M.; Plumlee, G.S.; Baumberger, K.L.; Backlin, A.R.; Gallegos, E.; Fisher, R.N. Trace elements in stormflow, ash, and burned soil following the 2009 Station Fire in Southern California. *PLoS ONE* **2016**, *11*, e0153372. [CrossRef] [PubMed]
61. Saidy, A.R.; Smernik, R.J.; Baldock, J.A.; Kaiser, K.; Sanderman, J. Microbial degradation of organic carbon sorbed to phyllosilicate clays with and without hydrous iron oxide coating. *Eur. J. Soil Sci.* **2015**, *66*, 83–94. [CrossRef]

3

Inherent Relationship between Flow Duration Curves at Different Time Scales: A Perspective on Monthly Flow Data Utilization in Daily Flow Duration Curve Estimation

Lei Ye [1], Wei Ding [1], Xiaofan Zeng [2], Zhuohang Xin [1,*], Jian Wu [1] and Chi Zhang [1]

1 School of Hydraulic Engineering, Dalian University of Technology, Dalian 116023, China; yelei@dlut.edu.cn (L.Y.); weiding@dlut.edu.cn (W.D.); 1455238909@mail.dlut.edu.cn (J.W.); czhang@dlut.edu.cn (C.Z.)

2 School of Hydropower and Information Engineering, Huazhong University of Science and Technology, Wuhan 430074, China; zengxiaofan@hust.edu.cn

* Correspondence: xinzh@dlut.edu.cn

Abstract: Modelling flow duration curves (FDCs) has long been a topic of interest since it is widely used in various hydrological applications. Most studies related to the estimation of FDCs in ungauged or partial gauged basins focus primarily on using climate and catchment characteristics to regionalize FDC at some single time scale. However, the relationship of FDCs at various time scales are rarely analyzed or studied. Here, we propose two methods, which are Modelled FDC Parameter comparison (M-FDC-P) and Empirical FDC Ratio comparison (E-FDC-R), to study the quantitative relationship between daily and monthly FDCs. One method M-FDC-P, selects a Kappa (KAP) distribution to represent the characteristics of the FDCs and then analyzes the relationship between KAP parameters of modelled FDCs at different time scales. Results indicate that three out of four parameters have strong correlations between FDCs at daily and monthly time scales. The other method, E-FDC-R, compares the quantitative relationship between daily and monthly empirical FDCs with given exceedance probabilities. The Power function is used for fitting the ratio-exceedance probability curves. In addition, the simulated daily FDC derived from monthly FDC can be very consistent with the observed daily flow records when the two parameters of power function are quantified precisely. These results clearly indicate that there are strong connections between daily and monthly FDCs, and monthly FDC can provide valuable information for daily FDC estimation. Since flow records at a large time scale are easier to obtain, daily FDC could be derived from monthly FDC by considering the inherent relationships between FDCs at different time scales, which is not sufficiently realized in previous studies.

Keywords: Kappa distribution; parameter relation; partial gauged basin; power function; ratio curve; ungauged basin

---

1. Introduction

The flow duration curve (FDC), which illustrates the percentage of time with which daily, monthly or some other time interval flows are equaled or exceeded over a historical period for a particular river basin, is recognized as one of the most important and widely used signatures of catchment runoff response [1]. The FDC is often interpreted as the complement of the cumulative distribution function of the flow, which provides a graphical representation of the relationship between the frequency and magnitude of flows, making it a compact signature of a catchment's functioning [2,3].

However, the modeling of FDCs has proven elusive, because of the complicated dynamics existing in flows, such as fractal and multifractal behavior [4], chaos-like dynamics [5–7], and periodicity in the mean, standard deviation, and skewness [8]. For example, at the daily time scale, daily flow is a complex time series with flow values ranging over many orders of magnitude [9]. It is very difficult to characterize the probability distribution that can approximate daily flow precisely.

At present, major studies related to the estimation of FDCs in ungauged or partial gauged basins focus primarily on statistical methods, which can be divided into three types [10,11]. The first one is the regression method, which involves establishing the relationship between flow with exceedance probabilities (or flow durations) and a set of climate and catchment characteristics [12–15]. The second one is the index method, which includes regionalization of parameters that represent the distribution function of FDC [16–18]. When no streamflow records are available, regional FDCs are used to synthesize FDCs [19]. The third one is the geostatistical method, which involves relying on kriging methods to produce predictions of hydrologic phenomena at ungauged sites with the combination of the information collected at neighboring gauging stations [20]. Over the past decade, geostatistical approaches predicting streamflow indices in ungauged basins have become increasingly popular [21–24].

However, as it can be seen, the above researches related with FDC mainly focus on a certain time scale, such as daily or annual. Except very few studies, the existing literature rarely concerns or analyzes the relationships among the FDCs at different time scales. Since FDCs at daily or monthly or annual time scale have different usages in the hydrology [9,25], and flow records at large time scale are easier to obtain, we are curious about the relationship between FDCs at different time scales. Are there any connections among the FDCs at different time scales? Is the information in FDC at large time scale implicit in small time scale? Can large scale FDCs provide some reference for small scale FDC?

To answer these questions, this paper comprehensively studies the correlations between FDCs at different time scales, especially for daily and monthly FDCs. The expected research results will be very important and useful for ungauged or partial gauged basins, which often just have flow records at larger time scale. To achieve this objective, the paper is structured as follows: Section 2 proposes the methodology to calculate the correlations between FDCs at different time scales. Section 3 describes the study area and data used. In Section 4, the variation characteristics of FDCs at different time scales and the correlations between different FDCs are revealed. Section 5 provides a detailed discussion of the results. The last section summarizes the conclusions of this study.

2. Methodology

In the present study, the correlations between FDCs at different time scales are studied by applying two types of methods. One method, called the Modelled FDC Parameter comparison (abbreviated as M-FDC-P), is to calculate the correlations between the parameters of modelled FDCs at different time scales. The other one, called the Empirical FDC Ratio comparison (abbreviated as E-FDC-R), is to compare the ratio relationship of different empirical FDCs with given exceedance probabilities.

2.1. M-FDC-P Method

M-FDC-P method first selects a specific probability distribution for modelling FDCs at different time scales and then compares the correlations between the probability distribution parameters of FDCs. By doing so, it is vivid to see to what degree the parameters are correlated, and which parameter is obviously strongly or weakly correlated.

Previous studies have noted that a complex distribution with at least four parameters is needed to approximate the probability distribution of the daily streamflow [2,16]. Blum et al. [9] demonstrated that four-parameter Kappa (KAP) distribution outperformed other three-parameter distributions and could provide a very good fit to the distribution of daily streamflow across most of the US. KAP distribution [18,25,26] is, therefore, selected as a probability distribution for modelling FDCs at different time scales, and its function is defined as follows:

$$F(x) = \left\{1 - h\left[1 - \frac{k}{\alpha}(x - \xi)^{1/k}\right]\right\}^{1/h} \tag{1}$$

where ξ is a location parameter, α is a scale parameter, and k and h are shape parameters. The corresponding quantile function is described as:

$$x(F) = \xi + \frac{\alpha}{k}\left\{1 - (\frac{1 - F^h}{h})\right\}^{1/h} \tag{2}$$

KAP distribution is a base function of many commonly used three-parameter probability distributions. When $k \neq 0$, the general Pareto, generalized extreme value and generalized logistic distributions are all special cases of KAP distribution for $h = 1$, $h = 0$ and $h = -1$, respectively.

KAP distribution parameters are usually estimated by the method of L-moments, in which the first p (number of the unknown parameters in the distribution, 4 for KAP distribution) sample L-moments or L-ratios is equaled to the corresponding population quantities. Estimates of L-moment ratios exhibit substantially less bias than ordinary moment ratio estimators and are resistant to the influence of data outliers [26]. Vogel and Fennessey [1], Hosking and Wallis [26] among others have described and summarized the advantages of L-moments, and therefore they are not reproduced herein. The detail description of the theory of L-moments and parameter estimation of KAP distribution can be found in [27,28].

When the probability distribution function is chosen, parameter is the only factor that affects the variation of the probability distribution. Finally, the correlation among KAP distribution parameters at different time scales are evaluated by the most widely used linear correlation coefficient to analyze the relationship between FDCs at different time scales.

2.2. E-FDC-R Method

The above M-FDC-P method uses theoretical probability distribution to model FDCs and then compares the relationship between parameters at different time scales, which may not be universally applicable due to the subjective choice of the probability distribution function. Therefore, the study also selects another way, called E-FDC-R, which does not consider specifying any probability distribution. The steps of the E-FDC-R method are presented as follows:

1. Estimation of the empirical FDCs. An empirical FDC is constructed by ranking flows at specific time scales from all recorded years and plotting them against an estimate of their exceedance probability, known as a plotting position [1]. The first step in empirical FDC construction is to sort flow data from highest to lowest. For the probability with which each flow is exceeded, the Weibull plotting position is then used, as it provides an unbiased estimate of exceedance probability, regardless of the underlying probability distribution of the ranked observations [1]. The Weibull plotting position is described as follows:

$$p = \frac{m}{N + 1} \times 100\% \tag{3}$$

   where p is the exceedance probability for the mth flow data, and N is the number of the total flow records. Plot p on x axis and the corresponding flow (the mth flow value) on y axis. The plotted dots and the x and y axis form the empirical FDC.
2. Calculation of the ratios between empirical FDCs. Firstly, sample a series of flow with pre-selected exceedance probabilities of empirical FDCs at different time scales, and then calculate the ratios of flow values at different time scales with given exceedance probabilities. Thus, the quantitative

relationship of FDCs at different time scales is achieved. It should be noted that the number of pre-selected exceedance probabilities needs to be large enough to sufficiently represent the ratio relation of FDCs. Secondly, the quantitative relationship is analyzed in order to find a certain function to represent the quantitative relationship between FDCs.

3. Evaluation of Modelled FDC. Once the specific function is obtained, the FDC at smaller time scale can be derived via the empirical FDC at larger time scale. To evaluate the performance of FDC at smaller time scale to reproduce observations, a measure of the standardized mean square error commonly referred to as Nash–Sutcliffe efficiency (NSE) is used. The description of NSE is introduced in [9], and hence not reproduced herein.

3. Study Area and Data

To keep the accuracy of the research data and the repeatability of the research, the international Model Parameter Estimation Experiment (MOPEX) data set is used in this study. The data set is described by Duan et al. [29] and can be downloaded from ftp://hydrology.nws.noaa.gov/. This data set includes mean areal precipitation, potential evaporation, daily streamflow, and daily maximum and minimum air temperature with an adequate number of precipitation gauges. As this study focuses on FDCs, only the streamflow for 219 catchments in the MOPEX data set is selected for analysis. Most of the streamflow data are included in the Hydroclimatic Data Network (HCDN) [30], which includes only gauges believed to be unaffected by upstream regulations [31] and therefore considered to represent near-natural hydrological conditions, without the need to worry that modifications to streamflow could have substantial impacts on FDCs [10,32]. The MOPEX data set consists of 56 years of daily streamflow from 1948 to 2003, which is long enough to construct FDCs. The catchment locations and boundaries are shown in Figure 1.

Figure 1. Locations and boundaries of 219 MOPEX catchments.

4. Results and Analysis

4.1. Empirical FDCs Variation with Different Time Scales

To see how the FDC variations with time scales, Figure 2 displays the empirical FDCs for a randomly selected catchment (No. 1064500) at time scales from 1 day to 365 days. It is evident from Figure 2a that the empirical FDCs change slowly with time scales, which means two FDCs could be very alike when their time scales are not much different, indicating that information of FDC at large time scale might be somewhat implicit in small time scales. FDC at a time scale of 1 day is the steepest (having largest slope) with obvious high flow values for low exceedance probabilities and low flow values for high exceedance probabilities, which then becomes flatter as the time scale gets larger. This is due to the fact that extreme flow events usually occur at the smaller time scale, and larger time scale flow will aggregate the difference, therefore resulting in flat FDC.

Figure 2b shows the empirical FDCs at time scales 1, 7, 15, 30, 90, and 365 days, corresponding to daily, weekly, half-monthly, monthly, seasonally, and yearly FDCs. The empirical FDCs at 1 day to 30 day time scales are very similar. Obvious differences occur when the time scale is larger than 90 days. Therefore, it is logical to investigate the possibilities of estimating daily FDC by using monthly flow data, since monthly flow data is far easier to obtain in the real world.

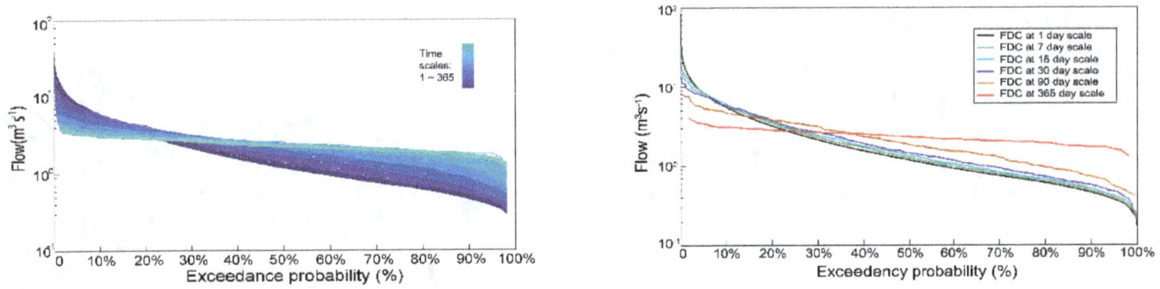

(a) (b)

Figure 2. Empirical flow duration curve (FDC) variations with time scales for a randomly selected catchment. (**a**) Time scales from 1 day to 365 days. (**b**) Time scales at 1, 7, 15, 30, 90, and 365 days.

4.2. Relationships of FDCs Derived via M-FDC-P

The daily flow data is averaged at different time scales and then fitted to a KAP distribution, and the relationships of the parameters of the KAP distribution is studied. As demonstrated earlier, the empirical FDCs at 1 to 30 day scales are very similar and might have strong correlations, therefore, this paper concentrates research interest on time scales from 1 to 30 days.

To assess how the KAP distribution parameters vary across time scales, Figure 3 presents the average of four KAP parameters of modelled FDCs for 219 catchments at time scales from 1 to 30 days. It can be seen that all the parameters change gradually with these time scales, some parameters even show a distinct changing trend. The maximum ξ (~ -0.4) occurs at a time scale of 1 day and then ξ decreases step by step, and finally stays steady around -0.8 after 19 days. For parameter α, it increases with time scales from 1 day to 30 days, with a minimum 1.1 at 1 day scale and a maximum around 2 at the 29 day scale. k is very similar to α; it increases along with all the time scales from negative to positive, and is very closed to zero at a time scale of 16 days. Parameter h is a little complicated compared to other parameters, for it first increases from 1 day to 11 day time scales and then decreases slightly until 30 day time scales.

The correlations of the four KAP distribution parameters for 219 catchments at different time scales are also studied. The linear correlation coefficients between four parameters at 1, 7, 15, and 30 day time scales are calculated in Table 1. For parameter ξ, all the correlation coefficients exceed 0.85, indicating very high correlations exist in parameter ξ at 1 to 30 days. The linear correlation coefficients between 1 day and other days become smaller as the time scale gets larger, so does the other time scales. It can also be seen in Table 1, except that the correlation coefficient for α between the largest time scale difference of 1 day and 30 day is 0.835; all the other correlation coefficients are higher or much closed to 0.9, which also indicates that high correlations exist in parameter α at 1 to 30 day. However, when compared to ξ, the correlations of α at different scales are a little weaker. When looking at parameter k, the linear correlation coefficients are dramatically lower than parameters ξ and α, with lowest correlation coefficients between 1 day and 30 day less than 0.5, indicating very weak correlation.

The correlation coefficients for parameter h between 1 day and other days become smaller as the time scale gets larger, just like other parameters. The linear correlation coefficients for h are the highest of all the parameters, with the lowest correlation coefficient between 1 day and 30 day still passing 0.94, indicating an extremely strong correlation.

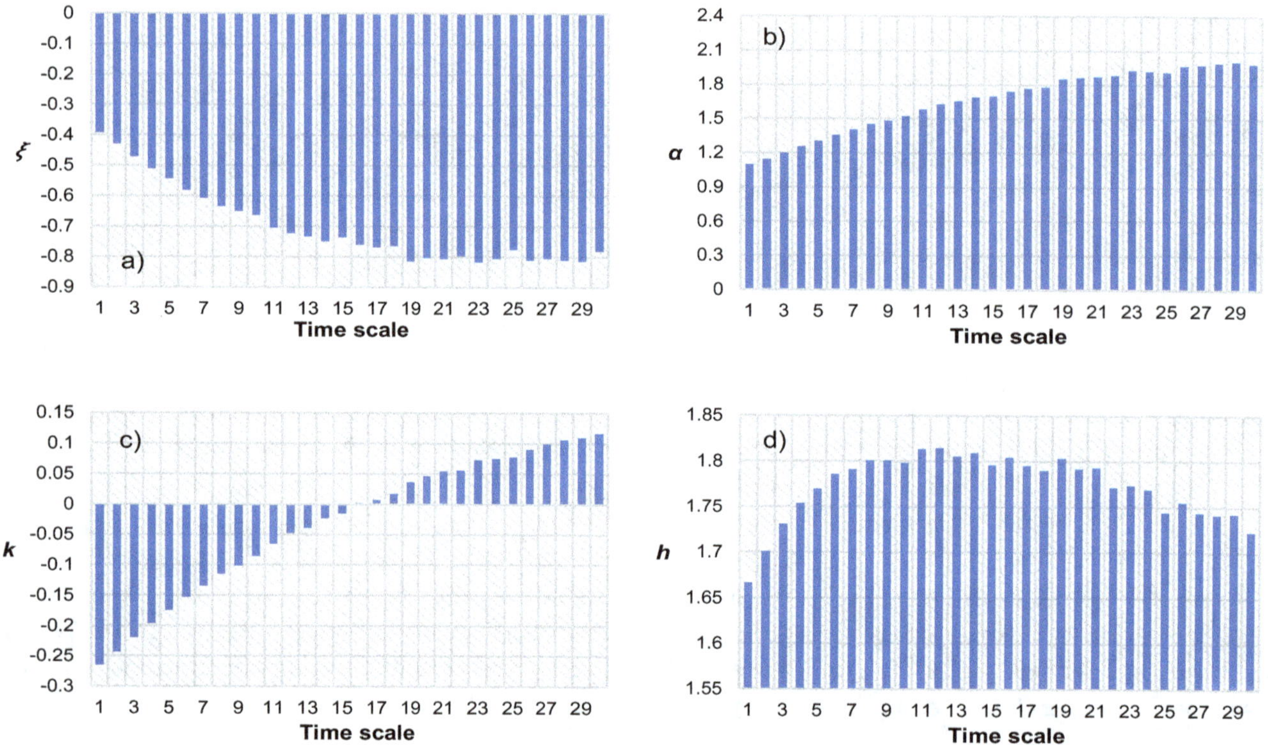

Figure 3. Variations of average four-parameter Kappa (KAP) parameters at time scales from 1 day to 30 day; (**a**) parameter ξ; (**b**); parameter α; (**c**) parameter k; (**d**) parameter h.

In addition, the relationships of the four parameters at different times are also studied by plotting corresponding parameter values for all the catchments. For illustration, the four parameters at the largest time scale difference of 1 day and 30 day are plotted in Figure 4. For parameter ξ, most of the points lie in or very near to the regression line. By comparing the linear correlation coefficients and the plot, it is easily to see the degree to which $\xi 1$ can be linear regressed by $\xi 30$. It can also be seen in Figure 4, that although most of the points lie in or very near to the regression line, there are more points far away from the regression line compared to parameter ξ. Meanwhile, these points are very centrally distributed, with a large proportion of points falling in the region bounded by $\alpha 1$ less than 2 and $\alpha 30$ less than 3. When looking at parameter k, the points are much dispersed distributed, and only a few proportion of the points lie in or near to the regression line. Further, these points do not follow a clear relationship, which indicates $k30$ is far less informative to estimate $k1$ and more information should be incorporated. For parameter h, it can be seen that except few points away from the regression line, most of the points lie in or very near to the regression line, which indicates that simple linear regression can have good performance for estimating h1 when only h30 is available.

Table 1. Linear correlation coefficients between four KAP parameters at different time scales.

Parameter	ξ				α				k				h			
Time Scale	**1 Day**	**7 Day**	**15 Day**	**30 Day**	**1 Day**	**7 Day**	**15 Day**	**30 Day**	**1 Day**	**7 Day**	**15 Day**	**30 Day**	**1 Day**	**7 Day**	**15 Day**	**30 Day**
1 day	1	0.959	0.901	0.858	1	0.953	0.893	0.835	1	0.867	0.654	0.475	1	0.987	0.967	0.940
7 day	0.959	1	0.982	0.954	0.953	1	0.981	0.941	0.867	1	0.931	0.817	0.987	1	0.992	0.973
15 day	0.901	0.982	1	0.990	0.893	0.981	1	0.985	0.654	0.931	1	0.962	0.967	0.992	1	0.992
30 day	0.858	0.954	0.990	1	0.835	0.941	0.985	1	0.475	0.817	0.962	1	0.940	0.973	0.992	1

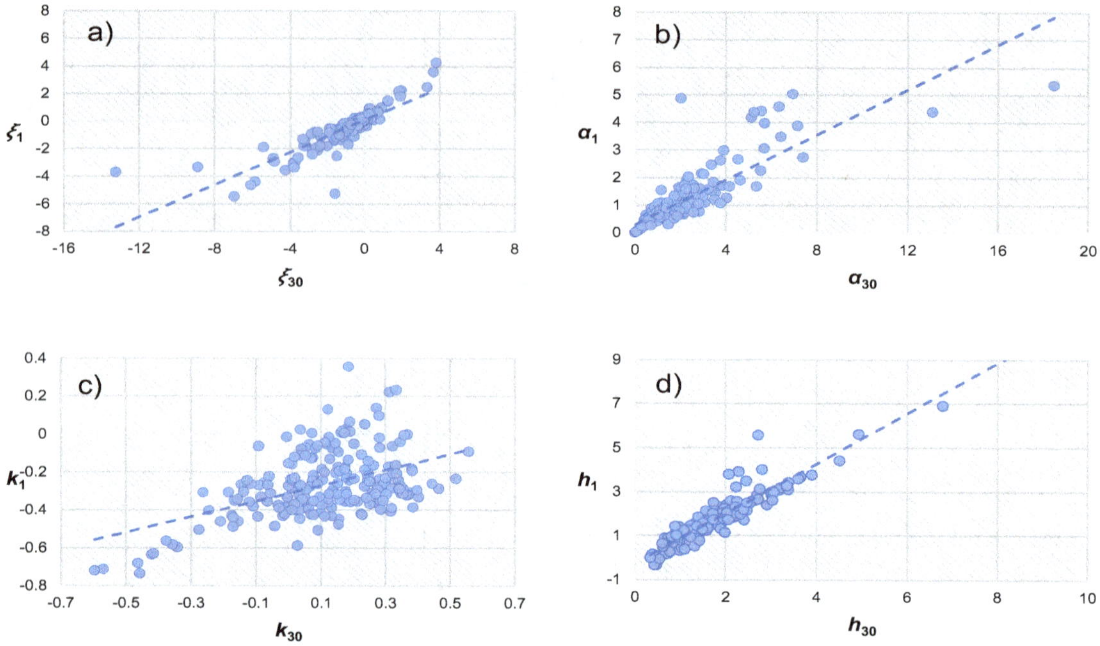

Figure 4. Scatter plots of four parameters at time scales 1 day and 30 day; (**a**) parameter ξ; (**b**); parameter α; (**c**) paramter k; (**d**) parameter h.

4.3. Relationships of FDCs Derived via E-FDC-R

Here we also focus research interest on the correlation between the FDCs at 1 day and 30 day. The flow values corresponding to 1 day and 30 day are sampled from the exceedance probabilities 0.5% to 99.5% with 0.5% as the sample interval. The total number of the sampled value is 199, which is large enough to analyze the quantitative relationship of FDCs at two time scales.

The ratios of FDC flow values at time scale 1 day to 30 day corresponding to sampled exceedance probabilities are shown in Figure 5. It is evident from Figure 5a that the ratio lines have distinct decreasing trend, these lines firstly decrease dramatically within 20% exceedance probability and slightly decrease thereafter. This is easy to understand that in flood season, maximum daily flow is certainly much higher than the monthly average, so the ratios at low exceedance probabilities are very high. While in non-flood season, minimum daily flow is also far lower than the monthly average, so the ratios at low exceedance probabilities are very low. Consequently, the ratio lines clearly show that daily FDC has higher flow values for low exceedance probability and lower flow values for high exceedance probability compared to monthly FDC, which is also consistent with Figure 2. It should also be noted that most of the curves reach 1 at less than 10% exceedance probability, consistent with [33], in which daily and monthly FDC curves normally cross (representing the ratio equals to 1) between the 1% and 10% exceedance probability in South Africa.

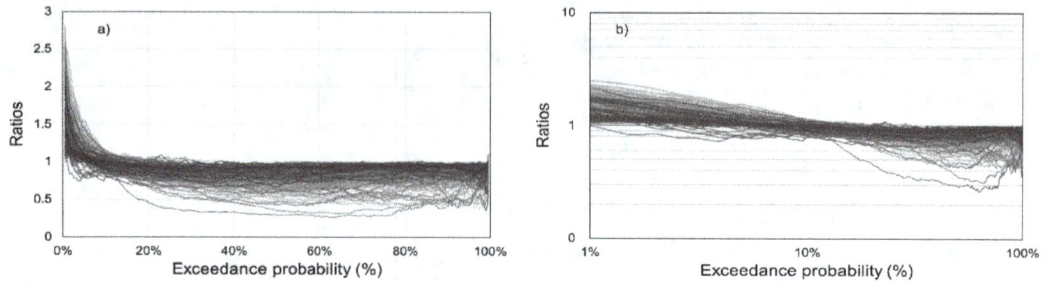

Figure 5. Ratios of daily to monthly FDCs at given exceedance probabilities; (**a**) conventional coordinates; (**b**) double logarithmic coordinates.

Seen from Figure 5a, the ratio relationship appears to be a power function, we apply the log-log transformation to the ratio relationship. It is evident from Figure 5b that the relationship between ratio value and exceedance probability is overall near-linear in the log-log plot, except a very small proportion in the very high exceedance probability. Here, the power function is used to fit the ratio curves as follows:

$$y = ax^{-b} \tag{4}$$

where x is the exceedance probability; y is the ratio value.

For each catchment, the parameter a and b are estimated and shown in Figure 6. The parameter a for all the catchments is positive and less than 1. This is due to the fact that the ratio is less than 1 for medium and has high exceedance probability. Despite for only two catchments, the parameter b for the rest of the catchments is negative; this is simply because the ratio curve is decreasing overall.

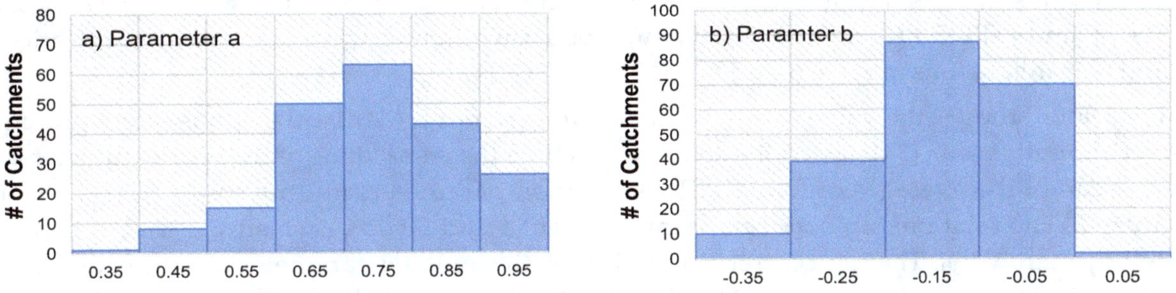

(a) **(b)**

Figure 6. Histograms of parameters (**a**,**b**) in the power function.

By applying the power function to fit the ratio of daily to monthly FDCs, the daily flow at given exceedance probability can be estimated based on the monthly FDC. The simulated daily FDC for a randomly selected catchment is presented in Figure 7. It is evident that an obvious improvement has been made for simulated daily FDC as it is much closer to daily FDC. Especially for the exceedance probability ranging from 10~70%, the simulated daily FDC and the observed daily FDC almost overlap.

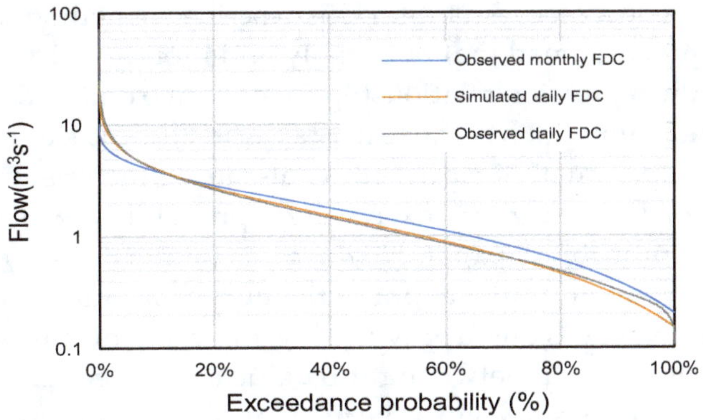

Figure 7. Simulated daily FDC based on monthly FDC.

After graphical comparison with observed daily FDC, the simulated daily FDC is modelled by KAP distribution and then evaluated with observed daily flow data. Figure 8 presents the NSE of modelled daily FDC. It is obvious that most of the catchments have very high NSE with median higher than 0.96. The results confirm again that power function is an appropriate relationship for the ratio of daily to monthly FDCs. After power transformation, the modelled daily FDC with good accuracy can be obtained based on monthly FDC.

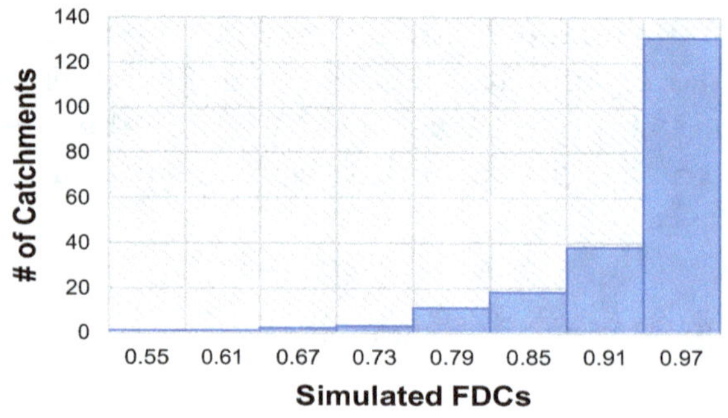

Figure 8. NSE of simulated daily FDCs based on monthly FDC.

5. Discussion

The present study aims to ascertain the relationships between FDCs at different time scales, based on M-FDC-P and R-FDC-P. Note that the two methods use different ways to examine the relationships, and some research results are discussed here.

In previous studies, the value of every single parameter of KAP distribution is rarely presented or discussed, though Parida [34] have done some research on the restrictions of the four parameters. In the present research, the averages of four KAP distribution parameters at time scales from 1 to 30 days are analyzed. The relationship between parameter h at 1 and 30 day is so strong that h_1 can be well estimated by merely h_{30}. The high correlation or insensitive of the h parameter is to be expected, as the generalized Pareto, a special case of KAP when $h = 1$, has been used to describe daily FDCs in many studies [16,18,25]. Even Blum et al. [9] illustrated that there are some regions in the United States such as New England, the Appalachian, and Valley and Ridge regions, for which the generalized Pareto distribution can provide a very good fit to daily FDCs. However, the relationship of the other three parameters are not following a very obvious linear trend or even a single trend (for parameter k). As stated before, even modeling daily FDC in daily gauged sites is very difficult, we are not expecting a single relationship can be found to estimate daily FDC parameters by using only monthly FDC parameters. Therefore, this paper proves that FDC parameters at different time scales are correlated and the relationships found in the paper can be used as an important supplement to the regional regression models to estimate daily FDC at ungauged or partial gauged sites.

As far as the E-FDC-R is concerned, Smakhtin [35] described a method to estimate daily FDCs from monthly flow data, which first assumed that both daily and monthly FDCs are linear in a log-normal space and then established the regression relationship for only two daily FDC indices based on monthly flow data. This method is very ambitious because it established the indices for 200 gauged hydrological stations without any specific parameters representing at-site characteristics. Sugiyama et al. [36] also stated that a distribution of FDC at given exceedance probability was usually represented with a straight line on log-normal probability paper. However, as for the ratio curves of flow values at given exceedance probability, they are very like exponential curves in our work, thus we use the power function to simulate the ratio curves and apply the log-log transformation to the ratio relationship. The results show that the relationship between ratio value and exceedance probability almost tends to be near-linear in the log-log axis, except a very small proportion in the very high exceedance probability. Future studies will consider further improving power function by using a piecewise function to describe the ratio curves for some kind of region (e.g., the stations with positive parameter b after figuring out a way to regionalize parameters in the power function).

6. Conclusions

This study provides a comprehensive assessment of relationships between FDCs at different time scales based on two methods including M-FDC-P and E-FDC-R. Selecting the MOPEX as the research data, the correlations of daily FDC and monthly FDC are deeply analyzed.

As for the M-FDC-P method, the Kappa (KAP) distribution is selected in this work to represent the streamflow distribution characteristics, due to the already existed existing research. The averages of all four KAP parameters tend to change gradually with time scales from 1 to 30 day. Meanwhile, three out of four parameters have linear correlations except the shape parameter k. The relationships found in the paper can be used as an important supplement to the regional regression models to estimate daily FDC at ungauged sites, when observed records with larger time scales are available.

As for the E-FDC-R method, it tries to quantify the relationship between daily and monthly FDCs, based on the ratio of FDCs rather than parameters of a specific distribution. Power function seems to be an effective functions for depicting the relationship between daily and monthly FDCs. The simulated daily FDC derived from monthly FDC can be very consistent with the observed daily flow records, when the two parameters in power function are quantified precisely. There are only two parameters in the power function that need to be regionalized by the site climate and catchment characteristics, which can reduce the uncertainties compared to other regional FDC methods that involve more parameters. Further, the parameter a in power function mainly controls the ratio for high exceedance probability or low flow data; and the parameter b mainly controls the shape of the ratio curve for low exceedance probability or high flow data.

Generally, results based on the above two methods clearly prove that there are strong connections between daily and monthly FDCs, which is not sufficiently realized in previous studies. If a specific probability distribution is applied for modelling FDCs at larger time scales, the gradual change characteristics of the distribution parameters for different FDCs can be considered to provide valuable information for daily FDC estimation. If there is no enough information to ascertain the probability distribution, the simple ratio relationships between different FDCs can help to assess the smaller time scale FDCs. Therefore, the inherent relationship between different FDCs will be very helpful for flow regulation and water resources management, especially in ungauged or partially gauged basins. Nonetheless, our future research will consider the regionalization of the distribution parameters to provide a complete daily FDC estimation using monthly flow data and other information collectively.

Author Contributions: L.Y. designed the study and wrote the manuscript; W.D. assisted with the results and discussion section; X.Z. assisted with the discussions and conclusions section; Z.X. and C.Z. provided feedback on the structure of the manuscript and reviewed the manuscript; J.W. performed the experiments and analyzed the data.

Acknowledgments: Special thanks are given to the two anonymous reviewers for their constructive comments that helped us to greatly improve the manuscript.

References

1. Vogel, R.M.; Fennessey, N.M. L moment diagrams should replace product moment diagrams. *Water Resour. Res.* **1993**, *29*, 1745–1752. [CrossRef]
2. LeBoutillier, D.W.; Waylen, P.R. A stochastic model of flow duration curves. *Water Resour. Res.* **1993**, *29*, 3535–3541. [CrossRef]
3. Cheng, L.; Yaeger, M.; Viglione, A.; Coopersmith, E. Exploring the physical controls of regional patterns of flow duration curves—Part 1: Insights from statistical analyses. *Hydrol. Earth Syst. Sci.* **2012**, *9*, 4435–4446. [CrossRef]
4. Zhang, Q.; Xu, C.Y.; Chen, Y.D.; Yu, Z. Multifractal detrended fluctuation analysis of streamflow series of the Yangtze River basin, China. *Hydrol. Process.* **2008**, *22*, 4997–5003. [CrossRef]
5. Sivakumar, B. Chaos theory in geophysics: Past, present and future. *Chaos Soliton Fract.* **2004**, *19*, 441–462. [CrossRef]
6. Wang, W.; Vrijling, J.K.; van Gelder, P.H.A.J.M.; Ma, J. Testing for nonlinearity of streamflow processes at different timescales. *J. Hydrol.* **2006**, *322*, 247–268. [CrossRef]

7. Tung, W.W.; Gao, J.; Hu, J.; Yang, L. Detecting chaos in heavy-noise environments. *Phys. Rev. E* **2011**, *83*, 046210. [CrossRef] [PubMed]
8. Bowers, M.C.; Tung, W.W.; Gao, J.B. On the distributions of seasonal river flows: Lognormal or power law? *Water Resour. Res.* **2012**, *48*, 5536. [CrossRef]
9. Blum, A.G.; Archfield, S.A.; Vogel, R.M. On the probability distribution of daily streamflow in the United States. *Hydrol. Earth Syst. Sci.* **2017**, *21*, 3093–3103. [CrossRef]
10. Castellarin, A.; Botter, G.; Hughes, D.A.; Liu, S.; Ouarda, T.B.M.J.; Parajka, J.; Post, D.A.; Sivapalan, M.; Spence, C.; Viglione, A.; et al. Prediction of flow duration curves ungauged basins. In *Runoff Prediction in Ungauged Basins: Synthesis Across Processes, Places and Scales*; Blöschl, G., Sivapalan, M., Wagener, T., Viglione, A., Savenije, H., Eds.; Cambridge University Press: Cambridge, UK, 2013; p. 465.
11. Hsu, N.S.; Huang, C.J. Estimation of Flow Duration Curve at Ungauged Locations in Taiwan. *J. Hydrol. Eng.* **2017**, *22*, 05017009. [CrossRef]
12. Chiang, S.M. Hydrologic Regionalization of Watersheds. II: Applications. *J. Water Res. Plan. Manag.* **2002**, *128*, 12–20. [CrossRef]
13. Mohamoud, Y. Prediction of daily flow duration curves and streamflow for ungauged catchments using regional flow duration curves. *Hydrol. Sci. J.* **2008**, *53*, 706–724. [CrossRef]
14. Hope, A.; Bart, R. Synthetic monthly flow duration curves for the Cape Floristic Region, South Africa. *Water SA* **2012**, *38*, 191–200. [CrossRef]
15. Shu, C.; Ouarda, T.B.M.J. Improved methods for daily streamflow estimates at ungauged sites. *Water Resour. Res.* **2012**, *48*, 2523. [CrossRef]
16. Castellarin, A.; Galeati, G.; Brandimarte, L.; Montanari, A.; Brath, A. Regional flow-duration curves: Reliability for ungauged basins. *Adv. Water Resour.* **2004**, *27*, 953–965. [CrossRef]
17. Li, M.; Shao, Q.; Zhang, L.; Chiew, F.H.S. A new regionalization approach and its application to predict flow duration curve in ungauged basins. *J. Hydrol.* **2010**, *389*, 137–145. [CrossRef]
18. Mendicino, G.; Senatore, A. Evaluation of parametric and statistical approaches for the regionalization of flow duration curves in intermittent regimes. *J. Hydrol.* **2013**, *480*, 19–32. [CrossRef]
19. Yu, P.S.; Yang, T.C.; Liu, C.W. A regional model of low flow for southern Taiwan. *Hydrol. Process.* **2010**, *16*, 2017–2034. [CrossRef]
20. Pugliese, A.; Farmer, W.H.; Castellarin, A.; Archfield, S.A.; Vogel, R.M. Regional flow duration curves: Geostatistical techniques versus multivariate regression. *Adv. Water Resour.* **2016**, *96*, 11–22. [CrossRef]
21. Chokmani, K.; Ouarda, T.B.M.J. Physiographical space-based kriging for regional flood frequency estimation at ungauged sites. *Water Resour. Res.* **2004**, *40*, 1–13. [CrossRef]
22. Skoien, J.O.; Merz, R.; Bloschl, G. Top-kriging-geostatistics on stream networks. *Hydrol. Earth Syst. Sci.* **2006**, *2*, 2253–2286. [CrossRef]
23. Castiglioni, S.; Castellarin, A.; Montanari, A. Prediction of low-flow indices in ungauged basins through physiographical space-based interpolation. *J. Hydrol.* **2009**, *378*, 272–280. [CrossRef]
24. Archfield, S.A.; Pugliese, A.; Castellarin, A.; Skøien, J.O. Topological and canonical kriging for design-flood prediction in ungauged catchments: An improvement over a traditional regional regression approach? *Hydrol. Earth Syst. Sci.* **2013**, *9*, 1575–1588. [CrossRef]
25. Castellarin, A.; Camorani, G.; Brath, A. Predicting annual and long-term flow-duration curves in ungauged basins. *Adv. Water Resour.* **2007**, *30*, 937–953. [CrossRef]
26. Hosking, J.R.M.; Wallis, J.R. *Regional Frequency Analysis: An Approach Based on L-Moments*; Cambridge University Press: Cambridge, UK, 1997; p. 244.
27. Hosking, J.R.M. L-moments: Analysis and estimation of distributions using linear combinations of order statistics. *J. R. Stat. Soc. B Methodol.* **1990**, *52*, 105–124.
28. Kjeldsen, T.R.; Ahn, H.; Prosdocimi, I. On the use of a four-parameter kappa distribution in regional frequency analysis. *Hydrol. Sci. J.* **2017**, *62*, 1–10. [CrossRef]
29. Duan, Q.; Schaakeb, J.; Andréassian, V.; Franks, S.; Gotetie, G.; Gupta, H.V.; Gusev, Y.M.; Habet, F.; Hall, A.; Hay, L.; et al. Model Parameter Estimation Experiment (MOPEX): An overview of science strategy and major results from the second and third workshops. *J. Hydrol.* **2006**, *320*, 3–17.
30. Slack, J.R.; Lumb, A.M.; Landwehr, J.M. Hydro-climatic data network (HCDN): A U.S. Geological Survey streamflow data set for the United States for the study of climate variation, 1874–1988. *J. Phys. Chem. C* **1992**, *113*, 2538–2544.

31. Wang, D.; Hejazi, M. Quantifying the relative contribution of the climate and direct human impacts on mean annual streamflow in the contiguous United States. *Water Resour. Res.* **2011**, *47*, 411. [CrossRef]
32. Kroll, C.N.; Croteau, K.E.; Vogel, R.M. Hypothesis tests for hydrologic alteration. *J. Hydrol.* **2015**, *530*, 117–126. [CrossRef]
33. Smakhtin, V.Y.; Watkins, D.A. *Low Flow Estimation in South Africa*; Report No 494/1/97; Water Research Commission: Pretoria, South Africa, 1997.
34. Parida, B.P. Modelling of Indian Summer Monsoon Rainfall Using a Four-parameter Kappa Distribution. *Int. J. Climatol.* **1999**, *19*, 1389–1398. [CrossRef]
35. Smakhtin, V.U. Estimating daily flow duration curves from monthly streamflow data. *Water SA* **2000**, *26*, 13–18.
36. Sugiyama, H.; Vudhivanich, V.; Whitaker, A.; Lorsirirat, K. Stochastic Flow Duration Curves for Evaluation of Flow Regimes in Rivers. *J. Am. Water Resour. Assoc.* **2007**, *39*, 47–58. [CrossRef]

4

Optimal Operation of Hydropower Reservoirs under Climate Change: The Case of Tekeze Reservoir, Eastern Nile

Fikru Fentaw Abera [1,2,*], Dereje Hailu Asfaw [1], Agizew Nigussie Engida [1] and Assefa M. Melesse [2]

1 School of Civil and Environmental Engineering, Addis Ababa Institute of Technology (AAiT), Addis Ababa 1000, Ethiopia; dereje_hasfaw@yahoo.com (D.H.A.); agiz70@yahoo.com (A.N.E.)
2 Department of Earth and Environment, Florida International University, Miami, FL 33199, USA; melessea@fiu.edu
* Correspondence: ffentawa@fiu.edu

Abstract: Optimal operation of reservoirs is very essential for water resource planning and management, but it is very challenging and complicated when dealing with climate change impacts. The objective of this paper was to assess existing and future hydropower operation at the Tekeze reservoir in the face of climate change. In this study, a calibrated and validated Soil and Water Assessment Tool (SWAT) was used to model runoff inflow into the Tekeze hydropower reservoir under present and future climate scenarios. Inflow to the reservoir was simulated using hydro-climatic data from an ensemble of downscaled climate data based on the Coordinated Regional climate Downscaling Experiment over African domain (CORDEX-Africa) with Coupled Intercomparison Project Phase 5 (CMIP5) simulations under Representative Concentration Pathway (RCP)4.5 and RCP8.5 climate scenarios. Observed and projected inflows to Tekeze hydropower reservoir were used as input to the US Army Corps of Engineer's Reservoir Evaluation System Perspective Reservoir Model (HEC-ResPRM), a reservoir operation model, to optimize hydropower reservoir release, storage and pool level. Results indicated that climate change has a clear impact on reservoir inflow and showed increase in annual and monthly inflow into the reservoir except in dry months from May to June under RCP4.5 and RCP8.5 climate scenarios. HEC-ResPRM optimal operation results showed an increase in Tekeze reservoir power storage potential up to 25% and 30% under RCP4.5 and RCP8.5 climate scenarios, respectively. This implies that Tekeze hydropower production will be affected by climate change. This analysis can be used by water resources planners and mangers to develop reservoir operation techniques considering climate change impact to increase power production.

Keywords: reservoir operation; optimization; SWAT; HEC-ResPRM; climate change; CORDEX-Africa; Tekeze basin

---

1. Introduction

Water resources reservoirs are important tools for integrated water resources development and management [1,2], but nowadays their operation and management is challenging due to various factors [3,4]. The reservoir operates to supply water for municipal consumption, hydropower production, irrigation and industrial needs, flood control, recreation, navigation or ecological requirements. Currently, due to water crisis the global freshwater supply to meet the needs of the different sectors is falling short [5–7]. Factors that contribute to this include population growth, urbanization, climate change, land use change, land degradation and poor water resources management [8,9]. Hence, to alleviate these problems and meet the freshwater and energy demand of communities, it will necessitate optimal operation of water resources reservoirs [10,11].

Various researchers studied the reservoir operation throughout the world to get optimum level of release and optimal volumes of storage considering inflows and needs [12–15]. Most research conducted in the reservoir operations have specific objectives like hydropower [12,14,15], flood control [13], irrigation [3] and environmental [16].

Water resources infrastructures have been designed and managed historically but these designs gave little attention to the effect of climate change and non-stationarity in hydrologic variables [17]. Evidence suggests that these hydrologic variables used for water resources planning and management previously assumed stationarity in time have changed by anthropogenic activities [18]. The increase in temperature, changes in precipitation and evapotranspiration rates due to climate change alters global hydrologic cycle [19]. The streamflow affected by intensity and frequency of precipitation leads to increase the intensity of floods and droughts. These changes affect water resources at local and regional levels [20]. The hydrological processes and water availability affected by a change in the patterns of precipitation and temperature impacts agriculture, industry, communities, hydropower and aquatic life [21]. Climate change impact on fresh water resources may change the mean annual streamflow, shift seasonal flows, increases floods and droughts and changes in sediment fluxes which affects reservoir operation [22–24].

Many researchers in different parts of the world have studied the impacts of climate variability and change on shifts in hydrological regimes and water resources (e.g., [25–28]). These studies assessed the current and future water resources availability and rainfall variability across the globe to support appropriate water resources planning and management. Different studies showed that Africa is highly vulnerable to climate change [29]. Climate change studies showed temperature increased and precipitation pattern changed throughout arid and semi-arid regions of Africa [30–32] and affected the hydrological processes that impacts reservoir operation. Most studies showed the impacts of climate change on African hydropower reservoirs [23,33–36]. Kim and Kaluarachchi [33] and Beyene et al. [34] projected that precipitation and temperature will be increased in the Nile River basin and have positive effect on hydropower production, but Yamba et al. [35] and Hamududu and Killingtveit [36] investigated that in the next 60 years hydropower production show a gradual reduction with large variability in the Zambezi River basin. In the Nile River basin, the rapidly growing hydropower based energy need, population growth, food insecurity and finite water resources will lead to competitions for water in the riparian countries and this will be aggravated by the climate change. Several studies have been conducted on the variability of precipitation and streamflow in the Nile River basin [37–41] that affects reservoir planning and management [3].

Most hydropower reservoir operators concern is existing hydrological variability without foreseeing climate change as a particular serious threat [23]. Hence reservoir operation need to incorporate plans to address hydrologic non-stationarity and uncertainty caused by climate change [20,26,42,43]. Due to this, ensembles of Global Circulation Models (GCMs), scenarios and regional climate models (RCMs) used as input to hydrological model to generate future streamflow [44,45] that can be used as an input for reservoir operations.

Sedimentation may cause serious impacts on reservoir operation and functionality by reducing reservoir storage capacity and shortening reservoir useful life for human benefits. Studies showed that Northern part of the Tekeze basin watersheds are vulnerable to sedimentation and/or soil erosion problems for the sustainable use of small reservoirs developed for irrigated agriculture and Tekeze reservoir [46,47]. This reservoir sedimentation problem may lead serious reduction in reservoir storage capacity, causing future hydropower generation problems. However, rate of sedimentation of Tekeze reservoir still remains unpredicted. More and wide knowledge is still needed to better understand and solve the sediment problem, and hence may improve future reservoir operation. But the focus of this research is to study potential climate change impact on hydropower reservoir operation and management by not varying sedimentation level.

Nowadays, reservoir operation techniques become increasingly important and researchers still searching the best technique. Many authors proposed and reviewed various reservoir operation models and methods [4,48–51]. Labadie [50] extensively reviewed and evaluated various optimization methods and reported that no universally approved algorithm for all reservoir operations. Rani and Moreira [4] investigated that optimization models usually require simulation models for verifying and testing planned operating policies. Dam managers use simulation models more relaxed than optimization models as simulation models are easier to interpret, apply and present to non-professionals [49,50]. But Optimization models give reliable results. In recent years, to overcome these problems, a combination of simulation and optimization models applied in reservoir operation. In this research, US Army Corps of Engineer's Reservoir Evaluation System Perspective Reservoir Models (HEC-ResPRM), a combination of simulation and optimization model is used.

In this study, Tekeze hydropower reservoir was chosen due to: (1) Tekeze basin shows high rainfall variability [37,52] which affects reservoir inflow; (2) Tekeze hydropower reservoir not designed by considering hydrological non-stationarity and climate change; (3) the reservoir has not been optimally operated and sometimes not fully functional during dry periods. Therefore, the objective of this research are to (1) assess impact of climate change on reservoir inflow using Soil and Water Assessment Tool (SWAT) and recent Coordinated Regional climate Downscaling Experiment over African domain (CORDEX-Africa) RCMs under Representative Concentration Pathway (RCP)4.5 and RCP8.5 climate scenarios, and (2) apply HEC-ResPRM optimization model to get optimal release, reservoir level and storage for optimal power production including in the face of climate change.

2. Material and Methods

2.1. Study Area

Tekeze basin, a tributary of Tekeze-Setit-Atbera river part of Eastern Nile (Figure 1) is geographically located from 11°40′ to 15°12′ N and 36°30′ to 39°50′ E. The surface area of the Tekeze reservoir watershed is 29,404 km^2. This basin has high mountainous areas in its sources in the central Ethiopian highlands up to 4517 m and low land areas near Ethio-Sudan border as low as 800 m with varying climate depending on altitude change. The rainfall increases with altitude from 600 mm to 1200 mm but it is a reverse for temperature which decreases from 26 °C to 10 °C. This basin has a mean annual inflow of 4.4 Billion cubic meters at Embamadre gauging station and annual potential evapotranspiration of 1778 mm.

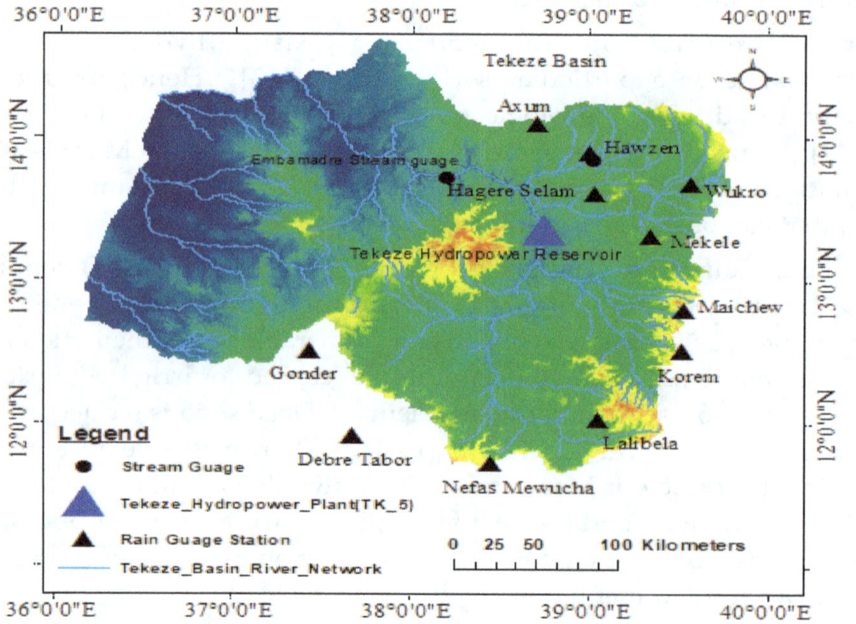

Figure 1. Location of Tekeze hydropower reservoir and weather stations.

Tekeze basin has a large elevation drop from its sources to low land areas near Ethio-Sudan border and offers a significant hydropower potentials of 5960 GWh/year. Tekeze single purpose hydropower reservoir located at 13°21′ N and 38°45′ E (Figure 1) is the second tallest double concrete arch dam in Africa next to Katse arch dam in Lesotho. The purpose of this reservoir is for hydropower production with total installed capacity of 300 MW in four 75 MW Francis turbines at underground power house. The reservoir has a total storage capacity of 9293 million cubic meters (MCM) of which 5293 MCM live storage at 1140 m above sea level (masl) and 4000 MCM below dead storage level (1096 masl). The reservoir also has 147 km^2 surface areas at full supply level with mean annual inflow of 3750 MCM.

2.2. Datasets Used

2.2.1. Historical and Future Hydrology

In this research, four hydrological data periods were analyzed. These were the reservoir inflow data of: (1) observed and RCP scenarios historical records (1994–2008); (2) the near future period (2011–2040), middle future period (2041–2070) and the far future periods (2071–2100). SWAT simulates historical (past) and all future reservoir inflows using precipitation and temperature projections from ensemble outputs of CORDEX-Africa RCMs downscaled from different GCMs from Coupled Intercomparison Project Phase 5 (CMIP5) simulations available in 0.44° resolution for Ethiopian domain under two recent representative concentration pathways (RCP4.5 and RCP8.5) climate scenarios. There are numerous weather stations in Tekeze basin. For this study, stations recording precipitation and temperature data that have long period of records with small data gaps were used. There are more than 20 streamflow gauged stations in the Tekeze basin but most of the stations are found in the small tributaries of Tekeze River which covers only small watershed areas. These stations, except Embamadre station, have large data gaps, short record periods and high amount of missing data. Observed streamflow at Embamadre station was collected from Ethiopian Ministry of Water, Irrigation and Electricity.

2.2.2. CORDEX-Africa

Currently, CORDEX-Africa initiated by World Climate Research Program (WCRP) provides an opportunity for the generation of high resolution regional climate projections over Africa that is used to assess future impacts of climate change at regional and local scales. In this study, results of CORDEX-Africa ensemble RCMs simulations for the past (1951–2005) and future (2006–2100) climate projections downscaled from different GCMs under RCP4.5 and RCP8.5 with spatial resolution of 0.44° is used. CORDEX-Africa RCMs generate an ensemble of high resolution historical and future climate projections at regional scale by downscaling different GCMs forced by RCPs based on the Coupled Intercomparison Project Phase 5 (CMIP5) [32,53]. CORDEX-Africa climate projections use RCP4.5 and RCP8.5 climate scenarios.

RCPs are new climate change scenarios established by CMIP5 [54,55], which can depict a wide variety of possible future climate scenarios. The fifth Assessment Report (AR5) scientific literature selects one mitigation scenario (RCP2.6), two medium stabilization scenarios (RCP4.5 and RCP6.0) and one high emission scenario. RCP2.6 scenario sees emissions peak early, then fall shown to be technically feasible. But one of RCP2.6 scenario key assumptions is the full participation of all developed and developing countries in the world in the short run to reduce all the main emitters, which is not possible in actual cases. Due to this, we decided to choose one medium scenario (RCP4.5) and high scenario (RCP8.5) covering entire range of radiative forcing. RCPs represent pathways of radiative forcing, not linked with exclusive socio-economic assumption in contrary to Special Report on Emission Scenarios (SRES). Any single radiative forcing pathway can result from a diverse range of socio-economic and technological development scenarios. RCP4.5 is a mid-range scenario that stabilizes radiative forcing at 4.5 W/m^2 (approximately 650 ppm CO_2-equivalent) in the year 2100 without exceeding this value, but this does not imply the climate system are stable [53,56]. Whereas RCP8.5 is upper bound of all RCP scenarios that stabilizes radiative forcing at 8.5 W/m^2 (greater than 1370 ppm CO_2-equivalent) in the year 2100 [53,57].

2.2.3. Reservoir Data

HEC-ResPRM, a reservoir operation model, requires data like back ground map of the watershed, reservoir outlet capacities, elevation-area-storage curve, current outflow-energy relationship, power production and flow time series to perform optimal operations. These physical data were used to develop model constraints and allow the model to calculate penalties. All these data were collected from Ethiopian Electric Power Corporation and Ministry of Water, Irrigation and Electricity.

2.3. Methods

2.3.1. Overview of SWAT

Soil and Water Assessment Tool (SWAT) was used to produce inflow projections and assess climate change impact on the streamflow used as an input for reservoir operation. The details of SWAT shown in Neitsch et al. [58]. It is a semi-distributed continuous widely used hydrological model in the Eastern Nile basins [59,60]. Since the objective of the study was to examine streamflow response to climate change, the land phase of the hydrologic cycle simulated by SWAT is based on the water balance equation:

$$SW_t = SW_o + \sum_{i=1}^{t}\left(R_{day} - Q_{surf} - E_a - W_{seep} - Q_{gw}\right) \quad (1)$$

in which SW_t is the final soil water content (mm), SW_o is the initial soil water content on day i (mm), t is the time (days), R_{day} is the amount of precipitation on day i (mm), Q_{surf} is the amount of surface runoff on day i (mm), E_a is the amount of evapotranspiration on day i (mm), W_{seep} is the amount of water entering the vadose zone from the soil profile on day i (mm), and Q_{gw} is the amount of return flow on day i (mm). Surface runoff volume was estimated using modified Soil Conservation Serves-Curve Number (SCS-CN) method.

In this study, SWAT was used together with the ArcSWAT interface, a geographic information system (GIS) based graphical user interface used to facilitate watershed delineation and initial parameterization. The SWAT model requires digital elevation model (DEM), land cover/land use information, soils and basic climate data. The land use/land cover data of the Tekeze basin includes agricultural land, shrub land (range grasses), mixed forests and pasture/grazing lands. SWAT subdivides the watershed in to hydrological response units (HRUs) with a homogeneous land use and soil properties based on topography and quantifies the relative impacts of soil, land use and climate change within each HRU. The 30 m × 30 m DEM, soil properties, land use/land cover and streamflow data were all collected from Ethiopian Ministry of Water, Irrigation and Electricity. Meteorological data of precipitation and temperature (1994–2008) were collected from National Meteorological Service Agency. Sensitivity analysis was done using Latin Hypercube sampling based on One Factor at a Time (LH-OAT) inbuilt in SWAT to identify sensitive parameters that influence model simulations. The sensitive parameters of this study mainly affecting model calibration were curve number (CN-2), soil available water capacity (Sol-AWC), alpha base flow recession constant (ALPHA-BF), soil evaporation compensation factor (ESCO), threshold water depth required for return flow to occur (GWQMN) and saturated hydraulic conductivity (SOL_K). Model calibration adjusts such high sensitive parameters to optimize the agreement between observed and simulated streamflow values at Embamadre station in Tekeze basin. Model performance was assessed using Nash-Sutcliffe efficiency (E_{NS}), coefficient of determination (R^2) and percent of bias (PBIAS). Finally, the historical and future RCP4.5 and RCP8.5 projections of precipitation and temperature were used as input into the calibrated and validated SWAT model to assess the impact of climate change on reservoir inflow.

2.3.2. HEC-ResPRM Optimization Model

In this study, the US Army Corps of Engineer's Reservoir Evaluation System Perspective Reservoir Model (HEC-ResPRM), a reservoir system operations optimizations software package developed to assist planners, operators and managers with reservoir operation plan and decision making, was used. It addresses a reservoir system operation problem of optimal long-term allocation of available water. HEC-ResPRM is a combination of simulation and optimization model when Perspective Reservoir Model (PRM) is integrated in to HEC-RES modeling platform. It is an implementation of HEC-PRM shared with HEC-ResSim in a graphical user interface for creating, running, sorting and analyzing optimization runs. HEC-ResPRM uses HEC's data storage system (HEC-DSS) to store and retrieve of input and output time series data.

It is a monthly network flow programming model and gives optimal values of release and storage by minimizing penalty functions [61,62]. Network flow programming is computationally efficient form of linear programming. A network solver finds optimal flow for the entire network simultaneously based on the unit cost associated with flow along each arc. Optimization problem represented by the network with cost associated with flow as follows:

$$Minimize: \sum_{t}^{n} C_t Q_t \text{ (For all nodes)} \quad (2)$$

$$Subject\ to: \sum Q_t - \sum a_t Q_t = 0 \text{ (For all nodes)} \quad (3)$$

$$L_t \leq Q_t \leq U_t \text{ (For all arcs)} \quad (4)$$

in which n is total number of network arcs; C_t is unit cost, weighting factor for flow along arc t; Q_t is flow along arc t; a_t is multiplier (gain) for arc t; L_t is lower bound on flow along arc t; and U_t is upper bound on flow along arc t. In this case, node represents a reservoir and river or channel junctions. Arcs represent inflow and outflow links in the reservoir system. Each arc has a minimum and maximum flow that it must carry in the reservoir system. The arcs (inflow and outflow links) may transfer water between two points in space (transferring water in channels) or in time (changing pool elevations in the reservoir). Also, flow is conserved in the reservoir (node). Equations (2) through (4) are special forms of linear programming problems solved using primal simplex method.

3. Results and Discussions

3.1. Climate Projections

Projected annual temperature and precipitation showed an increasing trend in 2020s, 2050s and 2080s over Tekeze basin under RCP4.5 and RCP8.5 climate scenarios. Projected mean annual temperature may increase up to 1.1 °C and 3.38 °C under RCP4.5 and RCP8.5 scenarios, respectively in all future time periods. Similarly, mean annual precipitation may increase up to 45% under both scenarios for all future time periods. Figure 2 shows future change rates of temperature in both scenarios for all future time periods. Mean monthly temperature will increase under both scenarios in all time periods except the months of January and February which showed a slightly decreasing trend in 2020s. Figure 3 shows future percentage changes of monthly precipitation amounts for different projected periods under RCP4.5 and RCP8.5 climate change scenarios. For RCP8.5 scenarios, the months of March, April and May would exhibit a decrease in precipitation amount compared to the baseline period whereas RCP4.5 scenario presented an increasing trend. The months of October through February would show an increase in precipitation compared to reference period for bothscenarios and projected periods considered.

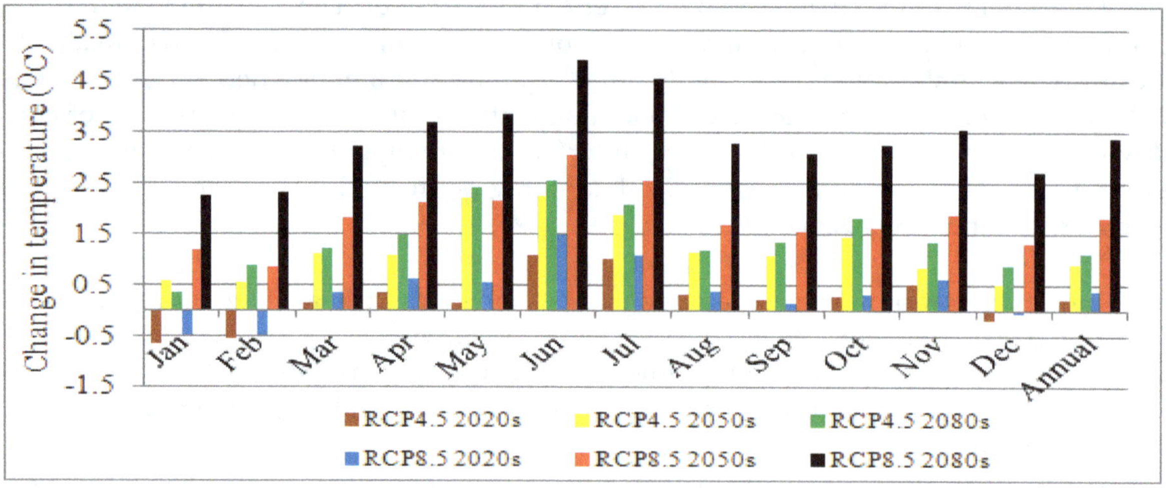

Figure 2. Rates of change of monthly mean temperature for the selected scenarios and projected periods.

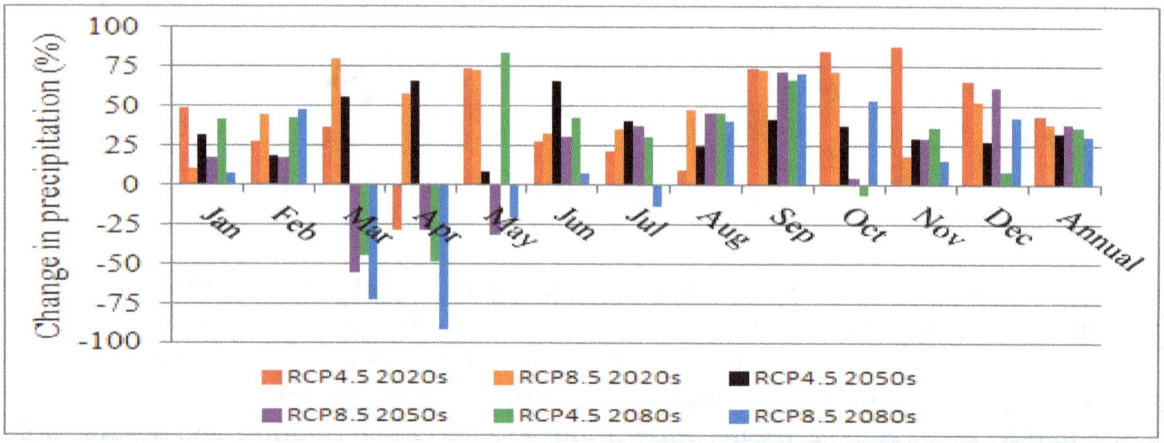

Figure 3. Change of monthly precipitation amount for the selected scenarios and projected periods.

3.2. Impacts of Climate Change on Streamflow

Impact of climate change on the streamflow at Embamadre station downstream of the reservoir was analyzed. Observed streamflow data from a period 1994–2002 was used for model calibration and from 2003–2008 was used for validation. Results in Figure 4 show that SWAT successfully simulated annual and monthly streamflow with a reasonable accuracy. The monthly result showed a good performance of SWAT, which indicated by the value of Nash-Sutcliffe efficiency (E_{NS}) 0.70 for calibration and 0.79 for validation and the coefficient of determination (R^2) 0.73 for calibration and 0.80 for validation. Similarly, percent of bias (PBIAS) value of 0.53% during calibration and 0.45% in validation periods showed good fit between observed and simulated streamflow.

Hence, the calibrated and validated SWAT forced to run for historical and future climate scenarios to generate future streamflow for both RCP4.5 and RCP8.5 climate scenarios. The effect of climate change on annual and monthly streamflow was also investigated as a percentage change with respect to the baseline period (1994–2008) under the two scenarios in three time periods 2020s (2011–2040), 2050s (2041–2070) and 2080s (2071–2100).

Figure 5 showed the percentage change of annual and monthly streamflow for both climate scenarios and the three time periods. Mean annual streamflow showed an increasing trend for both RCP4.5 and RCP8.5 for all time periods. Under RCP4.5, the mean annual percentage change of streamflow will increase by 49%, 39% and 47% in the 2020s, 2050s and 2080s, respectively. Similarly, for

RCP8.5, the mean annual percentage change of streamflow increases to 22%, 19% and 2% in the 2020s, 2050s and 2080s, respectively.

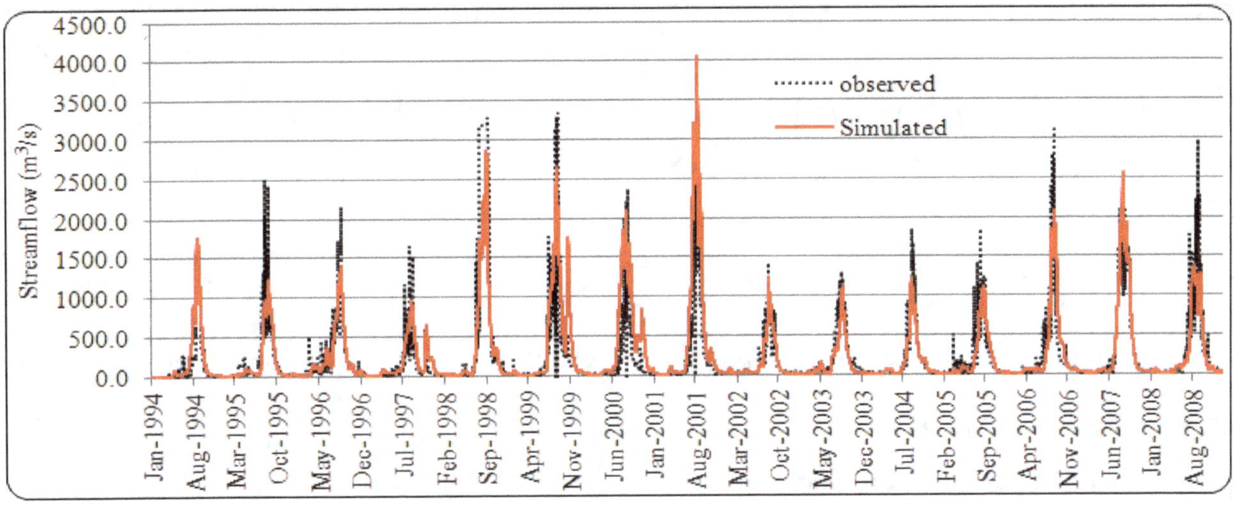

Figure 4. Observed and SWAT simulated monthly stream flow at Embamadre station.

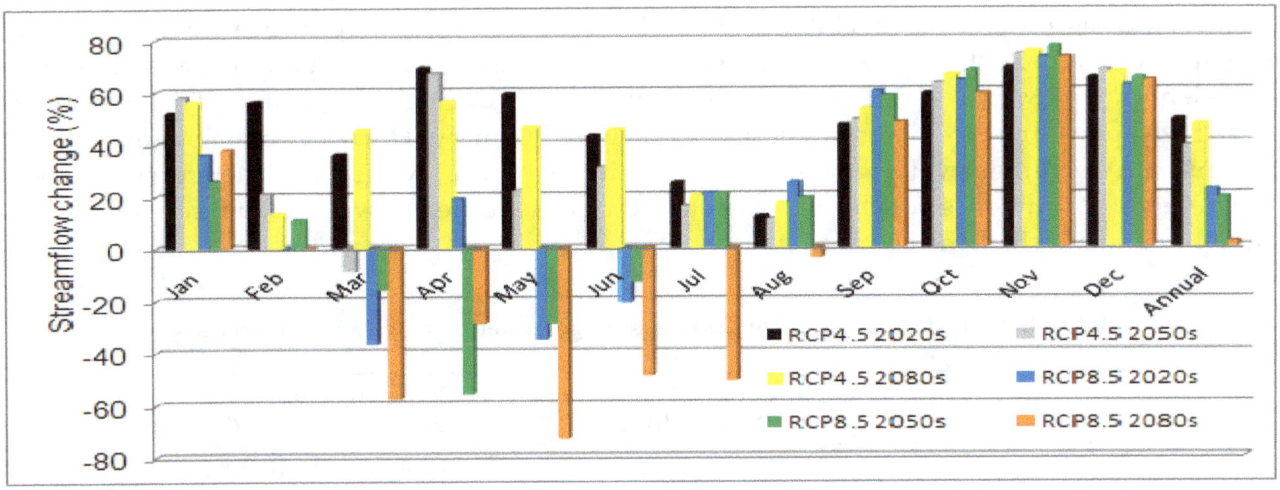

Figure 5. Mean monthly and annual streamflow changes under RCP4.5 and RCP8.5 climate scenarios.

The monthly streamflow change shows a mix of positive and negative trends. Under RCP4.5 climate scenario, change in average monthly streamflow ranges from 12 to 69% in 2020s and 13 to 67% in 2080s but in 2050s streamflow change shows mixed trend that decreases in March to May by up to 9% and increases on other months up to 39%. Mean monthly percentage change of streamflow under RCP8.5 climate scenario showed mixed trends in all time periods. Under RCP8.5, the mean monthly streamflow changes from −37 to 64%, −29 to 68% and −49 to 64% in 2020s, 2050s and 2080s, respectively. Individual month's trend showed that there was an increasing trend from August to February and a decreasing trend from March to July. Therefore, climate change will have a clear impact on the future streamflow an input of reservoir power production in Tekeze basin.

The changes and variability of monthly (inter-annual) streamflow will be much greater than the annual streamflow changes in both scenarios in all time periods. This result showed that it is important for the hydropower reservoir planners and managers to consider, the monthly streamflow variability and changes for future planning and operation of reservoirs.

The total mean annual historical (past) and future Tekeze hydropower reservoir inflow (m^3/s) trends under RCP4.5 and RCP8.5 climate scenarios for all time periods are shown in Figure 6.

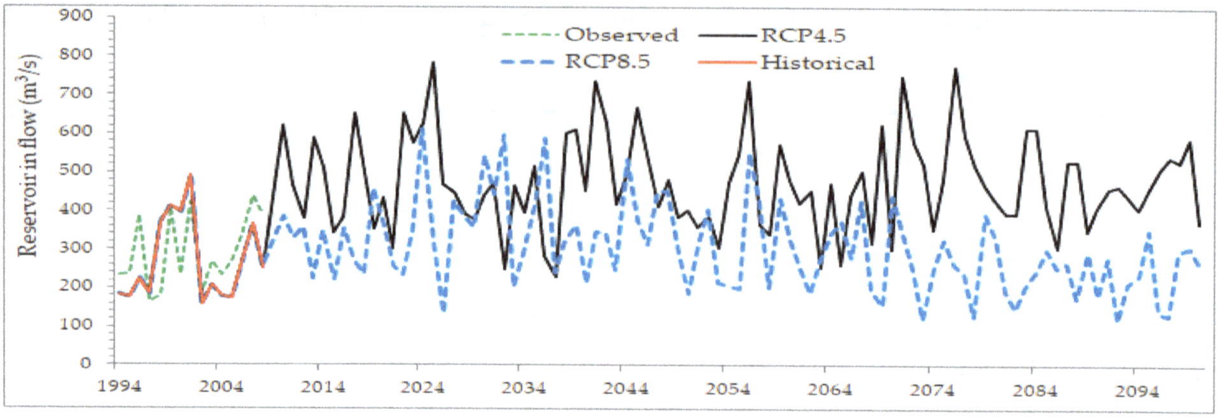

Figure 6. Annual Tekeze hydropower reservoir inflow trend for future time periods.

3.3. *Current Reservoir Operation*

HEC-ResPRM optimization model run under current baseline condition (2009–2017). This model optimized the current Tekeze hydropower reservoir operation. The current optimized value (Table 1) of HEC-ResPRM optimization model showed an increase in reservoir storage compared to current actual hydropower reservoir operation status. It is also indicated that the mean annual reservoir pool level increased up to 7.87 m (Table 2) that will store more water to produce power throughout the year. It contradicts the current actual Tekeze hydropower reservoir operation which produces insufficient power even very little or no power production during dry months. This implies that Tekeze hydropower reservoir was not optimally operated till now. The reservoir storage dropped to the minimum operating level and sometimes dries in the non-rainy months. Therefore, the current actual reservoir operation is not effective and should consider different well tested reservoir operation techniques under a changing climate.

Table 1. Mean annual optimized power storage under climate change scenarios.

Periods	Optimized Reservoir Storage (Mm^3)		Change in Optimized Reservoir Storage (%)	
	RCP4.5	RCP8.5	RCP4.5	RCP8.5
Current optimized	6639		24.0	
2020s	6688	6880	25.0	28.5
2050s	6669	6903	24.6	29.0
2080s	6665	6958	24.5	30.0

Table 2. Mean annual optimized pool level variation under RCP4.5 and RCP8.5 climate scenarios in three future time periods.

Time Periods	Optimized Pool Level (masl)		Pool Level Change (m)	
	RCP4.5	**RCP8.5**	**RCP4.5**	**RCP8.5**
Current optimized	1120.48		7.87	
2020s	1121.06	1123.27	8.45	10.66
2050s	1120.87	1123.37	8.26	10.76
2080s	1120.89	1123.85	8.28	11.24

3.4. *Reservoir Operation under Climate Change*

The future reservoir inflows generated by SWAT under RCP4.5 and RCP8.5 climate scenarios in three time periods 2020s (2011–2040), 2050s (2041–2070) and 2080s (2071–2100) with other reservoir data were used to run optimization model to get projected optimal reservoir outflow (release), storage and pool level results.

3.4.1. Projected Reservoir Inflow and Outflow

Climate change impacted inflow and outflow (release) hydrograph of Tekeze hydropower reservoir considered in this study are shown in Figure 7a,b. According to the inflow projections based on ensembles of CORDEX-Africa RCM climate model simulations, total inflows to Tekeze hydropower reservoir expected to increase under RCP4.5 and RCP8.5 climate scenarios for all future time periods. Figure 7a,b show that total monthly inflow under RCP4.5 is greater than the total monthly inflow projected under RCP8.5 climate scenarios. However, under RCP4.5 climate scenario, the reservoir inflow projections exhibit high fluctuations inter-annually as compared to RCP8.5 climate scenario and observed historical values. The highest inflow volumes under RCP4.5 were concentrated in the rainy months that spilled easily and affect the dry period reservoir storage level and or release.

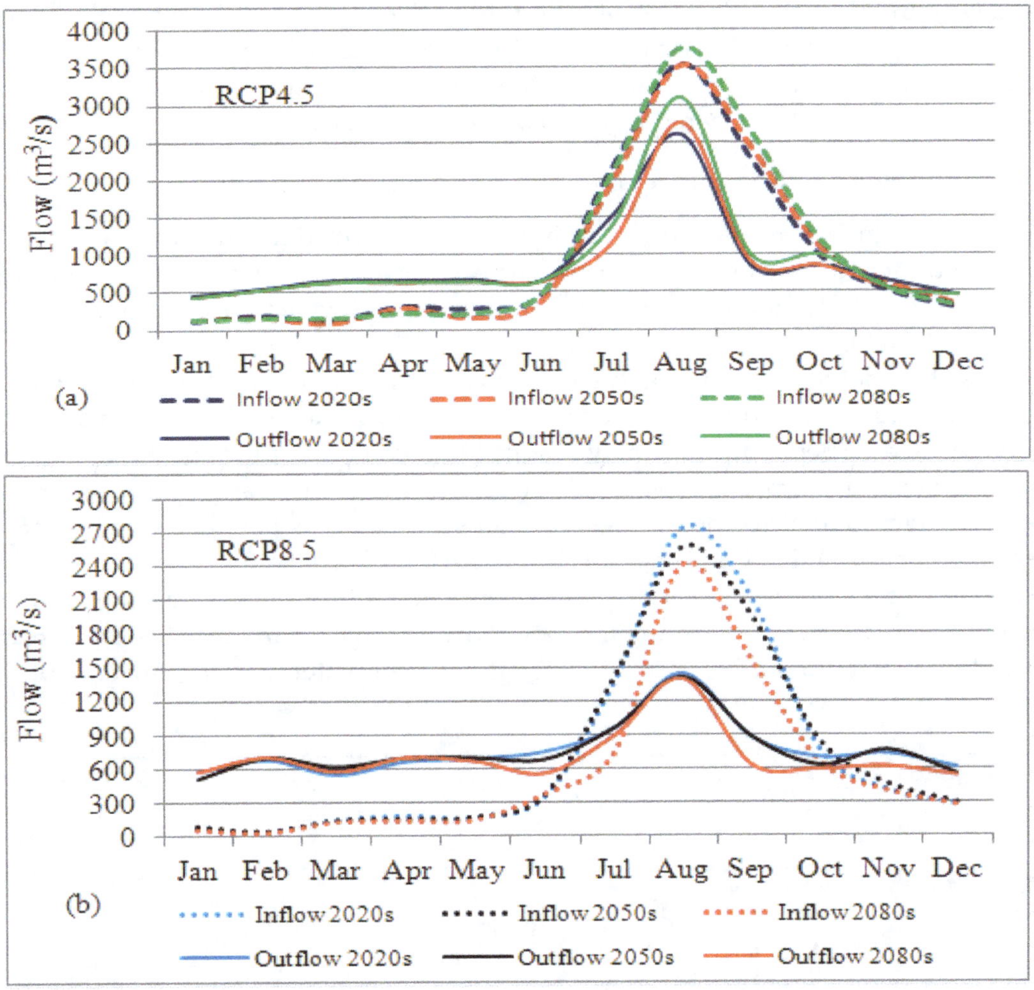

Figure 7. Mean monthly reservoir inflow and optimized outflow (release) for future time periods under: (**a**) RCP4.5 climate scenario; (**b**) RCP8.5 climate scenario.

There would be an increase in excess reservoir inflow during the rainy months of August through October under both RCP4.5 and RCP8.5 climate scenarios in all time periods. This increased spillage of available water inflow occurs because of the effect of climate change that increased the hydropower reservoir inflow under RCP4.5 and RCP8.5 future climate scenarios. According to the latest climate simulations, the overall inflow volume is predicted to be higher during rainy months and provided that the reservoir lack sufficient storage capacity to accommodate these high flows. As a result, Tekeze hydropower reservoir forced to spill water without generating hydropower. This indicates that the increased in overall reservoir inflow volume does not necessarily be advantageous to produce more power. Therefore, decision need to be taken on the amount of water to be released and or stored now and retained for future considering the variations in inflow and demands.

In this study, the reservoir outflow (release) was obtained by HEC-ResPRM optimization model under RCP4.5 and RCP8.5 climate scenarios for 2020s, 2050s and 2080s time periods. In all future time periods (Figure 7a,b) under the two climate scenarios, the reservoir release will be increased to produce more power due to an increased future reservoir inflow and optimum water storage using optimization model. Under RCP4.5 climate scenario average monthly reservoir outflow varies from 353 to 2590 m^3/s in 2020s, 435 to 2757 m^3/s in 2050s and 442 to 3090 m^3/s in 2080s. Similarly, average monthly reservoir outflow varies from 538 to 1445 m^3/s in 2020s, 514 to 1412 m^3/s in 2050s and 577 to 1396 m^3/s in 2080s under RCP8.5 scenario. In both scenarios, the minimum and maximum outflow value occurred during dry and wet periods, respectively. In all time periods, the optimum reservoir outflows (releases) under RCP8.5 climate scenario for the dry months of November through February were greater than the optimum releases under RCP4.5 climate scenario. These changes show that under RCP8.5, the optimized reservoir stored more water in wet months for dry period release and projected higher storage level compared to RCP4.5 climate scenario.

3.4.2. Optimum Reservoir Power Storage under Climate Change

HEC-ResPRM optimized result showed an increase in projected mean annual Tekeze hydropower reservoir storage under RCP4.5 and RCP8.5 climate scenarios. This increase was projected for three future time periods (Table 1) and the projected optimum stored water varies from 24 to 25% (RCP4.5) and 28.5 to 30% (RCP8.5).

HEC-ResPRM model result under both scenarios in current and all future time periods showed a minimum and maximum reservoir storage periods (Figure 8). Tekeze reservoir reached at maximum storage (reservoir filled) in September and stayed somewhat constant optimum storage up to November. During August to September, main rainy months, the reservoir is filled and optimization model keeps the maximum storage up to November. The reservoir storage tends to slightly be decreased starting from end of November until the beginning of February. After February, the reservoir storage decreased down to the optimization model capacity to store energy at a minimum flow and reached a minimum storage level in June to prepare and capture inflows in the wet main rainy months. In all future months, there will be a stored water to produce power which is always greater than the current optimized value.

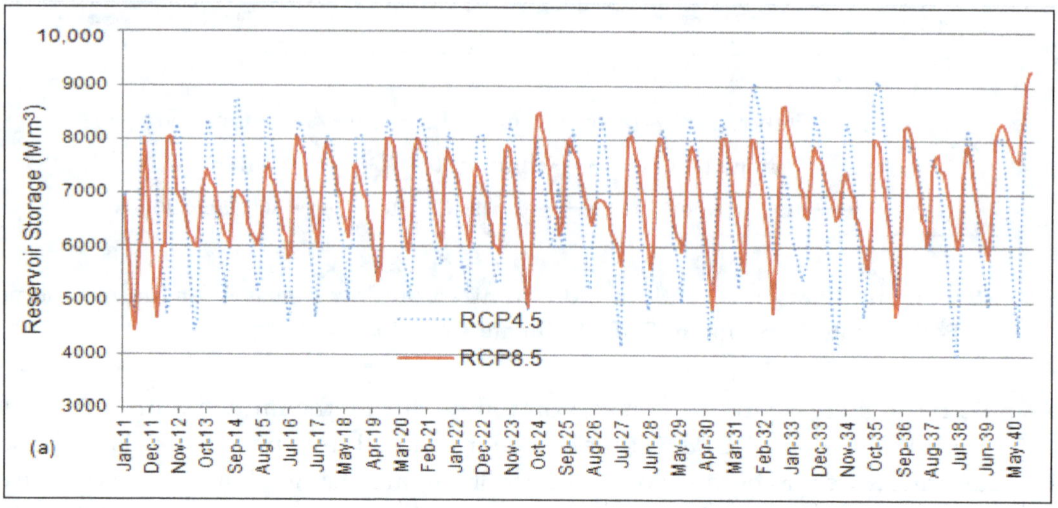

Figure 8. *Cont.*

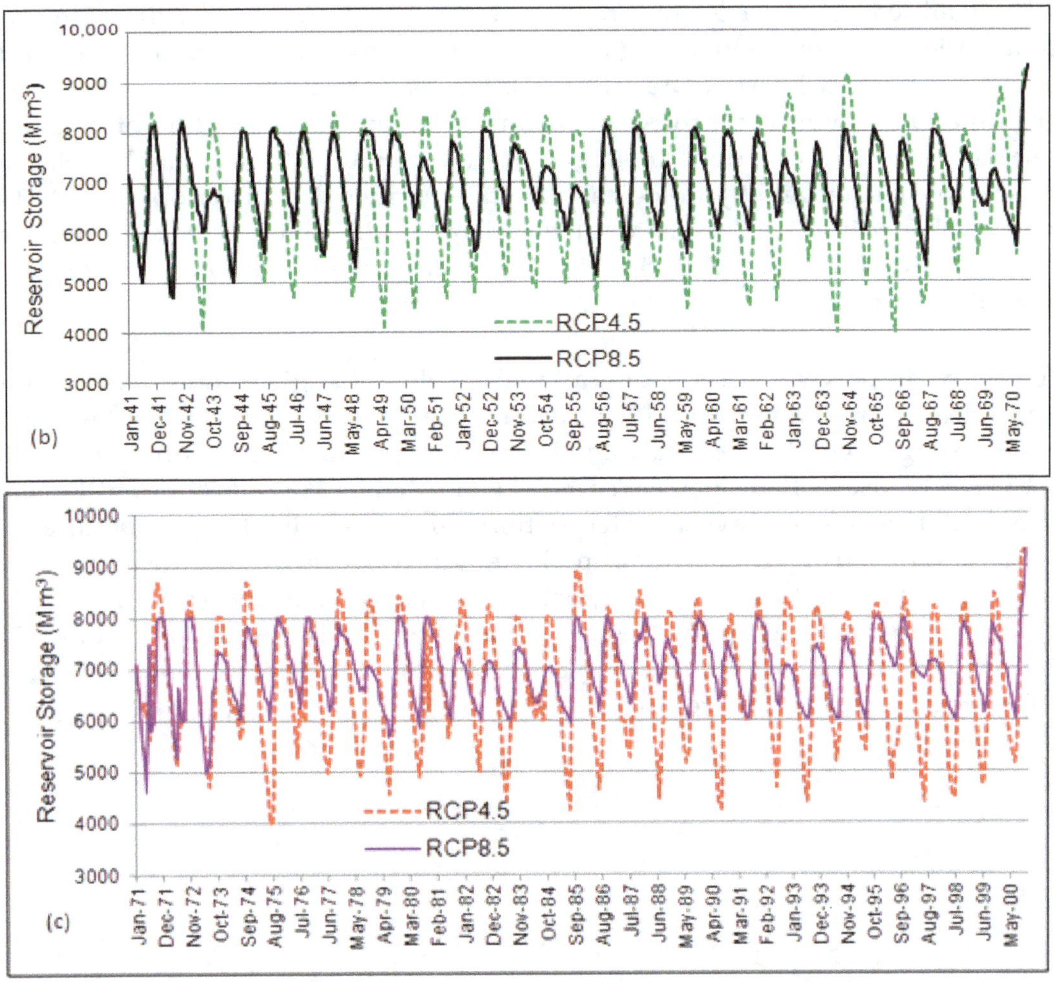

Figure 8. Monthly optimized reservoir storage variations under RCP4.5 and RCP8.5 climate scenarios for: (**a**) 2020s; (**b**) 2050s; (**c**) 2080s.

Figure 8 shows that more power is stored in both scenarios for the next 90 years as compared to current actual and current optimized value. The change in maximum optimum storage increased will be occurred in January and varies from 1693.4 to 1800.8 Mm^3 under RCP4.5 scenario and from 1731.9 to 1851.1 Mm^3 under RCP8.5 scenario in all time periods. The minimum optimal storage change increased will occur in July and varies from 392.2 to 424.7 Mm^3 under RCP4.5 scenario and 803.6 to 956.6 Mm^3 under RCP8.5 scenario in all time periods. This is due to climate change impact on the reservoir inflow and the capacity of the optimization model to operate the reservoir optimally. HEC-ResPRM optimization of future projections tends to make much greater seasonal use of reservoir storage than the current actual operations.

The monthly optimum stored water increases in all months for future time periods under both RCP scenarios as compared to the baseline period (base line varies from 4400 to 6500 Mm^3). Optimized monthly reservoir storage variations are shown in Figure 8. The mean monthly optimum reservoir storage in the future time periods varies for RCP4.5 from 5100 to 8300 Mm^3 in 2020s, 4700 to 8050 Mm^3 in 2050s and 5000 to 8100 Mm^3 in 2080s. It also varies for RCP8.5 from 4900 to 8100 Mm^3 in 2020s, 4850 to 8020 Mm^3 in 2080s and 750 to 7900 Mm^3 in 2080s.

3.4.3. Optimum Reservoir Pool Level (Elevation) under Climate Change

HEC-ResPRM optimization result indicates that Tekeze hydropower reservoir pool level will be increased under RCP4.5 and RCP8.5 scenarios in all time periods. This comparison made with the current mean annual reservoir operation pool level of 1112.61 masl from 2009–2017. In the three projected time periods, optimal pool level (Table 2) change varies from 8.26 to 8.45 m under RCP4.5 and 10.66–11.24 m under RCP8.5 climate scenarios. This is large elevation difference that will store more water in the rainy months for the dry season power production. The optimized pool levels under both scenarios in all time periods are larger than the current optimized pool level. This is due to the impact of climate change and hydrological non-stationarity on reservoir operation. The reservoir storage pool level change in RCP4.5 scenario is lower than RCP8.5 scenario due to increase in each year individual month's fluctuations in RCP4.5 scenarios because of future inflow variability that reduced the mean annual reservoir water storage level.

The Tekeze hydropower reservoir operational level is changing continually due to inflows occurred and releases are made to produce power. The start of dead storage level at 1096 masl (minimum live storage level) has been assumed for power production. Figure 9 shows the optimal reservoir pool level of Tekeze hydropower reservoir generated by HEC-ResPRM optimization model. These optimal pool level results have a similar pattern with the optimal reservoir storage variations and may be considered as rule curves for optimal operation of Tekeze hydropower reservoir under a given scenario and time period. The reservoir pool level stayed at high level every year from August to November when reservoirs filled during the rainy months of August through September. The drop of pool level in June caused due to optimization model constraint reservoir not emptied and a transition zone when the drawdown ends and reservoir refill start.

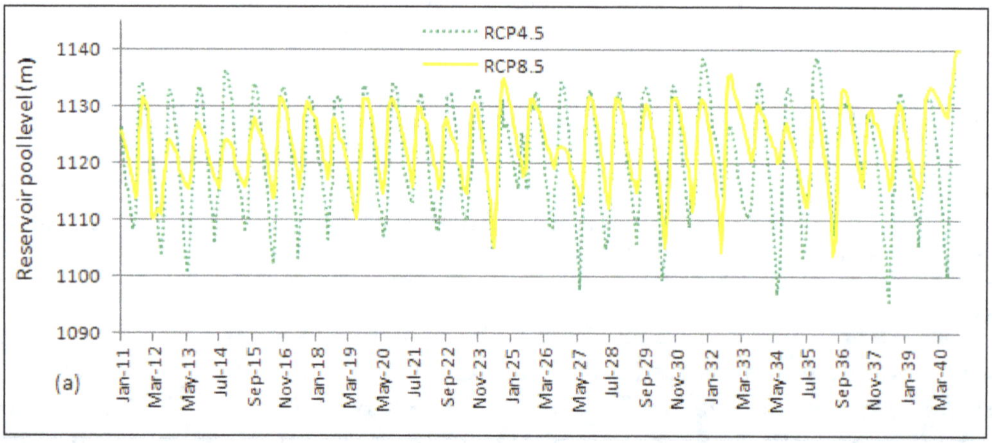

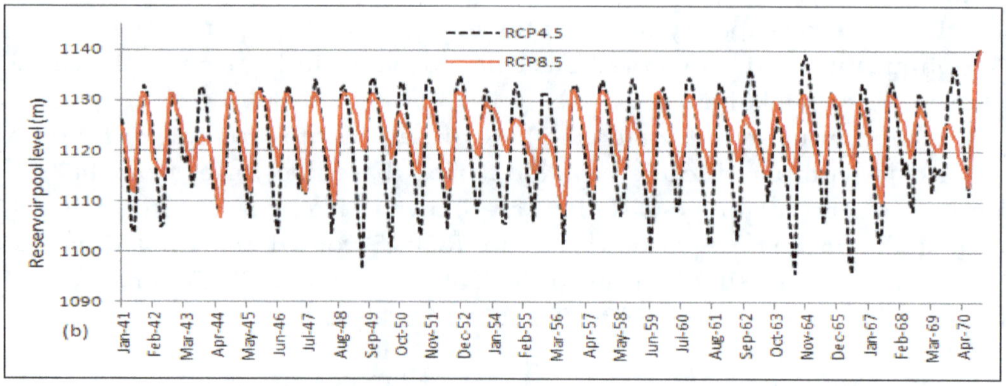

Figure 9. *Cont.*

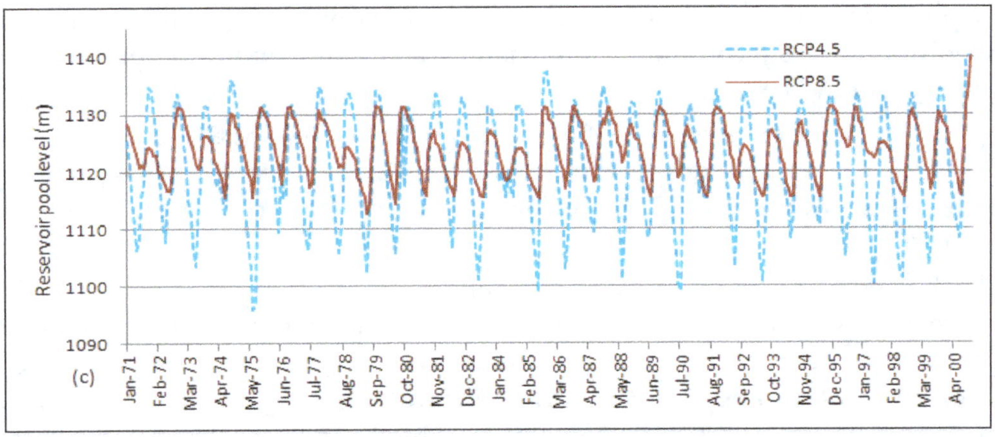

Figure 9. Monthly optimum pool level variations under RCP4.5 and RCP8.5 climate scenarios in three time periods: (**a**) 2020s; (**b**) 2050s; (**c**) 2080s.

There is no doubt that the hydropower system of Tekeze basin will be affected by climate change. With over all predicted increases in precipitation and streamflow, inflow to the reservoir anticipated to increase. Therefore, even though both RCP4.5 and RCP8.5 climate scenarios exhibited higher water inflow volume to Tekeze hydropower reservoir, this did not necessarily result in significantly more hydropower generation. Optimal operation of the reservoir using HEC-ResPRM considerably increases the power production by storing the higher inflow volume to inflow deficiency periods. Based on this research, the potential for hydropower production in the Tekeze hydropower reservoir is predicted to increase if appropriate reservoir operation techniques are used by operators and water managers to store and use the wet months flow to the dry months. Even if many optimization models and techniques have been developed in several fields of water resources system analysis such as hydropower reservoir operation around the world, the adaptation of such techniques and tools by water managers is slow. Researchers and scientists must accept the fact that the gap still exists between research studies and applications in practice. There need to be research on how to translate science to improve management operations of reservoirs for optimal results.

4. Conclusions

This study used a semi-distributed hydrological model (SWAT) and a reservoir optimization model (HEC-ResPRM) to evaluate the hydrological impacts of climate change on Tekeze hydropower reservoir operation in Tekeze basin part of Eastern Nile. We have evaluated climatic data (historical and future period) from the ensemble outputs of CORDEX-Africa RCMs under RCP4.5 and RCP8.5 climate scenarios for the periods of 2020s, 2050s, and 2080s. Calibrated SWAT model was used to generate climate change induced streamflow that was used as an input for optimal reservoir operation modeling. Analysis conducted on Tekeze hydropower reservoir inflows and outflow, reservoir storage volume and reservoir pool levels revealed the following:

1. This study found that the impact of climate change would increase in precipitation, temperature and streamflow in Tekeze basin under RCP4.5 and RCP8.5 climate scenarios over future periods which have an impact on current and future Tekeze hydropower reservoir operation.
2. Projected annual and inter-annual reservoir inflow showed increasing trend under both RCP4.5 and RCP8.5 climate scenarios.
3. HEC-ResPRM incorporates water storage, water surface elevation, release and power generation would provide better understanding of current and future conditions of Tekeze hydropower reservoir operation.

4. Current optimized power storage and pool level show more optimal results than the current actual operation, so it is recommended to change the current operating policy to produce more power throughout the year.
5. The projected increase of reservoir inflow under an ensemble of CORDEX-Africa RCP4.5 and RCP8.5 future climate scenarios lead to optimized reservoir power storage, pool level (head) and release that greatly exceed those historically observed, indicating a shift in current water system behavior.
6. The study showed that climate change clearly affects future reservoir planning and management in Tekeze basin. Therefore, water resources planners, managers and operators should consider climate change impacts in the design, planning and management of reservoir systems.
7. In practice, many reservoir system operators and water managers feel more comfortable to use pre-defined rule curves and simulation results which are easy to understand and operate as most optimal operating rules developed by scientists using sophisticated optimization models and algorithms are mathematically more complex. The use of a combination of simulation and optimization models may solve this problem.
8. This study has not considered the changes in land use/land cover due to socio-economic development in the future. Coupling climate models with projected changes in land use associated with climate change impacts and effect of climate change adaptation on erosion and sediment yield which is necessary to evaluate projected changes in runoff associated with future Tekeze River basin development. Hence, further studies are recommended to quantify future change in streamflow and sedimentation load in Tekeze hydropower reservoir as well as its implication on future hydropower generation.

Acknowledgments: The research presented has been financially supported by the Ethiopian Government via Addis Ababa University (PhD studentship for the first author). The first author would like to give his thanks to Florida International University in assisting for the two months research visit. We acknowledge the data support of Ethiopian Ministry of Water, Irrigation and Electricity, National Meteorological Service Agency and Electric Power Corporation.

Author Contributions: Fikru Fentaw was responsible for this current research article in the framework of his PhD program and initially wrote the manuscript. Dereje Hailu, Agizew Nigussie, and Assefa M. Melesse directed the study by helping to interpret the results, particularly Assefa M. Melesse who contributed more on editing and organizing of this research article.

References

1. Zhou, Y.; Guo, S.; Duan, W.; Chen, H.; Liu, P. Dynamic control of flood limited water level for cascade reservoirs. *J. Hydroelectr. Eng.* **2015**, *34*, 23–30.
2. Yazdi, J.; Moridi, A. Interactive reservoir-watershed modeling framework for integrated water quality management. *Water Resour. Manag.* **2017**, *31*, 2105–2125. [CrossRef]
3. Birhanu, K.; Alamirew, T.; Dinka, M.O.; Ayalew, S.; Aklog, D. Optimizing reservoir operation policy using chance constraint nonlinear programming for Koga Irrigation Dam, Ethiopia. *Water Resour. Manag.* **2014**, *28*, 4957–4970. [CrossRef]
4. Rani, D.; Moreira, M.M. Simulation–optimization modeling: A survey and potential application in reservoir systems operation. *Water Resour. Manag.* **2010**, *24*, 1107–1138. [CrossRef]
5. David, S.; Randolph, B.; Upali, A. Water scarcity in the Twenty-first century. *Int. J. Water Resour. Dev.* **1999**, *15*, 29–42. [CrossRef]
6. Jury, W.A.; Henry, J.V.J. The emerging global water crisis: Managing scarcity and conflict between water users. *Adv. Agron.* **2007**, *95*, 1–76.
7. Rijsberman, G. Water scarcity: Fact or fiction? *Agric. Water Manag.* **2006**, *80*, 5–22. [CrossRef]
8. Larson, K.L.; Polsky, C.; Gober, P.; Chang, H.; Shandas, V. Vulnerability of water systems to the effects of climate change and urbanization: A comparison of Phoenix, Arizona and Portland, Oregon (USA). *Environ. Manag.* **2013**, *52*, 179–195. [CrossRef] [PubMed]

9. Ghashghaie, M.; Marofi, S.; Marofi, H. Using system dynamics method to determine the effect of water demand priorities on downstream flow. *Water Resour. Manag.* **2014**, *28*. [CrossRef]
10. Guo, S.; Chen, J.; Li, Y.; Liu, P.; Li, T. Joint operation of the multi-reservoir system of the Three Gorges and the Qingjiang Cascade Reservoirs. *Energies* **2011**, *4*, 1036–1050. [CrossRef]
11. Zhou, Y.; Guo, S.; Liu, P.; Xu, C. Joint operation and dynamic control of flood limiting water levels for mixed cascade reservoir systems. *J. Hydrol.* **2014**, *519*, 248–257. [CrossRef]
12. Azizipour, M.; Ghalenoei, V.; Afshar, M.H.; Solis, S.S. Optimal operation of hydropower reservoir systems using weed optimization algorithm. *Water Resour. Manag.* **2016**, *30*, 3995–4009. [CrossRef]
13. He, Y.; Xu, Q.; Yang, S.; Liao, L. Reservoir flood control operation based on chaotic particle swarm optimization algorithm. *Appl. Math. Model.* **2014**, *38*, 4480–4492. [CrossRef]
14. Lu, B.; Li, K.; Zhang, H.; Wang, W.; Gu, H. Study on the optimal hydropower generation of Zhelin reservoir. *J. Hydro-Environ. Res.* **2013**, *7*, 270–278. [CrossRef]
15. Cheng, C.-T.; Wang, W.-C.; Xu, D.-M.; Chau, K.W. Optimizing hydropower reservoir operation using hybrid genetic algorithm and chaos. *Water Resour. Manag.* **2008**, *22*, 895–909. [CrossRef]
16. Yin, X.A.; Yang, Z.F. Development of a coupled reservoir operation and water diversion model: Balancing human and environmental flow requirements. *Ecol. Model.* **2011**, *222*, 224–231. [CrossRef]
17. Milly, P.; Betancourt, J.; Falkenmark, M.; Hirsch, R.M.; Kundzewicz, W.Z.; Lettenmaier, P.D.; Stouffer, R.J. Stationarity is dead: Whither water management? *Science* **2008**, *319*, 573–574. [CrossRef] [PubMed]
18. Stocker, D.; Qin, G.-K.; Plattner, L.V.; Allen, S.K. Technical summary. In *Climate Change 2013: The Physical Science Basis. Contribution of Working Group I to the Fifth Assessment Report of the Intergovernmental Panel on Climate Change*; Cambridge University Press: Cambridge, UK, 2013; pp. 31–116, ISBN 978-1-107-41532-4.
19. Huntington, T.G. Climate warming-induced intensification of the hydrologic cycle: An assessment of the published record and potential impacts on agriculture. In *Advances in Agronomy*; Sparks, D.L., Ed.; Academic Press: Cambridge, MA, USA, 2010; Volume 109, pp. 1–53.
20. Vicuña, S.; Garreaud, R.D.; McPhee, J. Climate change impacts on the hydrology of a snowmelt driven basin in semiarid Chile. *Clim. Chang.* **2011**, *105*, 469–488. [CrossRef]
21. Alazzy, A.; Lü, H.; Zhu, Y. Impact of climate change on evaluation of future water demand in the Euphrates and Aleppo basin, Syria. *Proc. Int. Assoc. Hydrol. Sci.* **2014**, *364*, 307–312. [CrossRef]
22. Zhang, C.; Zhu, X.; Fu, G.; Zhou, H.; Wang, H. The impacts of climate change on water diversion strategies for a water deficit reservoir. *J. Hydroinform.* **2014**, *16*, 872–889. [CrossRef]
23. Lumbroso, D.M.; Woolhouse, G.; Jones, L. A review of the consideration of climate change in the planning of hydropower schemes in sub-Saharan Africa. *Clim. Chang.* **2015**, *133*, 621–633. [CrossRef]
24. Lee, S.-Y.; Hamlet, A.F.; Grossman, E.E. Impacts of climate change on regulated streamflow, hydrologic extremes, hydropower production, and sediment discharge in the Skagit River Basin. *Northwest Sci.* **2016**, *90*, 23–43. [CrossRef]
25. Setegn, S.; Melesse, A.; Haiduk, A.; Webber, D.; Wang, X.; McClain, M. Modeling hydrological variability of fresh water resources in the Rio Cobre watershed, Jamaica. *Catena* **2014**, *120*, 81–90. [CrossRef]
26. Ehsani, N.; Vörösmarty, C.J.; Fekete, B.M.; Stakhiv, E.Z. Reservoir operations under climate change: Storage Capacity options to mitigate risk. *J. Hydrol.* **2017**, *555*, 435–446. [CrossRef]
27. Haile, A.T.; Akawka, A.L.; Berhanu, B.; Rientjes, T. Changes in water availability in the Upper Blue Nile Basin under the representative concentration pathways scenario. *Hydrol. Sci. J.* **2017**, *62*, 2139–2149. [CrossRef]
28. Zhu, X.; Zhang, C.; Qi, W.; Cai, W.; Zhao, X.; Wang, X. Multiple climate change scenarios and runoff response in Biliu River. *Water* **2018**, *10*, 126. [CrossRef]
29. Serdeczny, O.; Adams, S.; Baarsch, F.; Coumou, D.; Robinson, A.; Hare, W.; Schaeffer, M.; Perrette, M.; Reinhardt, J. Climate change impacts in Sub-Saharan Africa: From physical changes to their social repercussions. *Reg. Environ. Chang.* **2017**, *17*, 1585–1600. [CrossRef]
30. Conway, D.; Persechino, A.; Ardoin-Bardin, S.; Hamandawana, H.; Dieulin, C.; Mahé, G. Rainfall and water resources variability in Sub-Saharan Africa during the twentieth century. *J. Hydrometeorol.* **2009**, *10*, 41–59. [CrossRef]
31. Hales, S. *Contribution of Working Group II to the Fourth Assessment Report of the Intergovernmental Panel on Climate Change*; Climate Change 2007: Impacts, Adaptation and Vulnerability; Parry, M.L., Canziani, O.F., Palutikof, J.P., van der Linden, P.J., Hanson, C.E., Eds.; Cambridge University Press: Cambridge, UK, 2007.

32. Nikulin, G.; Jones, C.; Giorgi, F.; Asrar, G.; Büchner, M.; Cerezo-Mota, R.; Christensen, O.B.; Déqué, M.; Fernandez, J.; Hänsler, A.; et al. Precipitation climatology in an ensemble of CORDEX-Africa regional climate simulations. *J. Clim.* **2012**, *25*, 6057–6078. [CrossRef]
33. Kim, U.; Kaluarachchi, J.J. Climate change impacts on water resources in the Upper Blue Nile River Basin, Ethiopia. *JAWRA J. Am. Water Resour. Assoc.* **2009**, *45*, 1361–1378. [CrossRef]
34. Beyene, T.; Lettenmaier, D.P.; Kabat, P. Hydrologic impacts of climate change on the Nile River Basin: implications of the 2007 IPCC scenarios. *Clim. Chang.* **2010**, *100*, 433–461. [CrossRef]
35. Yamba, F.D.; Walimwipi, H.; Jain, S.; Zhou, P.; Cuamba, B.; Mzezewa, C. Climate change/variability implications on hydroelectricity generation in the Zambezi River Basin. *Mitig. Adapt. Strateg. Glob. Chang.* **2011**, *16*, 617–628. [CrossRef]
36. Hamududu, B.H.; Killingtveit, Å. Hydropower production in future climate scenarios; the case for the Zambezi River. *Energies* **2016**, *9*, 502. [CrossRef]
37. Abtew, W.; Melesse, A.M.; Dessalegne, T. Spatial, inter and intra-annual variability of the Upper Blue Nile Basin rainfall. *Hydrol. Process.* **2009**, *23*, 3075–3082. [CrossRef]
38. Mengistu, D.; Bewket, W.; Lal, R. Recent spatiotemporal temperature and rainfall variability and trends over the Upper Blue Nile River Basin, Ethiopia. *Int. J. Climatol.* **2014**, *34*, 2278–2292. [CrossRef]
39. Tarekegn, D.; Tadege, A. *Assessing the Impact of Climate Change on the Water Resources of the Lake Tana Basin Using the WATBAL Model*; CEEPA (the Centre for Environmental Economics and Policy in Africa): Pretoria, South Africa, 2006.
40. Melesse, A.M.; Loukas, A.G.; Senay, G.; Yitayew, M. Climate change, land-cover dynamics and ecohydrology of the Nile River Basin. *Hydrol. Process.* **2009**, *23*, 3651–3652. [CrossRef]
41. Setegn, S.G.; Rayner, D.; Melesse, A.M.; Dargahi, B.; Srinivasan, R. Impact of climate change on the hydroclimatology of Lake Tana Basin, Ethiopia. *Water Resour. Res.* **2011**, *47*, W04511. [CrossRef]
42. Jamali, S.; Abrishamchi, A.; Madani, K. Climate change and hydropower planning in the Middle East: implications for Iran's Karkheh hydropower systems. *J. Energy Eng.* **2013**, *139*, 153–160. [CrossRef]
43. Vonk, E.; Xu, Y.; Booij, M.; Zhang, X.; Augustijn, D.C.M. Adapting multireservoir operation to shifting patterns of water supply and demand. *Water Resour. Manag.* **2014**, *28*, 625–643. [CrossRef]
44. Wilby, R.L.; Harris, I. A framework for assessing uncertainties in climate change impacts: Low-flow scenarios for the River Thames, UK. *Water Resour. Res.* **2006**, *42*, W02419. [CrossRef]
45. Raje, D.; Mujumdar, P.P. Reservoir performance under uncertainty in hydrologic impacts of climate change. *Adv. Water Resour.* **2010**, *33*, 312–326. [CrossRef]
46. Haregeweyn, N.; Poesen, J.; Nyssen, J.; Gerared, G.; Verstraeten, G.; de Vente, J.; Deckers, L.; Moeversons, J.; Haile, M. Sediment yield variability in Northern Ethiopia: A quantitative analysis of its controlling factors. *Catena* **2008**, *75*, 65–76. [CrossRef]
47. Wolde, E. Identification and prioritization of sub watersheds for land and water management in Tekeze Dam watershed, Northern Ethiopia. *Int. Soil Water Conserv. Res.* **2016**, *4*, 30–38. [CrossRef]
48. Wurbs, R.A. Reservoir-system simulation and optimization models. *J. Water Resour. Plan. Manag.* **1993**, *119*, 455–472. [CrossRef]
49. Oliveira, R.; Loucks, D.P. Operating rules for multireservoir systems. *Water Resour. Res.* **1997**, *33*, 839–852. [CrossRef]
50. Labadie, J.W. Optimal operation of multireservoir systems: State-of-the-art review. *J. Water Resour. Plan. Manag.* **2004**, *130*, 93–111. [CrossRef]
51. Yeh, W.W.-G. Reservoir management and operations models: A state-of-the-art review. *Water Resour. Res.* **1985**, *21*, 1797–1818. [CrossRef]
52. Ayalew, L. The effect of seasonal rainfall on landslides in the highlands of Ethiopia. *Bull. Eng. Geol. Environ.* **1999**, *58*, 9–19. [CrossRef]
53. Van Vuuren, D.P.; Edmonds, J.; Kainuma, M.; Riahi, K.; Thomson, A.; Hibbard, K.; Hurtt, G.C.; Kram, T.; Krey, V.; Lamarque, J.-F.; et al. The representative concentration pathways: An overview. *Clim. Chang.* **2011**, *109*. [CrossRef]
54. Meinshausen, M.; Smith, S.J.; Calvin, K.; Daniel, J.S.; Kainuma, M.L.T.; Lamarque, J.-F.; Matsumoto, K.; Montzka, S.A.; Raper, S.C.B.; Riahi, K.; et al. The RCP greenhouse gas concentrations and their extensions from 1765 to 2300. *Clim. Chang.* **2011**, *109*, 213. [CrossRef]

55. Taylor, K.E.; Stouffer, R.J.; Meehl, G.A. An overview of CMIP5 and the Experiment design. *Bull. Am. Meteorol. Soc.* **2011**, *93*, 485–498. [CrossRef]
56. Thomson, A.M.; Calvin, K.V.; Smith, S.J.; Kyle, P.G.; Volke, A.; Patel, P.; Delgado-Arias, S.; Bond-Lamberty, B.A.; Wise, M.A.; Clarke, L.E.; et al. RCP4.5: A pathway for stabilization of radiative forcing by 2100. *Clim. Chang.* **2011**, *109*, 77–94. [CrossRef]
57. Riahi, K.; Rao, S.; Krey, V.; Cho, C.; Chirkov, V.; Fischer, G.; Kindermann, G.; Nakicenovic, N.; Rafaj, P. RCP8.5—A scenario of comparatively high greenhouse gas emissions. *Clim. Chang.* **2011**, *109*, 33. [CrossRef]
58. Neitsch, S.; Arnold, J.; Kiniry, J.R.; Williams, J.R.; King, K. SWAT theoretical documentation. *Grassland* **2005**, *494*, 234–235.
59. Van Griensven, A.; Ndomba, P.; Yalew, S.; Kilonzo, F. Critical review of SWAT applications in the Upper Nile Basin countries. *Hydrol. Earth Syst. Sci.* **2012**, *16*, 3371–3381. [CrossRef]
60. Dessie, M.; Verhoest, N.E.C.; Admasu, T.; Pauwels, V.R.N.; Poesen, J.; Adgo, E.; Deckers, J.; Nyssen, J. Effects of the floodplain on river discharge into Lake Tana (Ethiopia). *J. Hydrol.* **2014**, *519*, 699–710. [CrossRef]
61. Faber, B.A.; Harou, J.J. Multiobjective optimization with HEC Res-PRM? Application to the Upper Mississippi Reservoir system. *Oper. Reserv. Chang. Cond.* **2006**. [CrossRef]
62. Ostadrahimi, L.; Mariño, M.A.; Afshar, A. Multi-reservoir operation rules: Multi-swarm PSO-based optimization approach. *Water Resour. Manag.* **2012**, *26*, 407–427. [CrossRef]

5

Three-Dimensional Numerical Study of Free-Flow Sediment Flushing to Increase the Flushing Efficiency

Taymaz Esmaeili [1,*], Tetsuya Sumi [2], Sameh A. Kantoush [2], Yoji Kubota [3], Stefan Haun [4] and Nils Rüther [5]

1 Department of Civil Engineering, Gorgan Branch, Islamic Azad University, Gorgan 49147-39975, Iran
2 Disaster Prevention Research Institute (DPRI), Kyoto University, Uji, Kyoto 611-0011, Japan; sumi.tetsuya.2s@kyoto-u.ac.jp (T.S.); kantoush.samehahmed.2n@kyoto-u.ac.jp (S.A.K.)
3 Hydro-Soft Technology Institute Co., Ltd., Nakanoshima, Osaka 530-6126, Japan; kubotayj@hydro-soft.co.jp
4 Institute for Modelling Hydraulic and Environmental Systems, University of Stuttgart, 70569 Stuttgart, Germany; stefan.haun@iws.uni-stuttgart.de
5 Department of Civil and Environmental Engineering, Norwegian University of Science and Technology, S.P. Andersens vei 5, 7491 Trondheim, Norway; nils.ruther@ntnu.no
* Correspondence: t.esmaeili@gorganiau.ac.ir or taymaz.esmaeili@gmail.com

Abstract: The catchment of the Dashidaira reservoir located on the Kurobe River has high sediment yield. Because of the sufficient available amount of water in the catchment during flood events, the free-flow sediment flushing operation with full water-level drawdown is employed every year to preserve the effective storage capacity of the Dashidaira reservoir. This paper focuses first on the numerical simulation of a previously conducted free-flow flushing operation in the Dashidaira reservoir using the available in situ obtained data. Afterwards, to improve the flushing efficiency, the effects of water and discharge manipulation and the construction of an auxiliary channel on the total volume of the flushed sediment were studied. A fully 3D numerical model using the finite volume approach in combination with a wetting/drying algorithm was utilized to reproduce the flow velocity field and simulate the movable bed variations. The outcomes revealed that increasing the average free-flow discharge during the free-flow stage by approximately 56%, in the form of multiple discharge pulses, can enhance the flushing efficiency by up to 13%, and the construction of an auxiliary channel in the wide midstream of the reservoir can locally increase the sediment erosion from this area.

Keywords: reservoir flushing; numerical simulation; flushing efficiency; Kurobe River

---

1. Introduction

Dams interrupt the natural continuity of sediment transport through rivers, which results in sediment deposition in the reservoir behind the dam [1]. In a global context, sediment deposition is a challenging issue for the long-term utilization of dam reservoirs [2]. It is estimated that 0.5% of total storage volume of the reservoirs is lost annually in the world because of the sedimentation [3]. The loss of effective reservoir storage volume due to sediment deposition reduces the effective lifespan of dams and decreases reservoir functionality for flood control purposes, hydropower generation, irrigation and water supply, thereby generating a substantial economic loss [1,4,5]. Diverse measures, including sediment dredging, density current venting, bypassing, flushing, sluicing and upstream sediment trapping, have been used to control progressive sedimentation in reservoirs to prolong their life [1,6]. Among these measures, drawdown flushing (i.e., free-flow sediment flushing) plays a major role in reservoir storage capacity restoration and conservation since it is an efficient hydraulic technique for sediment removal [7]. Free-flow sediment flushing involves a complete lowering of the water level by opening low-level outlets to temporarily establish riverine flow (i.e., free flow) through the outlets. The accelerated flow erodes a channel through the deposits and flushes out both fine and coarse sediments [4,5,7]. However, only few experimental studies have focused on flushing channel formation [8], and detailed explanations of flushing channel formation and evolution in prototype reservoirs are scarce [9].

In Japan, the sediment yield coming from the catchments is generally high due to the geologically young mountains, steep slopes, flashy flow regimes, and frequent landslides, especially in the mountainous areas. Thus, a large amount of sediment is transported in Japanese rivers, and the sediment inputs in the reservoirs fed by these rivers are high. Kurobe River, located in the Toyama prefecture, is one of the most important rivers in Japan because of the cascade reservoir system along this river and the considerable amount of electricity generated by water from these reservoirs. Therefore, the reservoir owners are interested in implementing the applicable measures for increasing the sediment removal from these reservoirs especially from the specific areas, which encounters the excess deposition problems.

To optimize the sediment management strategies in reservoirs that feature conditions that change from shallow to deep areas and that also contain sandbars, numerical models that are more complicated than the simple one-dimensional (i.e., 1D) models should be used [10]. Advanced two-dimensional (i.e., 2D) numerical models have been employed to solve practical problems in rivers (e.g., simulation of morphological bed changes in river meanders) [11]. Nevertheless, 2D and quasi three-dimensional (i.e., 3D) numerical models are not able to directly simulate a complex 3D flow field that includes secondary currents. However, these phenomena contribute significantly to the natural sediment transport processes [12]. Thus, the application of 3D numerical models is essential if the velocity variation over the flow depth, i.e., helical flow, plays a major role in sediment transport (e.g., in channel bends). 3D numerical models were used by various researchers for simulating the flow field and sediment transportation in the dam reservoirs [13,14]. As for sediment flushing from reservoirs, a 3D numerical model has been shown to perform better than a 2D one in simulating the bed deformation, notably in channel bends [15]. Recently 3D numerical models were applied to investigate the sediment dynamics inside the reservoirs during a flushing operation at the both lab scale (e.g., [15,16]) and the full scale (e.g., [17–19]).

A Computational Fluid Dynamic (CFD) software package called SSIIM (Simulation of Sediment Movements In water Intakes with Multiblock Option) was utilized in this study. The SSIIM program implements a 3D numerical model of the flow field by solving the mass and momentum conservation equation in three dimensions using different turbulence closure approaches [20]. SSIIM was successfully applied to model the 3D flow field under various hydraulic and geometric boundary conditions [21]. SSIIM was also used for coupled computation of the flow and sediment field in physical model and prototype-scale studies by various researchers [22–28]. Recently, this CFD program was used with enhanced grid generation features to simulate the sedimentation and flushing channel evolution in reservoirs [19,29,30]. Nevertheless, it should be noted that simulation of a sediment-flushing event in a steep full-scale reservoir, such as the Dashidaira reservoir, is complex owing to the rapid variation of hydraulic boundary conditions and riverbed materials even in short distances. In addition, numerical studies focusing on potential measures for increasing the sediment flushing efficiency in a full-scale reservoir are scarce.

In the present study, the SSIIM model was employed to simulate the 2012 free-flow sediment flushing operation in the Dashidaira reservoir using the field-measured data. To do this, the sensitivity analysis of computed Total Volume of Flushed Sediments (i.e., TVFS) as a result of changes in the selected empirical parameters together with assessment of morphological bed changes was performed and subsequently model was calibrated. Because of the interest to increase the Flushing Efficiency (i.e., FE), when it is defined as the ratio of the flushed sediment volume to the used water volume, various scenarios, such as using additional artificial discharge during the free-flow state, increasing the water-level drawdown speed, and the construction of an auxiliary longitudinal channel, were numerically modeled to investigate how they affect the morphological bed changes in specific zones of the reservoir and whether they increase the Flushing Efficiency. To provide a more detailed and quantitative insight on morphological bed changes, the Bed Changes Index (i.e., BCI) is introduced and subsequently calculated bed changes were compared for each scenario with a reference case (e.g., calibrated case) to figure out the performance of each scenario. Moreover, the optimum relationship between the flushing efficiency and the amount of water used for flushing was established under the mentioned scenario conditions.

2. Study Case Description

2.1. Site Background

The Kurobe River originates from Washiba Mountain with an elevation of 3000 m, located in the northern Japanese Alps, and flows into Toyama bay in the Japan Sea (Figure 1a) [31]. The catchment area of the river is 682 km^2, and the length of the river is 85 km. The bed slope is steep and varies between 1% and 20%. In the catchment area of the Kurobe River, the average rainfall and total sediment yield are 4000 mm and 1.4×10^6 m^3/year, respectively, which are both among the highest in Japan [32]. Dashidaira dam, with a height of 76.7 m, was constructed in 1985 by Kansai Electric Power Company in the Kurobe River and has a power of 124 MW. The average bed slope of the reservoir is about 2.2% and also the gross and effective storage capacities are 9.01 and 1.66 million m^3, respectively [7,32]. The mean annual sediment load and the ratio of the total storage to the mean annual runoff in the Dashidaira reservoir are 0.62×10^6 m^3 and 0.00674, respectively [33]. The Dashidaira dam is one of the first dams in Japan constructed with sediment-flushing facilities. The first flushing operation in the Dashidaira reservoir was performed 6 years after the dam construction in 1991. Subsequently, the accumulated distorted sediments within 6 years were diffused to the downstream and estuary zone with many negative environmental implications. After that, the flushing operation is performed every year during the first major flood event in the rainy season to reduce negative environmental impacts on the downstream areas of the dam since aquatic animals have been adapted to the perturbation caused by floods. A flood not only provides enough energy to transport the flood-born sediments through the reservoir, but also scour the previously deposited sediments from the reservoir.

2.2. Field Data Organization

The bathymetric survey for measuring the bed levels is performed regularly by the reservoir owners before and after the annual flushing operation in the Dashidaira reservoir. The measured bed levels before the flushing operation in June 2012 and the locations of cross-sections A-A to L-L, which were used for further assessment of bed variations caused by the flushing, are shown in Figure 1b. The approximately 2-km length of the reservoir has been divided into three areas, namely, areas I, II and III, to analyze the study outcomes based on the generally similar types of bed materials in each area. In area I, the bed material is coarse (e.g., gravel), while in downstream half of area II and the area III, the bed material transitions to finer sediments (e.g., fine sand and even mud). Besides, temporal variation of the water level and discharge magnitudes are recorded at the dam site during the flushing. The inflow discharge and water level fluctuations during the 2012 flushing operation, in which the major sediment inflow was a wash load (i.e., transported without deposition in the reservoir), are illustrated in Figure 1c. After the start of the operation, the preliminary drawdown occurred from 8 to 24 h, and the free-flow state occurred from 24 to 38 h. The water level drawdown from the beginning of the operation to the free-flow state was approximately 25 m. Furthermore, based on the available onsite samples, seven sediment sizes ranging between 316 and 0.37 mm were considered to be representative grain sizes for the bed of the reservoir. Table 1 shows the average sediment size distribution in different cross-sections prior to the flushing operation. The large sediment sizes in cross-section L-L despite the size reduction in sections immediately upstream can be attributed to the last flushing operation process, in which closing the bottom outlets during the water level recovery stage causes the transported large sediments to deposit nearby the dam. During the 2012 flushing operation of the Dashidaira reservoir, the total volume of the eroded sediment from the reservoir was equal to 408,700 m^3.

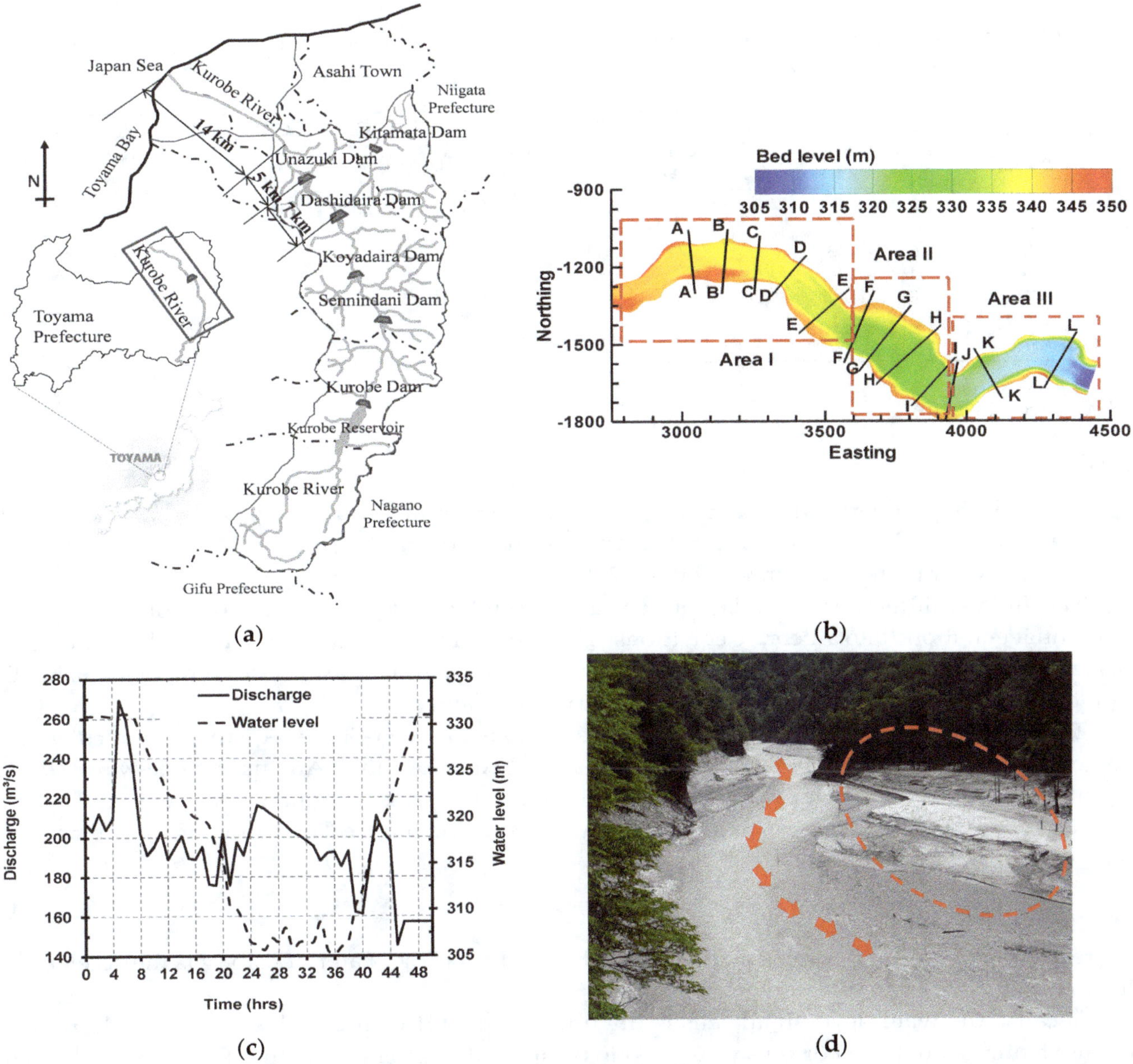

Figure 1. (**a**) Site map of the Kurobe River showing the location of the Dashidaira reservoir [31]; (**b**) Measured bed topography of the Dashidaira reservoir before the flushing operation in June 2012 and locations of the cross-sections A-L for further assessment of the bed variations in the upstream (i.e., area I), midstream (i.e., area II), and downstream (i.e., area III) portions of the reservoir; (**c**) Water level and discharge rates during the flushing operation in June 2012; (**d**) Onsite view of the dead zone area during the 2012 flushing operation (the dead zone area has been shown with ellipse dashed line and arrows show the flow direction).

The thalweg of the river channel exists close to the right bank of the wide middle area (i.e., area II) and subsequently fine sediments deposited along the left bank, indicated with an ellipse dashed line in Figure 1d, were not removed effectively during the flushing operation. Therefore, fine sediments have been accumulated and consolidated that may contain the degraded organic matters. The area highlighted by the ellipse dashed line is hereafter called the dead zone. Dam owners are interested in removing the accumulated sediments from this zone by annual flushing operation to prevent the formation of consolidated and distorted fine sediments and also to keep the effective storage capacity as much as possible.

Table 1. Average sediment size distribution in the specified cross-sections shown in Figure 1b. Cs. is an abbreviation for Cross-section.

Sediment Size (mm)	Cs. A-A (%)	Cs. B-B (%)	Cs. C-C (%)	Cs. D-D (%)	Cs. E-E (%)	Cs. F-F (%)	Cs. G-G (%)	Cs. H-H (%)	Cs. I-I (%)	Cs. J-J (%)	Cs. K-K (%)	Cs. L-L (%)
316	2	0	0	0	0	0	0	0	0	0	0	0
118.3	74	0	0	0	0	0	0	0	0	0	0	0
37.4	6	73	75	70	69	4	1	0	0	0	0	40
11.8	4	7	6	8	13	14	5	0	0	0	0	30
3.7	3	14	11	14	12	25	18	0	0	0	0	16
1.2	5	4	1	3	2	23	21	0	6	13	0	3
0.37	6	2	7	5	4	35	55	100	94	87	100	1

3. Numerical Model

3.1. Flow Field Modeling

The fully 3D numerical model SSIIM solves the continuity equation together with Reynolds-averaged Navier-Stokes equations in three dimensions and within non-orthogonal coordinates to compute the water motion for turbulent flow [20].

The finite-volume approach is applied as a discretization method to transform the partial differential equations into algebraic equations. The convection term in the Navier-Stokes equation is solved using the second-order upwind scheme [15,20]. The Reynolds stress term is modeled using the standard k-ε turbulence model with constant empirical values [34].

The unknown pressure field in the Reynolds-averaged Navier-Stokes equations is calculated by employing the semi-implicit method for pressure-linked equations [35]. An implicit free-water surface algorithm was used in the computations and was based on the pressure gradient between a cell and the neighbor cells. Then, the free-water surface is computed using the local elevation difference between a cell and the neighboring cells [19].

$$\frac{\partial p}{\partial x_i} = \rho g \frac{\partial z}{\partial x_i} \tag{1}$$

where p is the pressure, x_i is the special geometrical scale, ρ is the water density, g is the acceleration due to gravity and z is the water-level elevation.

Because the water level in all cells is unknown, an iterative method was used. Additionally, because a number of neighbor cells were used to compute the water elevation differences for each cell, different values appeared for the water elevation difference depending on the number of neighboring cells used in the computation. Therefore, a weighted average of these values was applied. The weighted average coefficient for each neighboring cell (a_i) is a function of the Froude number, the flow direction and the location of neighboring cells:

$$a_i = \begin{cases} min(2 - Fr; 1.0) & for\ w > -0.1\ and\ Fr < 2.0 \\ w^2(Fr - 1.0) & for\ w < -0.5\ and\ Fr > 2.0 \\ & 0.0 \end{cases} \tag{2}$$

with:

$$w = \frac{\vec{r} \times \vec{u}}{\left|\vec{r}\right| \times \left|\vec{u}\right|} \tag{3}$$

where Fr is the Froude number, w is the dot product of $\vec{r}$ and $\vec{u}$, is $\vec{r}$ the direction vector pointing from the center of a cell to the center of the neighbor cell aimed to take into account the upstream/downstream effect and $\vec{u}$ is the velocity vector of the cell. This coefficient is then used for discretizing the following equation [19]:

$$\sum_{i=1}^{8} a_i z_p = \sum_{i=1}^{8} a_i \left(z_i + \frac{1}{\rho g}(p_p - p_i) \right) \tag{4}$$

where z_p is the water level elevation in the cell, z_i is the water level elevation in the ith neighbor cell, p_p is the pressure in the cell and p_i is the pressure in the ith neighbor cell.

This implicit and iterative approach is a robust and stable method that can also be used in connection with estimates of the sediment transport and morphological bed changes under unsteady flow conditions. The use of an implicit discretization scheme allows large time-step sizes to be employed in the model [23].

In the SSIIM model, the grid is non-orthogonal, unstructured and adaptive, and it moves vertically with changes in the bed and free-water surface elevation. During the computations, only the water body is modeled. The water surface is recomputed after each time step, and the employed wetting/drying algorithm enables the model to have a varying number of grid cells (e.g., in the vertical and lateral directions) with respect to the water depth in the computational domain using Equation (5) [36]. The wetting/drying algorithm causes the cells to dry up and disappear from the computational domain if the water depth is smaller than a user-defined lower boundary. If the water depth becomes greater than the lower boundary in the dry area, this algorithm regenerates a cell. This approach makes it possible to have a dynamic grid that can move in the lateral direction, allowing changes in the bed and water level to be accurately modeled.

$$n = n_{max} \left(\frac{depth}{depth_{max}} \right)^p \tag{5}$$

where n is the number of grid cells in the vertical direction, n_{max} is the maximum number of grid cells in the vertical direction, and p is a user-defined parameter for the number of grid cells.

The Dirichlet boundary condition (logarithmic velocity distribution) was used for the water inflow, whereas the zero-gradient boundary condition was specified for the water outflow and the sediment concentration calculation. For the boundary condition at the bed and walls, where there is no water flux, the empirical wall law introduced by Schlichting (1979) was utilized [20,37]. Also, bed roughness in the form of dunes and ripples is taken into account using an empirical formula introduced by Van Rijn, which employs the characteristics of sediment size distribution and bed form height within the computational domain [20,38].

3.2. Sediment Transport Modeling

The sediment transport computation for simulating the morphological changes is divided into suspended sediment and bed load transport. The suspended sediment transport is calculated by solving the transient convection-diffusion equation.

$$\frac{\partial c}{\partial t} + U_j \frac{\partial c}{\partial x_j} + w \frac{\partial c}{\partial z} = \frac{\partial}{\partial x_j}(\Gamma_T \frac{\partial c}{\partial x_j}) \tag{6}$$

where U_j is the water velocity, w is the fall velocity of the sediments, c is the sediment concentration over time t and spatial geometries (i.e., x and z), and Γ_T is the turbulent diffusivity. To compute the equilibrium suspended sediment concentration, used as the boundary condition in the cells close to the bed, an empirical formula developed by Van Rijn is used [39]:

$$C_{bed} = 0.015\frac{d_i}{a}\frac{\left[\frac{\tau-\tau_{c,i}}{\tau_{c,i}}\right]^{1.5}}{\left(d_i\left[\frac{(\rho_s-\rho_w)g}{\rho_w \nu^2}\right]^{1/3}\right)^{0.3}} \tag{7}$$

where a is a reference level set equal to the roughness height, d_i is the diameter of the i-th fraction, ν is the kinematic viscosity, τ is the shear stress, $\tau_{c,i}$ is the critical shear stress for d_i, which was calculated from the Shield's curve, and ρ_w and ρ_s are the density of the water and sediment, respectively.

In the SSIIM model, the bed load can be simulated using the Van Rijn formula or alternatively by the Meyer-Peter-Müller (MPM) formula [40,41]. The MPM formula is appropriate for steep rivers, which mainly transport coarse sediments close to the bed:

$$q_{b,i} = \frac{1}{g}\left[\frac{\rho_w g r I - 0.047 g(\rho_s - \rho_w)d_{50}}{0.25\rho_w{}^{\frac{1}{3}}\left(\frac{\rho_s-\rho_w}{\rho_s}\right)^{\frac{2}{3}}}\right]^{\frac{3}{2}} \tag{8}$$

where $q_{b,i}$ is the sediment transport rate for the i-th fraction of the bed load per unit width, d_{50} is the characteristic sediment size (median sediment size), r is the hydraulic radius, and I is the slope of the energy line. The Van Rijn formula has been used to simulate a wide variety of sediment transport issues in both physical model and prototype scales:

$$\frac{q_{b,i}}{d_i{}^{1.5}\sqrt{\frac{(\rho_s-\rho_w)g}{\rho_w}}} = 0.053\frac{\left[\frac{\tau-\tau_{c,i}}{\tau_{c,i}}\right]^{1.5}}{d_i{}^{0.3}\left[\left(\frac{(\rho_s-\rho_w)g}{\rho_w \nu^2}\right)\right]^{0.1}} \tag{9}$$

The model accounts for side slope effects by utilizing a reduction function of the critical shear stress for incipient motion by introducing the formula of Brooks together with a sand slide algorithm [20,42]. The sand slide algorithm corrects the bed slope if it exceeds a defined critical angle of repose of the sediments during excessive erosion and thereby accounts for side bank erosion. In fact, the implemented sand slide algorithm acts as a limiter when erosion continues and the bed slope increases [26].

4. Numerical Simulations

4.1. Model Setup and Calibration

The computational grid was constructed based on the bathymetric surveys before the flushing (Figure 1b). The mesh cell sizes in the streamwise and transversal directions were 10–20 m and 5–10 m, respectively. The bed material density was assumed to be 2650 kg/m^3. The water levels and inflow discharge fluctuations, which were employed as the hydrodynamic boundary conditions in the simulations, are shown in Figure 1c. In addition, a non-uniform bed material size distribution with spatially varying fractions was introduced to the model by using the seven representative sediment sizes shown in Table 1. More specifically, the computational domain was divided into a number of small segments, and each segment had its own non-uniform grain size distribution. Because the wash load, by its definition, is transported without deposition through the reservoir, its effect on the simulation of the flushing process was neglected in the computations.

For calibration purposes, a reference case was first established, assuming general values for the empirical parameters in the SSIIM model. Due to the good performance of the MPM bed load sediment transport formula for flushing simulations of Alpine reservoirs revealed by Haun et al. with conditions almost similar to those of the Dashidaira reservoir (e.g., steep slopes and a wide variety of sediment size distributions) [29], this sediment transport formula was selected for the reference case and subsequently for the sensitivity analysis and model calibration. Then, the sensitivity of the computed TVFS to changes in the empirical parameters was investigated. The computed TVFS was compared with the measured TVFS (i.e., 408.7 $\times$ 10^3 m^3), and if it was larger than 75% compared to the measured one, it was considered a model case for further assessment. Then, the final simulated bed topography pattern was compared, both qualitatively and quantitatively, with the measurements. A qualitative assessment was performed to check whether the erosion in area I, where the coarser sediments exist, could be captured, and a quantitative assessment was conducted to measure the deviation of the simulated bed levels from the measured ones. Therefore, Mean Absolute Error (MAE) of the simulated bed levels after the flushing operation was calculated. Table 2 summarizes the results of the TVFS sensitivity analysis for the major selected empirical parameters. Also, variations in the MAE of the simulated bed levels for each case have been calculated. The TVFS increased with increasing the active layer thickness and water content but decreased with increasing the critical angle of repose. When a larger amount of bed materials can be eroded during one time step (i.e., a thicker active layer), a higher volume of erosion from the deposits is expected. A higher water content in the sediment deposits (e.g., 50%), which decreases the submerged density of the bed material, is assumed to lead to greater sediment entrainment. A higher critical angle of repose retains a steeper side bank of the flushing channel after each time step, resulting in further deepening of the channel, which is not an efficient approach for increasing the TVFS in the numerical model. In contrast, the lateral development of the flushing channel (i.e., channel widening), which is favorable for increasing the TVFS, can be achieved using a lower critical angle of repose. The roughness, active layer thickness, water content of the bed material, and critical angle of repose were after the calibration set to 0.5 m, 1 m, 40% and 32 degrees, respectively, because the application of these values can result in a reasonable TVFS and accuracy (i.e., 313.0 $\times$ 10^3 m^3, and 1.8 m) and can also satisfy the mentioned qualitative criteria.

Table 2. Sensitivity analysis of the TVFS in reference case with respect to the selected empirical parameters along with variation of the MAE of simulated bed levels for each case.

Parameter	Active Layer Thickness (m)			Water Content of the Bed Material(%)			Critical Angle of Repose (Degree)		
	0.3	**0.45**	**0.85**	**50**	**43**	**38**	**33**	**34**	**35**
TVFS ($\times 10^{-3}$ m^3)	261.6	290.5	299.8	369.1	306.2	316.8	311.0	302.7	284.5
MAE (m)	2.17	1.73	1.95	2.25	2.10	1.75	1.98	1.54	1.98

4.2. *Evaluation of the Flow Field and Morphological Bed Changes in the Reservoir*

Figure 2 shows the computational grid adjustment at the beginning (i.e., t = 10 h; Figure 2(a1)) and during the free-flow conditions with a low water head in the reservoir (i.e., t = 32 h; Figure 2(a2)), and corresponding surface water velocity field (i.e., Figure 2(b1,b2)). The cells with a smaller water head than a specified value were removed from the computational domain due to the employed wetting/drying algorithm. The grid adjustment also reveals that the flow was deflected to the right-hand side of area II during the free-flow conditions; consequently, the flushing channel location was close to the right bank. Figure 2(b1) shows that a complex flow field with a strong reverse flow pattern and water stagnant zone develops in the lower half of the reservoir. This can be attributed to the complex geometry of the computational domain, the variation in flow depths from shallow conditions in the upstream areas to deep conditions in the downstream areas, and the existing bed roughness of the computational domain. In addition, Figure 2(b2) shows the water flow concentration in the flushing channel when the water level is lowest in the reservoir. During the free-flow condition, as shown in Figure 2(b2), the velocities rise to approximately 4.5 m/s, and supercritical flows are likely to develop in several zones of the flushing channel. However, the 3D numerical model can capture the complex characteristics of the flow field in channel bends (i.e., erosion along the outside of the bend and deposition along the inside) and reproduced the non-symmetrical velocity profile over the width and the tilting the lateral water surface at the apex of the channel bend [43]. As can be observed from Figure 2(b1,b2), the wet area of the computational domain contains the water body and the corresponding surface velocity vectors so that it is distinguished from the dry area without surface velocity vectors. An approximately 2-m-thick deposit along the left bank of the dead zone, shown in Figure 1d, can be attributed to the reverse flows that developed beside the bank of the reservoir during the drawdown stage as shown in Figure 2(b1) and that transported suspended fine materials into the stagnant water zone.

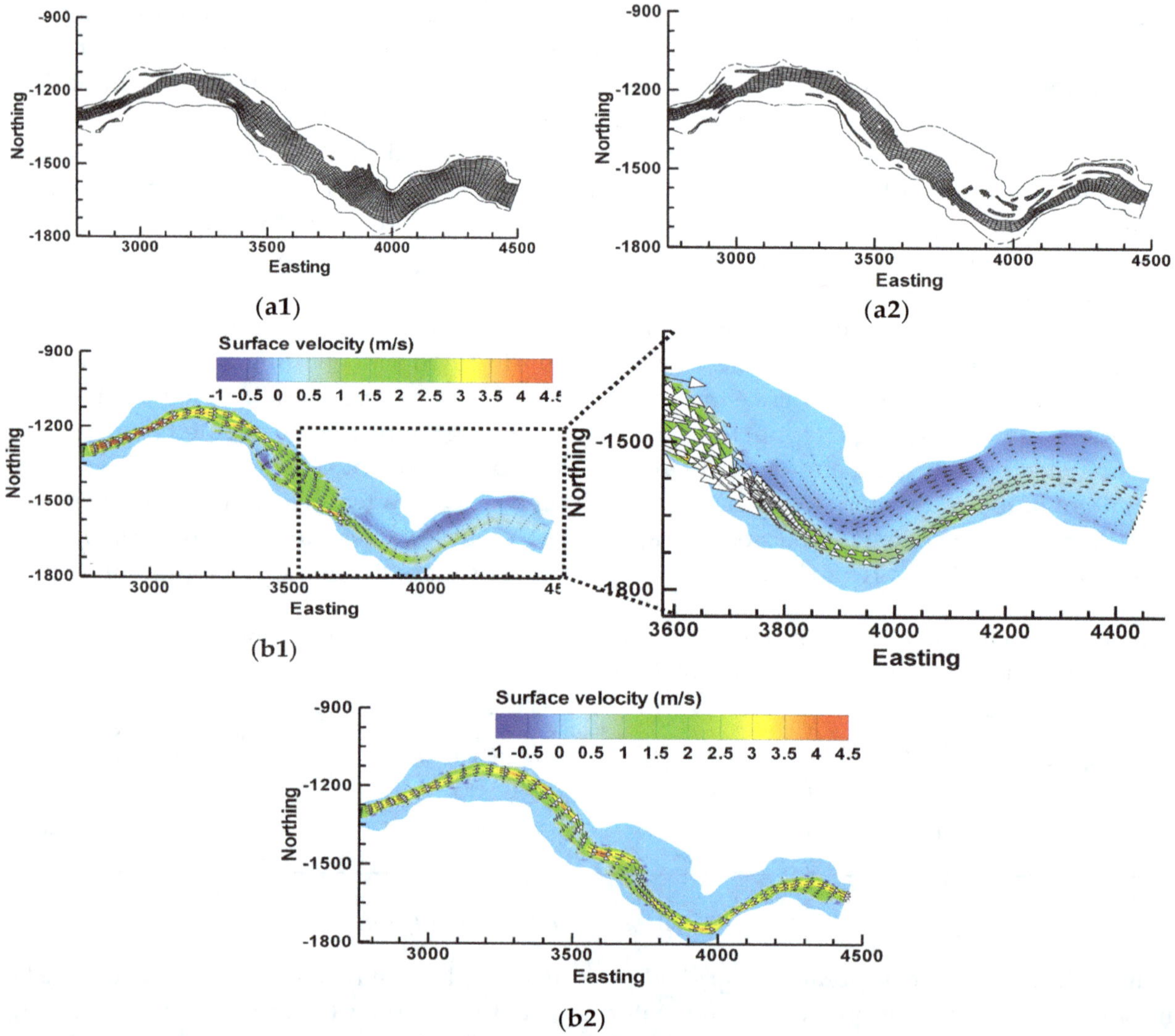

Figure 2. Computational grid (**a1**) at the beginning of the drawdown stage (t = 10 h) and (**a2**) during the free-flow condition in the Dashidaira reservoir (t = 32 h); corresponding surface velocity fields: (**b1**) at the beginning of the drawdown stage (t = 10 h) and (**b2**) during the free-flow condition (t = 32 h). The illustration on the right in (**b1**) shows the reverse flow domain and stagnant water zone.

Figure 3a illustrates the measured bed topography after the 2012 flushing operation in comparison with the simulated final bed topography after the flushing operation using the MPM and Van Rijn formulas (Figure 3b,c, respectively). To provide more quantitative insights into the simulated final bed topography after the flushing operation the BCI is defined as follows:

$$\text{BCI} = \frac{\sum_{i=1}^{n} (z_{i\_ms} - z_{i\_reference})}{n} \tag{10}$$

where z_{i_ms} is the measured or simulated bed level after the flushing operation at each grid node and $z_{i\_reference}$ is the measured or simulated reference bed level at the corresponding node, which is used for comparison purposes to provide information about erosion or deposition over a specific zone of the computational grid. Furthermore, n is the number of grid nodes considered for comparison purposes. Positive and negative values of BCI represent depositional and erosional conditions, respectively. In other words, BCI reveals the average change in the bed level of each target zone and readily indicates the dominant morphological process (i.e., erosion or deposition) in the zone compared to the reference case.

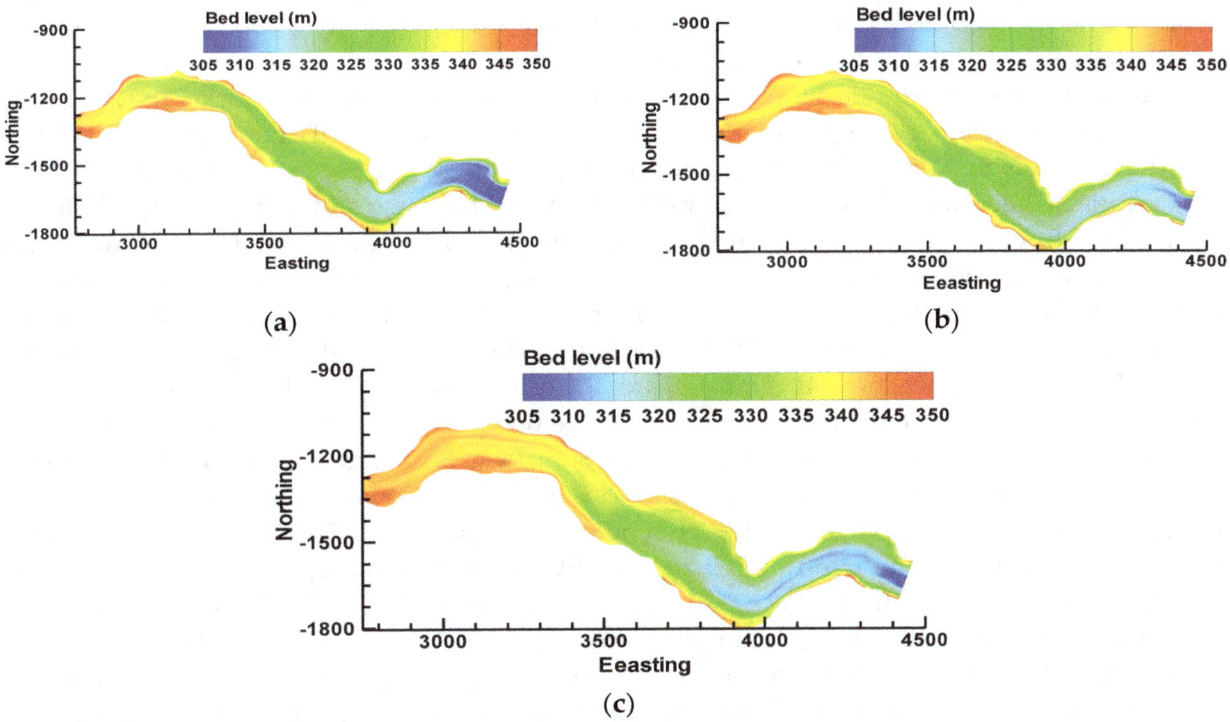

Figure 3. (**a**) Plan view of the measured bed topography after the 2012 flushing operation; Plan view of the simulated bed topography using (**b**) the MPM formula and (**c**) the Van Rijn formula.

BCI parameter was extracted in different areas of the reservoir using the measured bed levels after flushing and for calculations based on the MPM and Van Rijn bed load sediment transport formula when the measured bed levels before flushing (Figure 1b) are considered as reference case. In the upstream area I, the measurements reveal an average erosion value of 2.03 and simulated bed levels via MPM and Van Rijn formula in the calibrated case show average erosion value of 1.61 and 0.15 m, respectively. In this area, the performance of the MPM formula was better than Van Rijn formula in prediction of the morphological bed changes. In the wider midstream segment (i.e., area II), the measurements using the BCI parameter reveal an average erosion value of 0.67 m whereas simulated bed levels via MPM and Van Rijn formula show average erosion value of 1.59 and 3.11 m, respectively. The predicted morphological bed changes in this area obtained using the MPM formula were less overestimated than those obtained using the Van Rijn formula. In the area close to the dam (i.e., area III), the results of the simulations showed a narrower flushing channel compared to the measurements, regardless of the sediment transport formula used [43].

5. Discussion

5.1. Hydrodynamic Scenarios and Their Impacts on the Bed Morphology and Flushing Efficiency

5.1.1. Discharge Scenarios

One of the feasible scenarios in the Dashidaira reservoir is the introduction of an Additional inflow Discharge during the Free-flow stage of the flushing operation (i.e., the ADF scenario).

The additional discharge could be supplied from reservoirs located upstream of the Dashidaira reservoir. This additional inflow could enhance sediment entrainment by increasing the average flow velocity and bed shear stress in the Dashidaira reservoir. The ADF scenario is currently applicable, and preliminary tests for this scenario are being implemented in the Dashidaira reservoir.

Figure 4a shows the additional inflow that could be supplied in different ADF scenarios, and Figure 4b shows the discharge rates and original water level (i.e., Q and h, respectively) together with modified discharge rates under different ADF scenarios in the Dashidaira reservoir. Under the different ADF scenarios, the original boundary conditions of the water level (i.e., recorded during the 2012 flushing operation) have been retained, but the original boundary conditions of the discharge during the free-flow stage of the 2012 flushing operation change depending on which ADF scenario is used. For instance, ADF 110 indicates that 110 m^3/s of additional inflow discharge has been added to the original discharges during the free-flow stage of the 2012 flushing operation. With an original TVFS value equals to 313.0×10^3 m^3 in the reference case (i.e., resulting bed topography after simulation of 2012 flushing), the TVFS increased to 356.0×10^3 in the ADF 75 scenario, 396.1×10^3 in the ADF 110 scenario, and 425.0×10^3 m^3 in the ADF 170 scenario. The effects of introducing constant additional discharges under various ADF scenarios on the FE and the TVFS are illustrated in Figure 4c. The horizontal axis shows the ratio of average discharge during the free-flow stage using different ADF scenarios (i.e., Q_2) to the average discharge during the free-flow stage when no additional discharge is introduced in the reference case (i.e., Q_1). FE_2 and FE_1 are the flushing efficiencies when an ADF scenario is employed and when no additional discharge is employed in the reference case, respectively. The TVFS increases when the discharge increases during the free-flow stage. The FE values reached approximately −6.5% when the ADF 60 scenario was used. In this case, the increase in the flushed sediment volume was smaller than the increase in the used water volume according to the FE definition. Under the ADF 75, 90, and 110 scenarios, both the FE and TVFS increased with increasing average discharge during the free-flow stage. Then, the FE variation trended downward until a stable level was reached for the ADF 150 and ADF 170 scenarios. In contrast, the TVFS continued to increase. Under the given conditions, increasing the discharge magnitude during the free-flow condition can increase the TVFS, but this increase is not proportional to the discharge increase that causes the decrease in the FE for some cases. According to the diagram shown in Figure 4c, when the average discharge during the free-flow stage increased by approximately 56% under the ADF 110 scenario, (i.e., $Q_2/Q_1 = 1.56$), the FE increased approximately by 5%. Under these conditions, the total used water volume for the flushing operation increased by approximately 21%. In Table 3 the average bed level changes compared to the reference case using the BCI parameter in the upstream, midstream and downstream areas have been revealed for the ADF 75, ADF 110, and ADF 170 scenarios. As shown in Figure 4c and Table 3, introducing additional discharge increases the erosion in all areas by between 3% and 36% depending on the additional discharge, but the effect in the areas close to the dam (i.e., areas II and III) is more pronounced. Instead of adding a constant discharge to the original discharge values during the free-flow condition in the reference case, another scenario (i.e., the PDF scenario) using the same additional water volume over a shorter duration and in the form of discharge pulses was tested.

The PDF scenario was introduced to determine whether changing the characteristics of the additional inflowing water (e.g., the discharge intensity) markedly affects the quantity of flushed sediments and the bed changes in specific zones of the reservoir. The concept of the PDF scenario has been illustrated schematically in Figure 4d. It should be noted that before the flushing operation, area III was mainly covered with fine materials. During the simulation of the flushing operation, eroded coarser materials from area I were deposited in the lower parts of area III due to the reduced bed shear stress. Thus, introducing an additional discharge during the free-flow condition can contribute to flushing the deposited sediments out of this area.

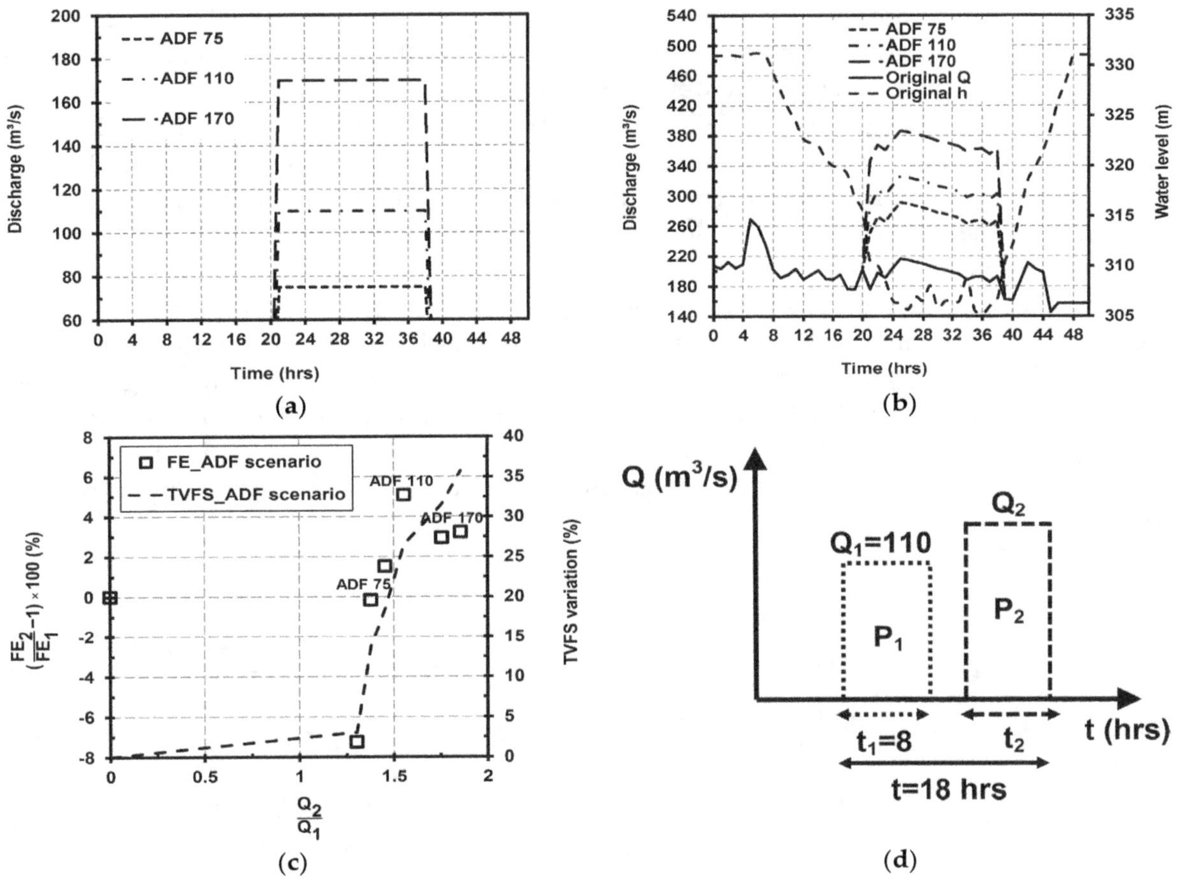

Figure 4. (**a**) Constant additional discharge rates used for different ADF scenarios in the Dashidaira reservoir. These additional rates are added to the original discharge rates shown in Figure 1c; (**b**) Modified hydrodynamic boundary conditions for different ADF scenarios in the Dashidaira reservoir; (**c**) Non-dimensional curves showing the relationships among TVFS, FE, and the water discharge used under different ADF scenarios compared to the reference case; (**d**) Schematic figure illustrating the PDF scenario with a variable discharge pulse in the second half of the free-flow stage.

Because the best FE correlation was found for the ADF 110 scenario as shown in Figure 4c, the total volume of the additional water used during the 18 h of the free-flow stage was calculated, and this additional water volume was introduced into the model in the form of two discharge pulses, the first one in the first half of the free-flow stage (i.e., P1) and the second one (i.e., P2) in the second half. More specifically, in the PDF scenario, a constant 110 m^3/s discharge pulse within 8 h in the first half of the free-flow flushing (i.e., P1 110 8) and a second discharge pulse with variable magnitude and duration in the second half (i.e., P2 Q_2 t_2) were introduced for further assessments.

In Figure 5, the bed changes with and without the introduction of the additional discharge (i.e., the reference case, ADF 110, and ADF 170) have been plotted at the cross-sections A-A, E-E, F-F, H-H, K-K, and L-L. As can be observed from Figure 5(f1,f2), instead of erosion, deposition occurs in cross section L-L. At the end of free-flow stage during the flushing process, bottom outlets are closed and water level starts to increase beside the dam while it is still low in the upstream portions of the reservoir and free-flow condition exists. In such a condition, the induced bed shear stress is reduced noticeably in areas close to the dam. Therefore, the eroded large size sediments are still transported towards downstream but they are deposited close to the dam without chance to flush. Table 3 reveals that the introduction of an additional discharge during the free-flow condition, in ADF scenarios, has a marginal effect on the erosion of coarser sediments in area I. This pattern can be attributed to the major role of the water-level drawdown stage in the initial development and evolution of the flushing channel, as revealed by experimental and numerical model studies by Esmaeili et al. and Kantoush and Schleiss [30,44]. Notably, from the reservoir entrance up to the cross-section A-A, the BCI values showed 0.04, 0.06, and 0.10 m of erosion under ADF 75, 110, and 170 scenarios compared to the simulated bed levels after 2012 flushing operation. Because the BCI values indicate overall deposition in area I, the eroded coarser bed materials from the area upstream of section A-A are likely deposited again after passing the cross-section A-A as a result of the increasing flow depth and the consequent velocity reduction. This process prevents the smaller grain size bed materials underneath from being eroded. However, if the deposition of coarser materials occurs in the main flushing channel, the erosion will increase, even with a smaller additional discharge (e.g., in the ADF 75 scenario) due to an increase in the bed shear stresses in the channel. In addition, the inflowing discharge during the free-flow condition is concentrated in the central flushing channel, which mainly contributes to a slow widening and deepening of the existing channel. This widening and deepening process is more effective in areas covered by finer sediments (e.g., areas II and III). Thus, use of the 170 m^3/s of additional inflow during the free-flow stage may result in a further increase in the flushing channel width and depth in the lower part of area II and throughout area III.

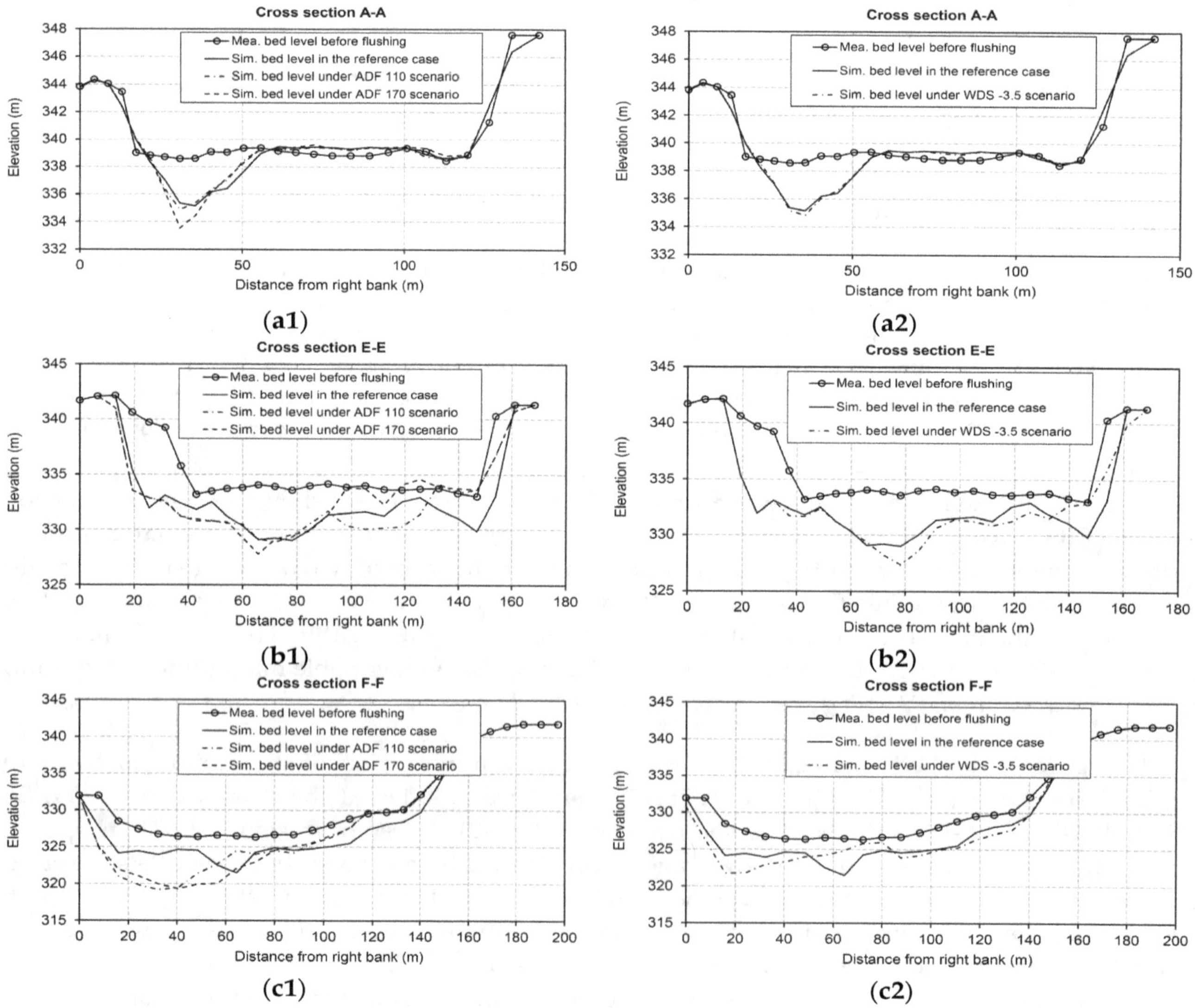

Figure 5. *Cont.*

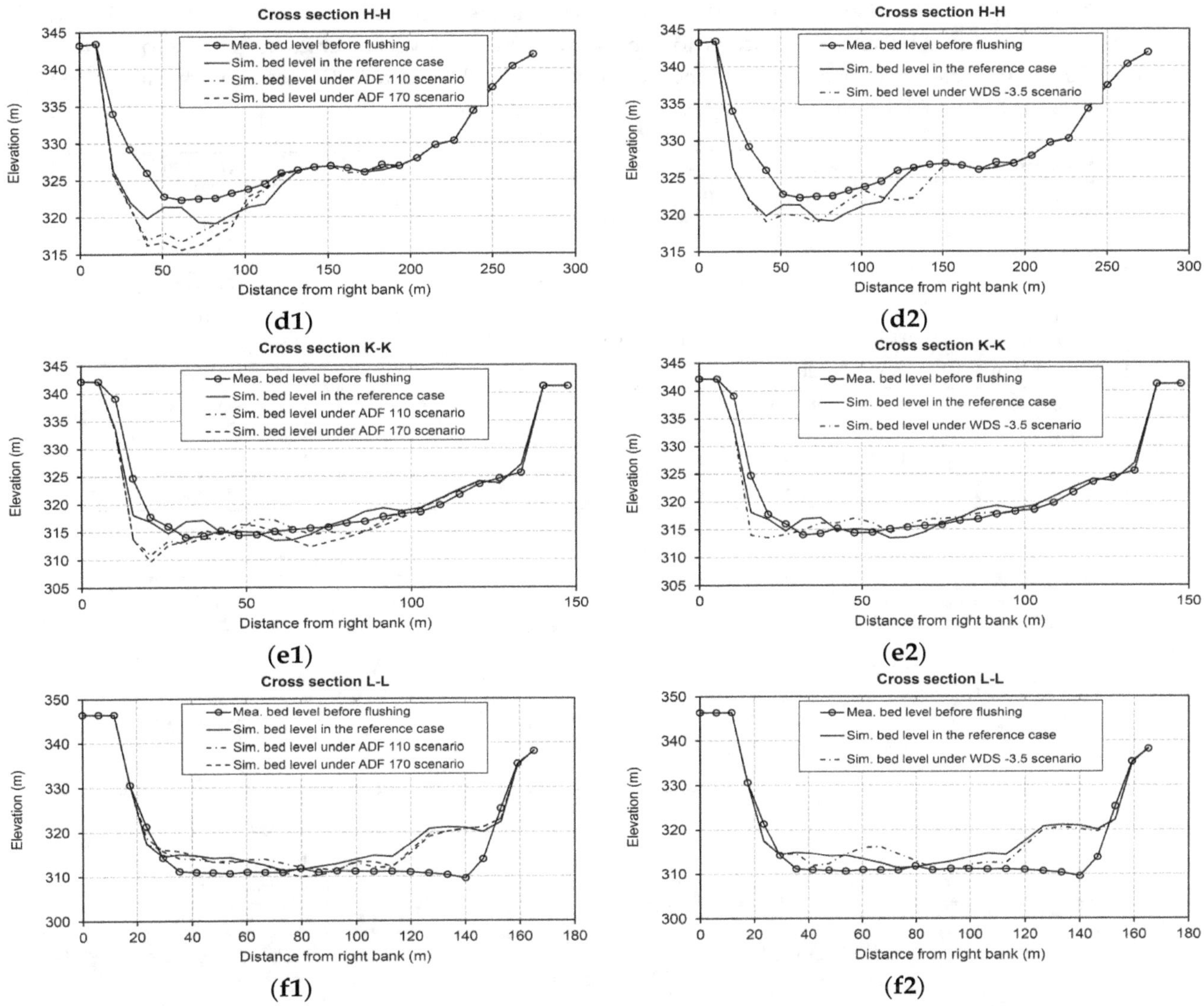

Figure 5. Measured bed levels before flushing along with the simulated bed levels after flushing in the reference case and under the ADF 110, ADF 170 and WDS −3.5 scenarios at (**a1,a2**) cross section A-A; (**b1,b2**) cross section E-E; (**c1,c2**) cross section F-F; (**d1,d2**) cross section H-H; (**e1,e2**) cross section K-K; and (**f1,f2**) cross-section L-L. Locations of the cross sections can be found in Figure 1b. Mea. and Sim. are abbreviations for Simulated and Measured, respectively.

Also, the resulting BCI parameter for the PDF scenario has been shown in Table 3. Using the PDF scenario not only enhanced the TVFS by increasing the erosion from areas II and III but also increased the erosion of coarser material from area I when the second discharge pulse was sufficiently high (i.e., PDF P1 110 8-P2 183.5 6). These increases can be attributed to the transport of already eroded coarser materials from area I to farther downstream areas due to a higher induced bed shear stress. Increasing the TVFS under the PDF scenarios, can increase the FE about 13% compared to the reference case. Nevertheless, due to the further deposition of sediments eroded from the head of the reservoir in the dam vicinity, the bed level changes appear to be marginal compared to the ADF 110 scenario when the second discharge pulse is high. Figure 6 shows the bed changes in different cross-sections in area I under the PDF P1 110 8-P2 183.5 6 scenario compared to the bed changes using the ADF 110 scenario to quantitatively show the advantageous performance of the PDF scenario in eroding the coarser bed materials in area I. However, the use of ADF or PDF scenario does not affect the main flushing channel location close to the right bank in area II. Consequently, the deposited bed materials close to the left bank in area II (i.e., the dead zone) still cannot be effectively removed.

Table 3. Average bed level changes in different areas of the reservoir under different ADF, PDF, and WDS scenarios. For ADF and WDS scenarios, the simulated bed levels after 2012 flushing operation were used as the reference case to extract the BCI parameter, whereas for PDF scenario the simulated bed levels under the ADF 110 scenario were used as the reference case.

Scenario		**ADF 75**			**ADF 110**			**ADF 170**	
Area	I	II	III	I	II	III	I	II	III
BCI (m)	0.32	−0.47	−0.54	0.06	−0.49	−0.55	0.20	−0.76	−0.90
TVFS ($\times 10^{-3}$ m^3)		356.0			396.1			425.0	
Scenario		**PDF P1 110 8-P2 137.5 8**			**PDF P1 110 8-P2 157 7**			**PDF P1 110 8-P2 183.5 6**	
Area	I	II	III	I	II	III	I	II	III
BCI (m)	−0.01	−0.09	−0.21	0.02	−0.04	−0.20	−0.09	0.01	0.04
TVFS ($\times 10^{-3}$ m^3)		426.2			417.3			410.9	
Scenario		**WDS −0.5**			**WDS −2.5**			**WDS −3.5**	
Area	I	II	III	I	II	III	I	II	III
BCI (m)	0.03	−0.14	0.02	0.10	−0.13	−0.23	−0.08	−0.33	−0.22
TVFS ($\times 10^{-3}$ m^3)		322.8			331.9			378.9	

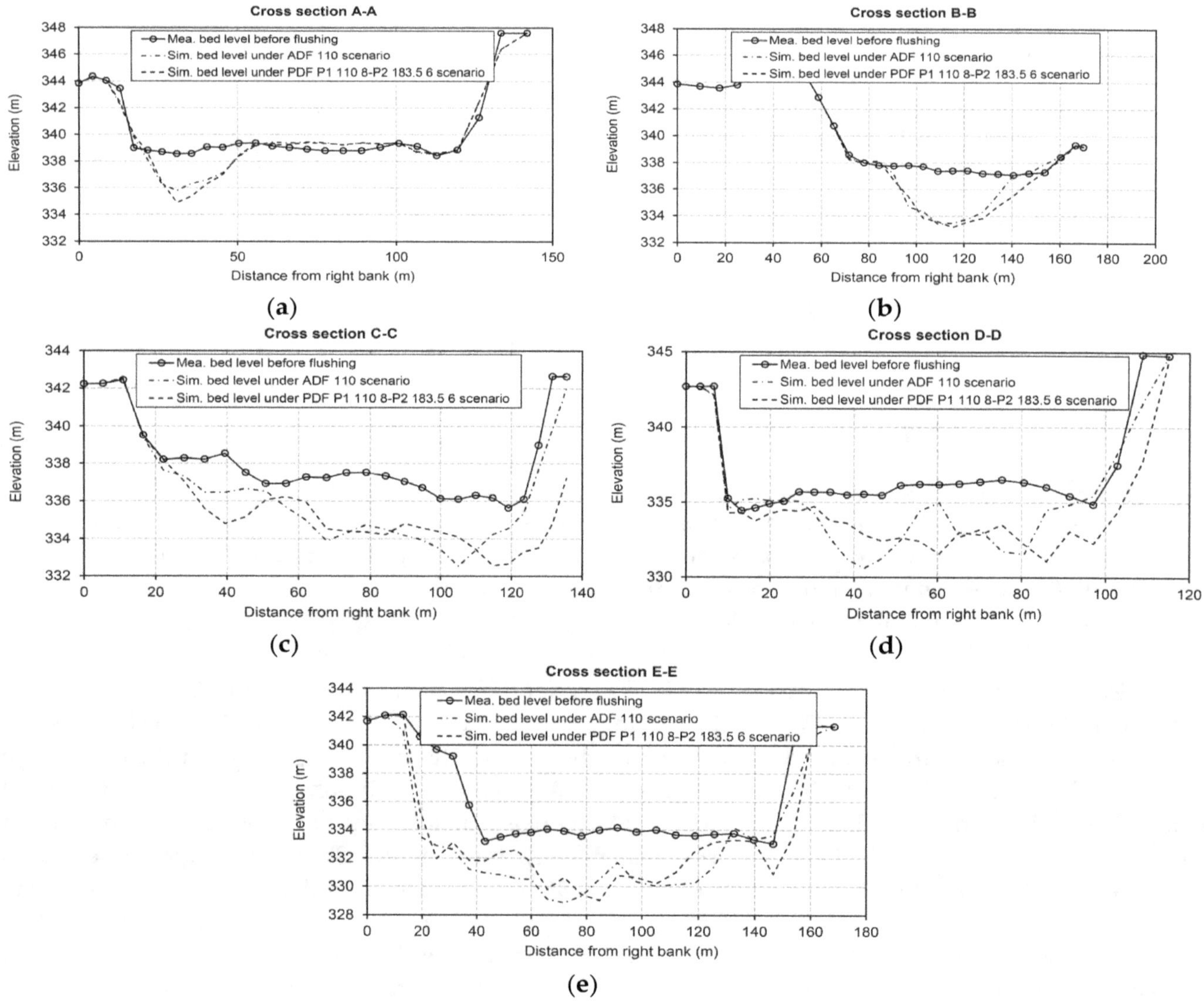

Figure 6. Measured bed levels before flushing and the simulated bed levels after flushing under the ADF 110 and PDF P1 110 8-P2 183.5 6 scenarios at (**a**) cross section A-A; (**b**) cross section B-B; (**c**) cross section C-C; (**d**) cross section D-D; and (**e**) cross section E-E. Locations of the sections can be found in Figure 1b.

5.1.2. Water Level Scenarios

Dominant role of the water-level drawdown stage in scouring the flushing channel has been already highlighted in physical model studies [30,44]. Therefore, another feasible scenario for the Dashidaira reservoir is increasing the Water-level Drawdown Speed (i.e., the WDS scenario). A target limb during the second half of the drawdown stage (i.e., between t = 12 and 20 h after starting the flushing) in the water-level variation curve is selected. Then, the original water-level drawdown rate is increased significantly for 1 hour at the beginning of the target limb (i.e., between t = 12 and 13 h after starting the flushing). Afterwards, the original drawdown rates remain unchanged during the remaining time (i.e., from t = 13 until t = 20 h). As a result, introducing an extra drop in the water level (e.g., 0.5, 2.5, and 3.5 m) for one hour while the original discharge rates remain unchanged causes the water velocity to increase abruptly, which acts as a strong driving force over the bed materials. For example, the WDS −0.5 scenario represents an extra 0.5-m drop in the water level at the beginning of the relevant limb (i.e., between t = 12 and 20 h). This scenario can be created via gate operations during a flushing event (e.g., a temporary increase in the gate opening speed) or with modification of the bottom outlet geometry to increase the discharge capacity of the bottom outlets, which can contribute to a faster drawdown process. Figure 7a illustrates the original water level and discharge rates, together with water-level modifications, during the target limb for three different WDS scenarios in the Dashidaira reservoir.

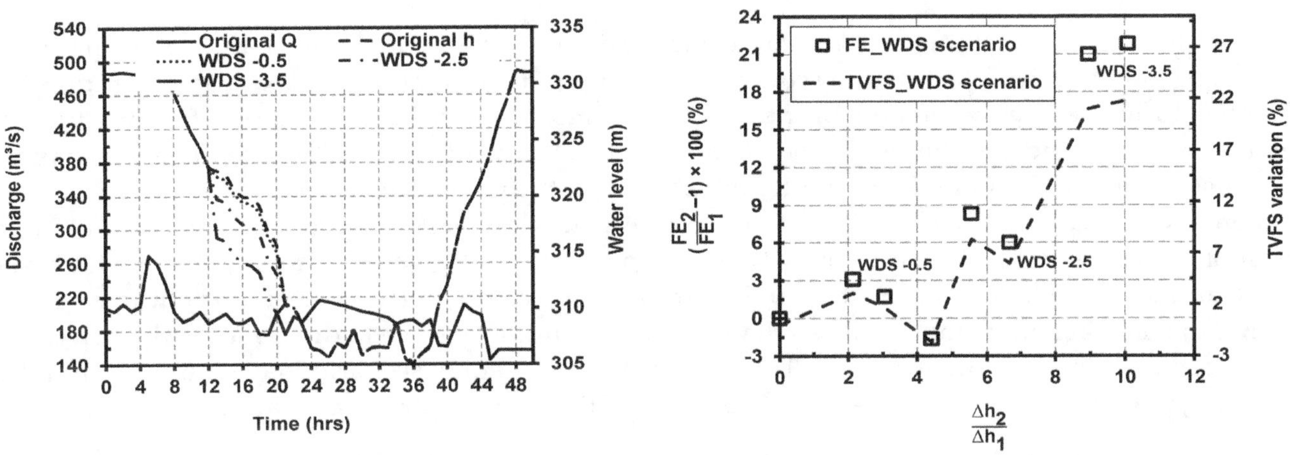

(a) (b)

Figure 7. (**a**) Utilized water levels and discharge rates for different WDS scenarios; (**b**) Non-dimensional curves showing the relationship among TVFS, FE, and Δh under different WDS scenarios compared to the reference case.

The effects of introducing a faster drawdown of the water level on the FE and the TVFS under various WDS scenarios are shown in Figure 7b. The horizontal axis shows the ratio of the extra imposed water level drop (i.e., Δh_2) to the original one (i.e., Δh_1) at the beginning of the target limb during the drawdown stage. On the left vertical axis, FE_2 represents the flushing efficiency when a WDS scenario is employed, and FE_1 represents the flushing efficiency when the original water level of the 2012 flushing operation is applied. All calculated and presented values are relative to the reference case (i.e., the 2012 flushing operation). As shown in Figure 7b, the FE variations are overall directly related to the variations in $\Delta h_2/\Delta h_1$. However, removing the coarser material from the far upstream area of the reservoir requires a high extra drop in the water level (i.e., a high $\Delta h_2/\Delta h_1$). Moreover, in some cases, increasing $\Delta h_2/\Delta h_1$ results in lower TVFS and FE values because coarser eroded material moves from upstream areas to downstream areas and is deposited on finer materials in the deeper areas. If the driving forces produced by the extra drop in the water level are not strong enough to remobilize the newly deposited coarser sediments overlying finer sediments, the erosion of finer fractions may be lower, resulting in lower TVFS values and consequently lower FE_2 values. Table 3 also shows the average BCI values in the upstream, midstream and downstream areas of the Dashidaira reservoir after the application of the WDS −0.5, WDS −2.5, and WDS −3.5 scenarios. As shown in Table 3, compared to the reference case with the TVFS of 313.0×10^3 m^3, the TVFS increases slightly when the magnitude of the extra drop in the water level is small (e.g., 0.5 m in the WDS −0.5 scenario). When the extra drop in the water level is larger (e.g., 2.5 m), the TVFS increases but not remarkably. Although the water-level decrease can enhance the relative roughness (i.e., the ratio of the roughness height to the water depth), this increased roughness is not high enough to lead to the erosion of the coarser materials in the upstream areas. Thus, the main effect is limited to the finer materials in the WDS −0.5 and WDS −2.5 scenarios. Due to the larger extra water level drop in the WDS −3.5 scenario, higher bed erosion occurs over the entire reservoir, including the upstream areas covered with coarser materials. In Figure 5, the bed changes in different cross-sections located in areas I, II, and III under the WDS −3.5 scenario are compared to the bed levels in the reference simulations.

Although a marked drop in the water level enhances the sediment erosion from the entire reservoir area and increases the FE more than 20% as shown in Figure 7b, the flushing channel location remains close to the right bank in area II, and the deposits along the left bank (i.e., the dead zone) are not effectively removed. However, constraints imposed by the gate facilities for safe and quick opening, and existing risks regarding the side bank failures are necessary to be assessed carefully.

5.2. Auxiliary Channel Scenario

According to the literature, construction of longitudinal channels in the wide areas of the reservoir has been proposed for affecting the scouring pattern [4]. Thus, a potential scenario for improving the erosion in the dead zone area of the Dashidaira reservoir is the application of a longitudinal auxiliary (i.e., secondary) channel in the dead zone to act as a side channel beside the main channel. In this scenario, the flushing flow is partly diverted from the main channel into the auxiliary channel and enters the main channel again at a confluence further downstream of the diversion point. Thus, the morphological processes occur along two flushing channels in area II. In contrast to the discharge and water-level scenarios, which affect the hydrodynamic characteristics within the entire reservoir during the flushing operation, auxiliary channel scenario influences the hydrodynamic characteristics in a specific segment of the reservoir. Thus, the auxiliary channel scenario may be more useful for resolving the issue of ineffective sediment scouring within the dead zone area in the reservoir.

The presence of two flushing channels in area II is better than just one in terms of several aspects. First, erosion occurs in two flushing channels instead of one, which means that flushing occurs over a longer channel (i.e., larger area). Consequently, higher TVFS and FE values are expected. When a longitudinal auxiliary channel is not used, the flushing channel merely develops along the river thalweg (i.e., pre-existing flushing channel). The flushing channel attracts the entire flow, and the channel mainly deepens, which is not an effective approach for increasing the TVFS and FE. Second, a remarkable amount of previously unerodible sediments can be eroded from the dead zone following the construction of a longitudinal auxiliary channel.

The schematic illustration showing the location of the applied auxiliary channel and also the surface velocity vectors in different stages of the flushing using an auxiliary channel are shown in Figure 8. The location of the entrance of auxiliary channel was selected in a place where a portion of the flushing flow can be easily diverted into the channel (Figure 8a). In addition, in the numerical model, side banks of the auxiliary channel were implemented along the existing grid lines in the selected area of the dead zone. The depth and length of the channel were set to be 2.5 and 403 m, respectively. As shown in Figure 8b, the auxiliary channel deflects a portion of the water flow towards the dead zone, and the remaining flow continues its original path close to the right bank of area II along the original thalweg of the main flushing channel. At the beginning of the drawdown stage (i.e., $t = 10$ h) in Figure 8c, the incoming flow is mainly deflected towards the auxiliary channel. At the middle of the drawdown stage (i.e., $t = 16$ h), the flow bifurcates completely, with a major portion of the incoming flow being deflected towards the main channel and a smaller portion being deflected towards the auxiliary channel, as shown in Figure 8d. After erosion of the existing sandbar, between the main and auxiliary channel, and the consequent water exchange between the channels due to the small water level fluctuations during the free-flow stage, the water flow is distributed over a wider flushing channel. These conditions continue during the free-flow stage (i.e., t = 32 h), as shown in Figure 8e. More specifically, using a longitudinal auxiliary channel along the dead zone develops another thalweg away from the main flushing channel thalweg by attracting a portion of the incoming flow and preventing the entire flow from heading towards the right bank of area II.

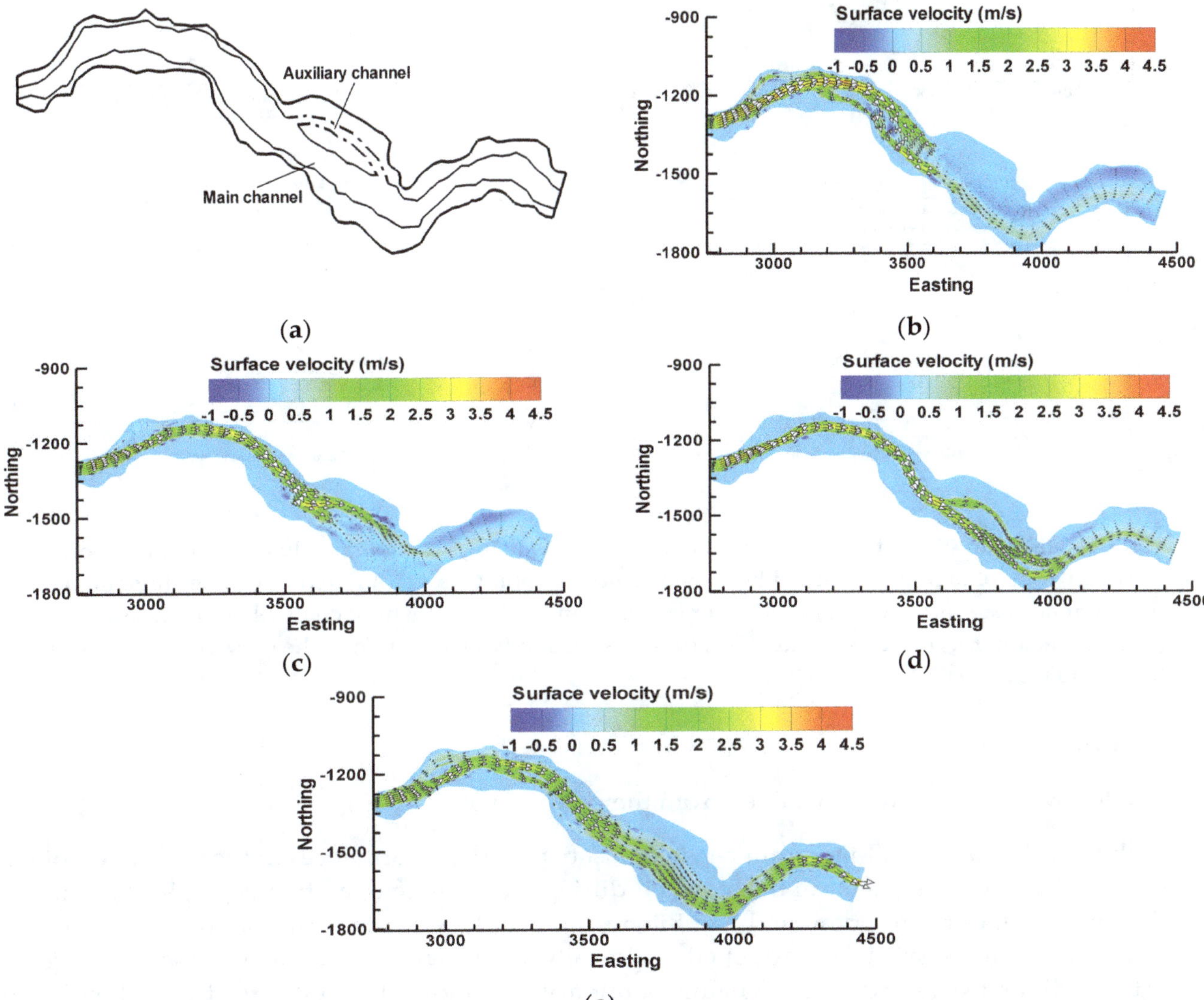

Figure 8. (**a**) Schematic illustration of the concept of longitudinal auxiliary flushing channel in the dead zone area of the Dashidaira reservoir; The surface velocity field in different stages using an auxiliary flushing channel: (**b**) before onset of the drawdown stage (i.e., t = 5 h); (**c**) at the beginning of the drawdown stage (i.e., t = 10 h); (**d**) at the middle of the drawdown stage (i.e., t = 16 h); and (**e**) during the free-flow condition (i.e., t = 32 h).

Figure 9a illustrates the plan view of the final bed morphology in the scenario with an auxiliary longitudinal channel in the dead zone of area II. Utilizing this scenario increases the TVFS to 340.0×10^3 m^3, compared to the 313.0×10^3 m^3 of the reference case. In Figure 9b–d, the final bed levels in different cross-sections of area II are shown after the flushing operation and are compared to the bed levels in the reference case. One can clearly see that the bed levels decreased noticeably along the thalweg of the auxiliary flushing channel and meanwhile the channel widened. Both deepening and widening process of the auxiliary channel contributes in increasing the TVFS and subsequently the flushing efficiency.

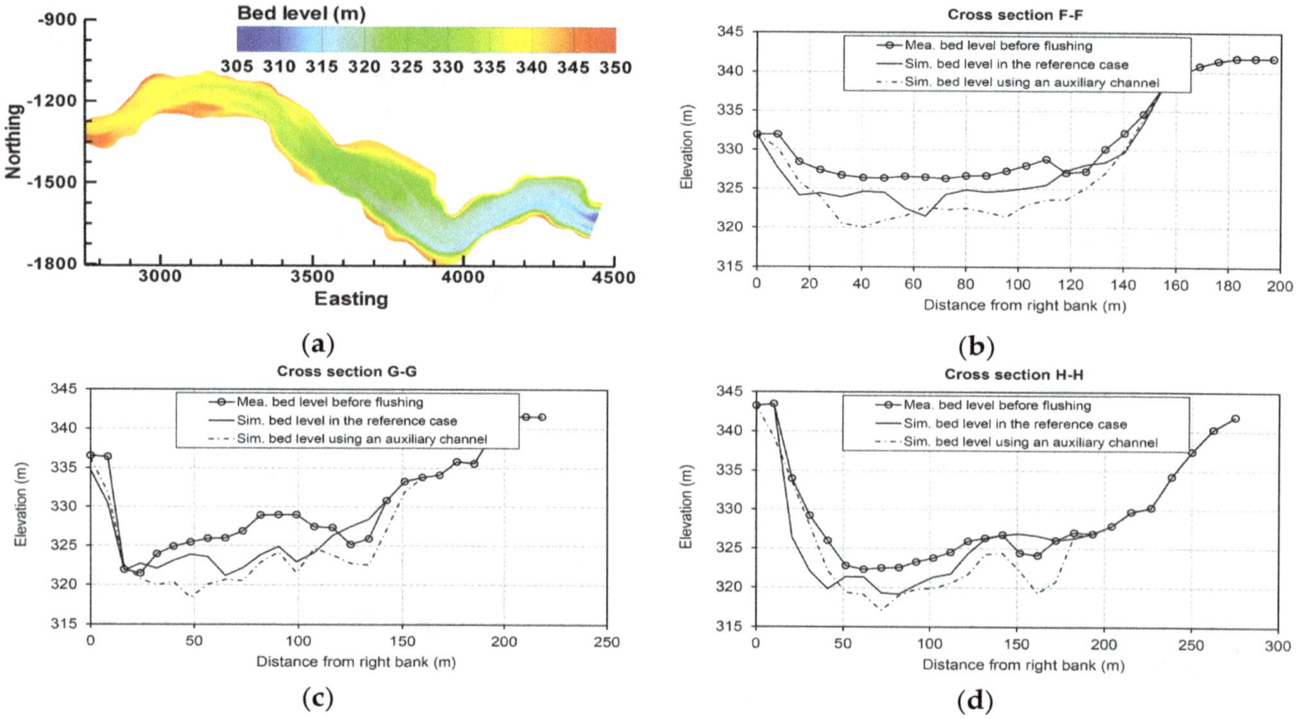

Figure 9. (**a**) Plan view of the final simulated bed levels using an auxiliary flushing channel along the dead zone of area II. Measured bed levels before flushing and simulated ones after flushing in the reference case along with simulated ones using an auxiliary flushing channel, at the location of (**b**) cross section F-F; (**c**) cross section G-G; (**d**) cross section H-H. Location of the cross sections can be found in Figure 1b.

6. Conclusions

The following results were obtained from the present work:

- Both the MPM and Van Rijn formulas yielded satisfactory performances in the simulation of bed changes in specific segments of the reservoir during the flushing operation (e.g., MPM formula in the upper half of the reservoir and Van Rijn formula in the vicinity of the Dam). These sediment transport formulas have been developed empirically to calculate the sediment transport for a given set of sediment sizes and hydrodynamic boundary conditions. However, the bed sediment size distribution, bed roughness, and hydrodynamic boundary conditions change dynamically during the free-flow flushing process. Such significant changes cannot be handled by empirical sediment transport formulas due to their inherent limitations. Nevertheless, the MPM bed load sediment transport formula qualitatively and quantitatively performed better than the Van Rijn formula for the entire reservoir. The MPM formula was able to achieve TVFS values that were more than 75% that of the measured TVFS values. Due to the application of the empirical formulas, the alluvial roughness also could not be estimated appropriately, which further magnifies the mentioned inability of the sediment transport formulas to accurately represent the morphological bed changes.
- For the Dashidaira reservoir, introducing an artificial additional discharge during the free-flow stage is practically feasible since this discharge can be supplied from upstream reservoirs. In addition, because this additional discharge is introduced when the flushing gates are fully

opened and the water level is low, this discharge can be passed through the bottom outlets if its value is less than the maximum capacity of the outlets. Additional discharge has two major effects: first, it increases the induced bed shear stress and bed erosion and supplies an additional driving force to transport eroded sediments farther downstream in the reservoir and flush them out from the reservoir; second, it causes the water level to increase in the downstream river channel, which can be beneficial from an environmental point of view because it washes away fine materials from the downstream channel terraces (thereby preventing river channel clogging). However, it was found that introducing the extra discharge in the form of two discharge pulses with a larger discharge pulse in the second half of the free-flow stage more efficiently increases the FE ratio and the bed degradation in the upstream areas covered with coarser materials. The numerical outcomes showed that introducing approximately 21% more water from upstream reservoirs (i.e., an approximately 56% increase in the average free-flow discharge) can enhance the FE by approximately 5−13% compared to the reference case (i.e., the 2012 flushing operation), depending on how this additional discharge is delivered.

- The construction of an auxiliary longitudinal flushing channel in the dead zone area of the Dashidaira reservoir causes a portion of the flushing flow to deviate from the main channel into the auxiliary channel and to enter the main channel again via a confluence downstream of the diversion point. The non-diverted flow continues along its original path along the thalweg of the main flushing channel and the diverted flow towards the auxiliary channel scour the deposited sediments from the targeted dead zone in the reservoir. The flushing processes associated with the auxiliary longitudinal channel result in a flushing channel that is overall longer and wider. Hence, the FE is higher by as much as approximately 9% compared to the reference case.

Acknowledgments: The first author was funded by the Ministry of Education, Culture, Sports, Science, and Technology of Japan (MEXT) through the HSE program for Asian Megacities. The support of the Disaster Prevention Research Institute (DPRI) of Kyoto University for open access publication is acknowledged. Authors would like to give their thanks to Kansai Electric Power Co., Inc., NEWJEC Inc., and also Shuhei Minami for providing the field data used in this study framework.

Author Contributions: Taymaz Esmaeili analyzed and processed the data, performed the numerical model tests and prepared the first and final draft of this manuscript. Tetsuya Sumi, Sameh A. Kantoush, Stefan Haun, and Nils Rüther provided their theoretical and practical insights and expertise on the outcomes and also contributed to preparation of the manuscript with reading and corrections. Yoji Kubota contributed to preparation and interpretation of a part of input and output data in the numerical model development stage.

References

1. Kondolf, G.M.; Gao, Y.; Annandale, G.W.; Morris, G.L.; Jiang, E.; Zhang, J.; Cao, Y.; Carling, P.; Fu, K.; Guo, Q.; et al. Sustainable sediment management in reservoirs and regulated rivers: Experiences from five continents. *Earth's Future* **2014**, *2*, 256–280. [CrossRef]
2. Morris, G.L. Chapter 5: Sediment management and sustainable use of reservoirs. In *Modern Water Resources Engineering*; Wang, L.K., Yang, C.T., Eds.; Humana Press: Totowa, NJ, USA, 2014; Volume 15, pp. 279–337.
3. White, R. *Evacuation of Sediments from Reservoir*; Thomas Telford: London, UK, 2001.
4. Morris, G.L.; Fan, J. *Reservoir Sedimentation Handbook: Design and Management of Dams, Reservoirs and Watersheds for Sustainable Use*; McGraw-Hill: New York, NY, USA, 1998.
5. Shen, H.W. Flushing sediment through reservoirs. *J. Hydraul. Res.* **1999**, *37*, 743–757. [CrossRef]
6. Sumi, T.; Kantoush, S.A. Integrated management of reservoir sediment routing by flushing, replenishing, and bypassing sediments in Japanese river basins. In Proceedings of the 8th International Symposium on Ecohydraulics, Seoul, Korea, 12–16 September 2010.
7. Liu, J.; Minami, S.; Otsuki, H.; Liu, B.; Ashida, K. Prediction of concerted sediment flushing. *J. Hydraul. Eng.* **2004**, *130*, 1089–1096. [CrossRef]
8. Kantoush, S.A. Experimental Study on the Influence of the Geometry of Shallow Reservoirs on Flow Patterns and Sedimentation by Suspended Sediments. Ph.D. Thesis, École Polytechnique Fédérale de Lausanne, Lausanne, Switzerland, 2008.
9. Sloff, C.J.; Jagers, H.R.A.; Kitamura, Y. Study on the channel development in a wide reservoir. In Proceedings of the 2nd River Flow conference, Napoli, Italy, 23–25 June 2004.

10. Fukuoka, S.; Sumi, T.; Horiuchi, S. Sediment management on the arase dam removal project. In Proceedings of the 12th International Symposium on River Sedimentation, Kyoto, Japan, 2–5 September 2013.
11. Asahi, K.; Shimizu, Y.; Nelson, J.; Parker, G. Numerical simulation of river meandering with self-evolving banks. *J. Geophysic. Res.* **2013**, *118*, 2208–2229. [CrossRef]
12. Fukuoka, S.; Uchida, T. Toward integrated multi-scale simulations of flow and sediment transport in rivers. *J. Jpn. Soc. Civ. Eng.* **2013**, *69*, II_1–II_10. [CrossRef]
13. Fang, H.; Rodi, W. Three-dimensional calculation of flow and suspended sediment transport in the neighbourhoods of the dam for the Three Gorges Project (TGP) reservoir in the Yangtze River. *J. Hydraul. Res.* **2003**, *41*, 379–394. [CrossRef]
14. Lu, Y.J.; Wang, Z.Y. 3D numerical simulation for water flows and sediment deposition in dam areas of the Three Gorges project. *J. Hydraul. Eng.* **2009**, *135*, 755–769. [CrossRef]
15. Haun, S.; Olsen, N.R.B. Three-dimensional numerical modelling of the flushing process of the Kali Gandaki hydropower reservoir. *Lakes Reserv.* **2012**, *17*, 25–33. [CrossRef]
16. Khosronejad, A.; Rennie, C.D.; Salehi Neyshabouri, A.A.; Gholam, I. Three-dimensional numerical modeling of reservoir sediment release. *J. Hydraul. Res.* **2008**, *46*, 209–223. [CrossRef]
17. Harb, G.; Dorfmann, C.; Schneider, J.; Badura, H. Numerical analysis of sediment transport processes during a flushing event of an Alpine reservoir. In Proceedings of the 7th River Flow conference, Lausanne, Switzerland, 3–5 September 2014.
18. Gallerano, F.; Cannata, G. Compatibility of Reservoir Sediment Flushing and River Protection. *J. Hydraul. Eng.* **2011**, *137*, 1111–1125. [CrossRef]
19. Haun, S.; Olsen, N.R.B. Three-dimensional numerical modelling of reservoir flushing in a prototype scale. *Int. J. River Basin Manag.* **2012**, *10*, 341–349. [CrossRef]
20. Olsen, N.R.B. *A Three Dimensional Numerical Model for Simulation of Sediment Movement in Water Intakes with Multiblock Option*; Department of Hydraulic and Environmental Engineering, The Norwegian University of Science and Technology: Trondheim, Norway, 2014.
21. Esmaeili, T.; Sumi, T.; Kantoush, S.A.; Haun, S.; Rüther, N. Three-dimensional numerical modelling of flow field in shallow reservoirs. *Proc. ICE-Water Manag.* **2015**, *169*, 229–244. [CrossRef]
22. Dehghani, A.A.; Esmaeili, T.; Chang, W.Y.; Dehghani, N. 3D numerical simulation of local scouring under hydrographs. *Proc. ICE-Water Manag.* **2013**, *166*, 120–131. [CrossRef]
23. Haun, S.; Kjærås, H.; Løvfall, S.; Olsen, N.R.B. Three-dimensional measurements and numerical modelling of suspended sediments in a hydropower reservoir. *J. Hydrol.* **2013**, *479*, 180–188. [CrossRef]
24. Esmaeili, T.; Dehghani, A.A.; Zahiri, A.R.; Suzuki, K. 3D Numerical simulation of scouring around bridge piers (Case Study: Bridge 524 crosses the Tanana River). *World Acad. Sci. Eng. Technol.* **2009**, *58*, 1028–1032.
25. Fischer-Antze, T.; Olsen, N.R.B.; Gutknecht, D. Three-dimensional CFD modeling of morphological bed changes in the Danube River. *Water Resour. Res.* **2008**, *44*. [CrossRef]
26. Rüther, N.; Olsen, N.R.B. Modelling free-forming meander evolution in a laboratory channel using three-dimensional computational fluid dynamics. *Geomorphology* **2007**, *89*, 308–319. [CrossRef]
27. Ruether, N.; Singh, J.M.; Olsen, N.R.B.; Atkinson, E. 3-D computation of sediment transport at water intakes. *Proc. ICE-Water Manag.* **2005**, *158*, 1–7. [CrossRef]
28. Olsen, N.R.B.; Kjellesvig, H.M. Three-dimensional numerical modelling of bed changes in a sand trap. *J. Hydraul. Res.* **1999**, *37*, 189–198. [CrossRef]
29. Haun, S.; Dorfmann, C.; Harb, G.; Olsen, N.R.B. Numerical modelling of the reservoir flushing of the bodendorf reservoir, Austria. In Proceedings of the 2nd IAHR European Congress, Munich, Germany, 27–29 June 2012.
30. Esmaeili, T.; Sumi, T.; Kantoush, S.A. Experimental and numerical study of flushing channel formation in shallow reservoirs. *J. Jpn. Soc. Civ. Eng.* **2014**, *70*, I_19–I_24. [CrossRef]
31. Kantoush, S.A.; Sumi, T.; Suzuki, T.; Murasaki, M. Impacts of sediment flushing on channel evolution and morphological processes: Case study of the Kurobe River, Japan. In Proceedings of the 5th River Flow Conference, Braunschweig, Germany, 8–10 June 2010.
32. Minami, S.; Noguchi, K.; Otsuki, H.; Fukuri, H.; Shimahara, N.; Mizuta, J.; Takeuchi, M. Coordinated sediment flushing and effect verification of fine sediment discharge operation in Kurobe River. In Proceedings of the ICOLD Conference, Kyoto, Japan, 6–8 June 2012.
33. Sumi, T. Evaluation of efficiency of reservoir sediment flushing in Kurobe River. In Proceedings of the 4th International Conference on Scour and Erosion, Tokyo, Japan, 5–7 November 2008.

34. Launder, B.E.; Spalding, D.B. *Lectures in Mathematical Models of Turbulence*; Academic Press: London, UK, 1972.
35. Patankar, S. *Numerical Heat Transfer and Fluid Flow*; Taylor & Francis: New York, NY, USA, 1980.
36. Olsen, N.R.B.; Haun, S. Free surface algorithms for 3-D numerical modeling of flushing. In Proceedings of the 5th River Flow Conference, Braunschweig, Germany, 8–10 June 2010.
37. Schlichting, H. *Boundary Layer Theory*; McGraw-Hill: New York, NY, USA, 1979.
38. Van Rijn, L.C. Sediment transport, part III: Bed forms and alluvial roughness. *J. Hydraul. Eng.* **1984**, *110*, 1733–1754. [CrossRef]
39. Van Rijn, L.C. Sediment transport, part II: Suspended load Transport. *J. Hydraul. Eng.* **1984**, *110*, 1613–1641. [CrossRef]
40. Van Rijn, L.C. Sediment transport, part I: Bed load transport. *J. Hydraul. Eng.* **1984**, *110*, 1431–1456. [CrossRef]
41. Meyer-Peter, E.; Müller, R. Formulas for bed load transport. In Proceedings of the 2nd Meeting of the International Association for Hydraulic Structures Research, IAHR, Stockholm, Sweden, 7–9 June 1948.
42. Brooks, H. Boundary shear stress in curved trapezoidal channels. *J. Hydraul. Div. Am. Soc. Civ. Eng.* **1963**, *89*, 327–333.
43. Esmaeili, T.; Sumi, T.; Kantoush, S.A.; Kubota, Y.; Haun, S. Numerical study on flushing channel evolution, case study of Dashidaira reservoir, Kurobe river. *J. Jpn. Soc. Civ. Eng.* **2015**, *71*, I_115–I_120. [CrossRef]
44. Kantoush, S.A.; Schleiss, A.J. Channel formation in large shallow reservoirs with different geometries during flushing. *J. Environ. Technol.* **2009**, *30*, 855–863. [CrossRef] [PubMed]

6

Long-Term Scheduling of Large-Scale Cascade Hydropower Stations using Improved Differential Evolution Algorithm

Xiaohao Wen [1,2], Jianzhong Zhou [1,3,*], Zhongzheng He [1,3] and Chao Wang [4]

1 School of Hydropower and Information Engineering, Huazhong University of Science and Technology, Wuhan 430074, China; wen_xiaohao@ctg.com.cn (X.W.); hezz_hae@hust.edu.cn (Z.H.)
2 China Three Gorges Corporation, Beijing 100038, China
3 Hubei Key Laboratory of Digital Valley Science and Technology, Wuhan 430074, China
4 China Institute of Water Resources and Hydropower Research, Beijing 100038, China; wangchao@iwhr.com
* Correspondence: jz.zhou@mail.hust.edu.cn

Abstract: Long-term scheduling of large cascade hydropower stations (LSLCHS) is a complex problem of high dimension, nonlinearity, coupling and complex constraint. In view of the above problem, we present an improved differential evolution (iLSHADE) algorithm based on LSHADE, a state-of-the-art evolutionary algorithm. iLSHADE uses new mutation strategies "current to pbest/2-rand" to obtain wider search range and accelerate convergence with the preventing individual repeated failure evolution (PIRFE) strategy. The handling of complicated constraints strategy of ε-constrained method is presented to handle outflow, water level and output constraints in the cascade reservoir operation. Numerical experiments of 10 benchmark functions have been done, showing that iLSHADE has stable convergence and high efficiency. Furthermore, we demonstrate the performance of the iLSHADE algorithm by comparing it with other improved differential evolution algorithms for LSLCHS in four large hydropower stations of the Jinsha River. With the applications of iLSHADE in reservoir operation, LSLCHS can obtain more power generation benefit than other alternatives in dry, normal, and wet years. The results of numerical experiments and case studies show that the iLSHADE has a distinct optimization effect and good stability, and it is a valid and reliable tool to solve LSLCHS problem.

Keywords: energy; hydropower stations; differential evolution algorithm; optimal scheduling; ε-constrained method

1. Introduction

Hydropower has a significant share on the total energy consumption as it is renewable, clean, and cheap. Therefore, many countries have been working on the development of the utility of hydropower [1], and many hydropower plants have been put into operation in the past few decades [2–6]. Large cascade hydropower stations (LHS) play an increasingly important role in energy production. Many scholars have conducted a lot of research on the water resources management of LHS. Zhou et al. [7] proposed a joint optimal refill rules for cascade reservoirs to solve the conflict between the flood control and refill operation. The energy storage operation chart combined with discriminant coefficient method was put forward by Jiang [8], which was successfully applied to cascade reservoirs of Li Xianjiang River in southwest China. Regarding the input (e.g., inflow) imprecision and uncertainties, Chen et al. [9–12] analyzed the influence of the uncertainty in water resources management and the distribution of flood forecasting error. Djebou et al. [13,14] presented the interactions between these hydrologic factors that interplay at the watershed scale using the entropy-based index. Aiming to determine the optimal strategy that hedges the risk of energy shortfall, Xu et al. [15] develops a multi-objective stochastic programming model for informing hedging decisions for hydropower operations. Due to complex hydrodynamic relation, various complex constraints and diversified interest demand, the long-term scheduling of large cascade hydropower stations (LSLCHS) has developed into a multi-dimensional, non-convex and non-linear optimization problem. Correspondingly, optimization of LSLCHS has become a fairly challenging theoretical and practical problem, which is urgent for optimization techniques and constraints treatment [16–18]. Over many years, to solve this problem, various optimization algorithms have been applied. Usually there are two categories of methods: traditional optimizers and modern heuristic algorithms. Traditional optimizers include linear programming (LP) [19,20], nonlinear programming (NLP) [21,22], dynamic programs (DP) [23,24], progressive optimality algorithm (POA) [25,26], etc. These algorithms have rigorous mathematical foundations but low convergence efficiency. They suffer from curse of dimensionality. Modern heuristic algorithms use intelligent strategies to guide search to better areas, such as particle swarm optimization (PSO) [27,28], genetic algorithm (GA) [29], cultural algorithm (CA) [30], binary artificial sheep algorithm (BASA) [31], ant colony optimization (ACO) [18,32], etc. Compared with traditional optimizers, modern meta-heuristics are significantly more flexible and have high search efficiency as the meta-heuristics are inspired by different nature principles from biology, ethology, or physics. However, the common heuristic algorithms have some disadvantages such as premature convergence because of local fast convergence, and bad local search capability owing to many global searches. Moreover, they lack effective measures to handle complex constraints, making it difficult to be applied to solve complex optimal problems with high dimensions such as LSLCHS.

Differential evolution (DE) is a simple yet practical modern heuristic algorithm for global optimization over continuous spaces introduced by Price and Storn [33]. The DE algorithm has been used in many practical cases [34,35] and gradually become more popular. Similar to all other modern heuristic algorithms, the evolutionary process of DE uses mutations, crossover, and selection operators at each generation to reach the global optimum. The performance of DE basically depends on the mutation strategy, the crossover operator. Besides, the intrinsic control parameters (population size NP, scaling factor F, the crossover rate Cr) play a vital role in balancing the diversity of population and convergence speed of the algorithm. Therefore, Brest et al. [36] proposed a self-adaptive DE (jDE), in which both F and Cr are applied at random with probability $\tau 1$ and $\tau 2$. SaDE is proposed by Qin et al. [37] adaptively adjusts its trial vector generation strategies and control parameters simultaneously by learning from the previous search. JADE [38] is a well-known, effective DE variant which employs a control parameter adaptation mechanism and puts forward mutation strategy "current-to-pbest/1", differential evolution with composite trial vector generation strategies, control parameters (CoDE) [39], differential evolution with ensemble of parameters and mutation strategies (EPSDE) [40]. Success-History-based Adaptive DE (SHADE) [41] is an improved version of JADE which uses a different parameter adaptation mechanism. LSHADE [42] further extends SHADE with Linear Population Size Reduction (LPSR), which continually decreases the population size according to a linear function. In addition, LSHADE is the best ranked DE algorithm on CEC2014 Competition on Real-Parameter Single Objective Optimization [43] (see http://www.ntu.edu.sg/home/EPNSugan/index_files/CEC2014/-CEC2014.htm).

To avoid premature convergence and to accelerate convergence, we present an improved version of the LSHADE algorithm in this paper, called iLSHADE. The main improvement of iLSHADE is that a new mutation strategy "current to pbest/2-rand" is put forward for wider search range to improve search ability and prevent individual repeated failure evolution (PIRFE) strategy applied in the population evolution process. Finally, iLSHADE is applied to LSLCHS in Jinsha River combined with the improved constraints handling technique. Results of the study demonstrates its superiority in dealing with LSLCHS problem.

The remainder of this paper is organized as follows: Section 2 introduces the formulation of LSLCHS problem. In Section 3, a brief view of DE framework and improvement strategies of iLSHADE is presented. Section 4 presents numerical simulation experiment of iLSHADE. In Section 5, implementation of iLSHADE in LSLCHS is shown in detail. In Section 6, iLSHADE is applied to solve LSLCHS, and the results are analyzed. Finally, conclusions are summarized in Section 7.

2. Optimization Model

The primary objective of LSLCHS problem is to maximize the total power generation of LHS over the whole operation periods, while subjecting to kinds of equality and inequality constraints. The objective formula and constraints are described as follows.

2.1. Objective Function

$$\text{obj} = max\sum_{i=1}^{M}\sum_{t=1}^{T} N_{it}\Delta t,\ N_{it} = A_i H_{it} Q_{it} \tag{1}$$

where obj is the total power generation of LHS over the whole operation periods, M is the number of hydro plants; T is the whole periods; A_i is output coefficient of the i-th hydro plant; Δt shows interval of scheduling term; N_{it}, H_{it} and Q_{it} denote output, pure water head and water discharge through hydro-turbine of the i-th hydro plant in the t-th period, respectively. Moreover, H_{it} is calculated by upstream water level, trail water level and head loss shown in formula (8).

2.2. Constraints

In the process of long-term optimal dispatch, various complex equality and inequality constraints, such as water level, output, and hydraulic connection, should be considered for restricting the total power generation optimization. The constraints of LHS are described as follows:

1. Water balance constraint.

$$\begin{aligned} V_{i,t+1} &= V_{i,t} + (I_{i,t} - Q_{i,t} - S_{i,t})\Delta t, \\ I_{i,t} &= q_{i,t} + Q_{i-1,t} + S_{i-1,t} \end{aligned} \tag{2}$$

$V_{i,t}$ is reservoir storage of the i-th hydropower station at the beginning of period t, $I_{i,t}$ is inflow, $q_{i,t}$ stands for local inflow and $S_{i,t}$ is deserted outflow.

2. Hydraulic connection.

$$Z_{i,t}^{down} = \begin{cases} F(Q_{i,t} + S_{i,t}) & \text{without backwater effect,} \\ F(Q_{i,t} + S_{i,t},\ Z_{i+1,t}) & \text{with backwater effect.} \end{cases} \tag{3}$$

where $Z_{i,t}$ stands for upstream water level, $Z_{i,t}^{down}$ is trail water level. Function F represents the hydraulic connection between upstream and downstream hydropower stations. Generally, the trail water level is a function of outflow. However, when the hydropower station is located at the backwater region of its downstream hydropower station, the upstream water level of the downstream hydropower station must be taken into consideration in the function.

3. Water level constraint.

$$Z_{i,t}^{min} \le Z_{i,t} \le Z_{i,t}^{max} \tag{4}$$

$$|Z_{i,t} - Z_{i,t+1}| \le \Delta Z_i \tag{5}$$

$Z_{i,t}^{min}$ and $Z_{i,t}^{max}$ are the upper and lower water level limits and ΔZ_i is the maximum amplitude of water level variation.

4. Power generating constraint.

$$N_{i,t}^{min} \le N_{i,t} \le N_{i,t}^{max}(H_{i,t}) \tag{6}$$

$N_{i,t}^{max}(H_{i,t})$ represents the maximum output. The maximum output is a function of pure water head. $N_{i,t}^{min}$ is the lower output limit, which is generally called Guaranteed output.

5. Outflow constraint.

$$Q_{i,t}^{min} \leq Q_{i,t} + S_{i,t} \leq Q_{i,t}^{max} \tag{7}$$

$Q_{i,t}^{max}$ is the maximum outflow limit and $Q_{i,t}^{min}$ is the minimum outflow limit.

6. Water head equation.

$$H_{i,t} = (Z_{i,t} + Z_{i,t+1})/2 - Z_{i,t}^{\text{down}} - H_{i,t}^{loss}(Q_{i,t}) \tag{8}$$

$H_{i,t}$ stands for the pure water head. $Z_{i,t}^{\text{down}}$ is trail water level described in formula (3). $H_{i,t}^{loss}(Q_{i,t})$ represents water head loss, which is a function of outflow through hydro-turbines.

7. Boundary condition.

$$Z_{i,0} = Z_i^{begin},\ Z_{i,T} = Z_i^{end} \tag{9}$$

where Z_i^{begin} and Z_i^{end} are initial water level and terminal water level of the i-th hydro plant, respectively.

3. Overview of iLSHADE

3.1. DE

DE is a group-based evolutionary algorithm. It is used to solve the following continuous domain global optimization problem:

$$\begin{array}{l} minimize\ f(\vec{x}),\ \vec{x} = (x_1, \ldots, x_D) \\ x_i \in [\overline{x}_i, \underline{x}_i]\ \ \forall i \in \{1, \ldots, D\},\ \ -\infty < \overline{x}_i < \underline{x}_i < +\infty \end{array} \tag{10}$$

where $f(\vec{x})$ is continuous fitness evaluation function, D is the dimension of the problem. DE has three control parameters that need to be set before the calculation: F is scaling factor, CR is crossover control parameter, and NP is population size. The framework of DE is as follows (Figure 1):

Set the generation number $G = 0$
Sample NP $\vec{x}_{i,G}, \ldots, \vec{x}_{NP,G}$ form an initial population
while the termination criteria are not meet **do**
  for $i = 1$ to NP **do**
    Mutation: generate a mutant individual $\vec{v}_{i,G}$ using a DE mutation strategy
    Repair: correct $\vec{v}_{i,G}$ to make it feasible if $\vec{v}_{i,G}$ is not feasible,
    Crossover: mix $\vec{x}_{i,G}$ and $\vec{v}_{i,G}$ to generate a trial individual $\vec{u}_{i,G}$
    Selection: If $f(\vec{u}_{i,G}) \leq f(\vec{x}_{i,G})$, set $\vec{x}_{i,G+1} = \vec{u}_{i,G}$, otherwise, set $\vec{x}_{i,G+1} = \vec{x}_{i,G}$
  end for
  $G = G + 1$
end while
Output: the individual with the smallest objective function value in population

Figure 1. Pseudocode of DE.

The mutation strategy in original DE is "rand/1", which is expressed in formula (11):

$$\vec{v}_{i,G} = \vec{x}_{r1,G} + F \cdot (\vec{x}_{r2,G} - \vec{x}_{r3,G}) \tag{11}$$

Other common DE mutation strategies are as follows:

- "rand/2":

$$\vec{v}_{i,G} = \vec{x}_{r1,G} + F \cdot (\vec{x}_{r2,G} - \vec{x}_{r3,G}) + F \cdot (\vec{x}_{r4,G} - \vec{x}_{r5,G}) \tag{12}$$

- "best/1":

$$\vec{v}_{i,G} = \vec{x}_{best,G} + F \cdot (\vec{x}_{r1,G} - \vec{x}_{r2,G}) \tag{13}$$

- "best/2":

$$\vec{v}_{i,G} = \vec{x}_{i,G} + F \cdot (\vec{x}_{best,G} - \vec{x}_{i,G}) + F \cdot (\vec{x}_{r1,G} - \vec{x}_{r2,G}) \tag{14}$$

- "current to best/1":

$$\vec{v}_{i,G} = \vec{x}_{i,G} + F \cdot (\vec{x}_{best,G} - \vec{x}_{i,G}) + F \cdot (\vec{x}_{r1,G} - \vec{x}_{r2,G}) \tag{15}$$

where the indexes $r1 - r5$ represent the random and mutually different integers generated within the range $\{1, \text{NP}\}$, and also different from index i. $\vec{x}_{best,G}$ is the best individual in a current generation. Each strategy has a different ability to maintain the diversity of the population, which may increase/reduce the rate of convergence in the process of evolution.

3.2. *iLSHADE*

An improved LSHADE (iLSHADE) with new mutation strategy "current to best/2-rand" and the PIRFE strategy is proposed. The details of these strategies and algorithm procedure are shown below.

3.2.1. Mutation Strategy "Current to pbest/2-rand"

The mutation strategy "current to pbest/1" was proposed by in the framework of JADE (which is expressed in formula (16)).

$$\vec{v}_{i,G} = \vec{x}_{i,G} + F_i \cdot (\vec{x}_{pbest,G} - \vec{x}_{i,G}) + F_i \cdot (\vec{x}_{r1,G} - \vec{x}_{r2,G}) \tag{16}$$

In Equation (16), the individual $\vec{x}_{pbest,G}$ is randomly selected from the top $N \times p (p \in [0,1])$ members in the G-th generation. "current to pbest/1" depends on the control parameter p to balance exploitation and exploration (small p behaves more greedily). $\vec{x}_{\mathrm{i},G}$ and $\vec{x}_{r1,G}$ are selected from P in the same way as in Equation (12), while $\vec{x}_{r2,G}$ is randomly chosen from the union $\mathrm{P} \cup \mathrm{A}$, of the current population and the archive. We present an improved mutation strategy "current to pbest/2-rand" to improve the search range based on mutation strategy "current to pbest/1", which is expressed as follows:

$$\vec{v}_{i,G} = \vec{x}_{i,G} + F_i \cdot (\vec{x}_{pbest,G} - \vec{x}_{i,G}) + F_i \cdot [(\vec{x}_{r1,G} - \vec{x}_{r2,G}) \cdot \text{rand}_i + (\vec{x}_{r3,G} - \vec{x}_{r4,G}) \cdot (1 - \text{rand}_i)] \tag{17}$$

where rand_i is a uniformly distributed random number between [0,1]. $\vec{x}_{\mathrm{i},G}$, $\vec{x}_{r1,G}$ and $\vec{x}_{r3,G}$ are selected randomly and different within the range {1, NP} from P in the same way as in formula (16), while $\vec{x}_{r2,G}$ and $\vec{x}_{r4,G}$ is randomly chosen from the union, $\mathrm{P} \cup \mathrm{A}$, of the current population and the archive. The two mutation strategies are illustrated in Figure 2.

As seen in Figure 2a, $\vec{v}_{i,G}$ is the mutation individual generated for individual $\vec{x}_{i,G}$. According to the principle of vector addition, the position of $\vec{v}_{i,G}$ changes with the associated mutation factor F_i, and its position only exists on this "Search Line". Mutation strategy "current to pbest/2-rand" uses rand_i and linear combination of $(\vec{x}_{r1,G} - \vec{x}_{r2,G}) \cdot \text{rand}_i + (\vec{x}_{r3,G} - \vec{x}_{r4,G}) \cdot (1 - \text{rand}_i)$ to expand the search range. By the varying value rand_i, $\vec{v}_{i,G}$ can search anywhere in the shaded triangle area with the change of the F_i and rand_i (see Figure 2b). Obviously, the search range of "current to pbest/2-rand" is much larger than that of "current to pbest/1".

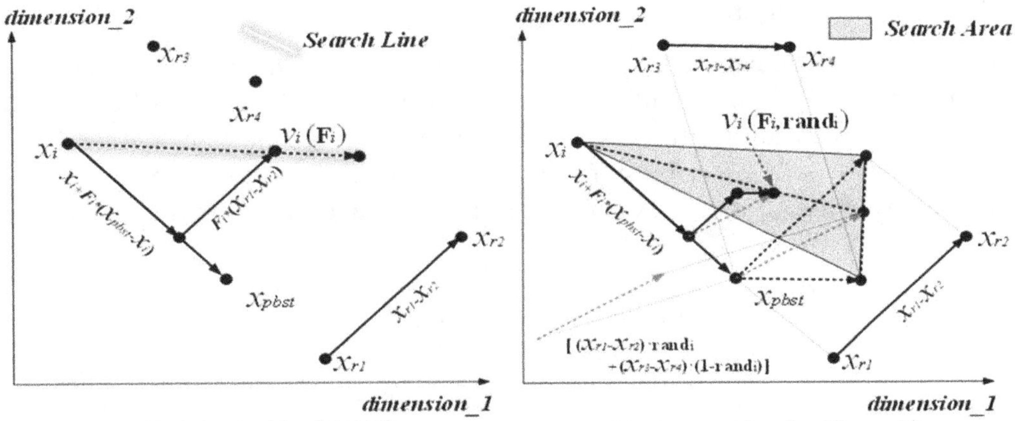

(a) "current to pbest/1" (b) "current to pbest/2-rand"

Figure 2. Illustration of the DE mutation strategy in two dimensions.

3.2.2. The PIRFE Strategy

The PIRFE strategy is proposed to avoid individuals in a local optimum lead useless evolution. When an individual falls into a local optimal, it is extremely difficult to produce an effective mutation to jump out of the local optimal. The individual failing evolution number (IFEN) is defined to record the number of individual failing evolution, and it is initialized to zero. $\vec{x}_{i,G}$ generates $\vec{u}_{i,G}$ through mutation and crossover. If $\vec{x}_{i,G+1}$ produces a failing evolution and $\vec{x}_{i,G+1}$ is better than $\vec{u}_{i,G+1}$, $IFEN_{i,G+1}$ is equal to $IFEN_{i,G}$ plus one, otherwise $IFEN_{i,G+1}$ will reset to zero. $IFEN_{i,G+1}$ is illustrated as follows:

$$IFEN_{i,G+1} = \begin{cases} 0 & \text{if } \vec{u}_{i,G+1} \text{ is better than } \vec{x}_{i,G+1} \\ IFEN_{i,G}+1 & \text{otherwise} \end{cases} \quad (18)$$

If $\vec{x}_i$ produces failing evolutions for $IFEN_i$ generations, and $IFEN_i$ is more than limit evolution generations (see formula (18)), $\vec{x}_i$ falls into local optimal. The LEG represents the largest evolution generation allowing failure evolutions. In the next evolutionary process, we should avoid involving $\vec{x}_i$. $\vec{x}_i$ should be replaced by a new individual randomly selected in population. The framework of the PIRFE strategy is as follows (Figure 3):

```
Initialization- IFEN
   IFEN_{i,G} is initialized to zero, (i ∈ {1,···,NP}, G = 1) and set LG
Update-IFEN IFEN_{i,G}
   #DE Selection
   If u_{i,G} is better than x_{i,G}
      IFEN_{i,G+1} = 0
   Else
      IFEN_{i,G+1} = IFEN_{i,G} + 1
   End
Mutation-PIRFE
   If IFMN_{i,G} ≥ LEG
      random select x_{j,G} instead of x_{i,G} in "current to pbest/2-rand",
and j ≠ i, pbest, r1, r2, r3, r4, generate v_{i,G}.
   Else
      generate v_{i,G} using mutation strategy "current to pbest/2-rand"
   End
```

Figure 3. Pseudocode of PIRFE strategy.

3.2.3. Control Parameters Assignments

SHADE maintains a historical memory with H entries for both DE control parameters CR, F, M_{CR} and M_F. The scaling factor $F \in [0,\ 1]$ controls the magnitude of the differential mutation operator and $CR \in [0,\ 1]$ is the crossover rate. In the beginning, the contents of $M_{CR,k}$, $M_{F,k}$ $(k = 1, \cdots, H)$ are all initialized to 0.5. In each generation G, the control parameters CR_i and F_i used by each individual x_i are generated by randomly selecting an index r_i from $[1,\ H]$, and then applying the formulas (19) and (20):

$$CR_i = \begin{cases} 0 & \text{if } M_{CR,r_i} = \perp \\ \text{randn}_i(M_{CR,r_i}, 0.1) & \text{otherwise} \end{cases} \tag{19}$$

$$F_i = \text{randc}_i(M_{F,r_i}, 0.1) \tag{20}$$

In case a value for CR_i outside of [0, 1] is generated, it is replaced by the limit value (0 or 1) closest to the generated value. When $F_i > 1$, is truncated to 1, and when $F_i \leq 0$, formula (20) is repeatedly applied to generate a valid value. These manners are determined according to the procedure for JADE [38]. In formula (19), if M_{CR,r_i} has been assigned the "terminal value" $\perp$, CR_i is set to 0.

In each generation, in formula (17), CR_i and F_i values that succeed in generating a trial individual $\vec{u}_{i,G}$ better than the parent individual $\vec{x}_{i,G}$ are recorded as S_{CR}, S_F. At the end of the generation, the contents of memory are updated as follows:

$$M_{CR,k,G+1} = \begin{cases} \perp & \text{if } M_{CR,k,G} = \perp \text{ or } \max(S_{CR}) = 0 \\ \text{mean}_{WA}(S_{CR}) & \text{if } S_{CR} \neq \varnothing \\ M_{CR,k,G} & \text{otherwise} \end{cases} \tag{21}$$

$$M_{F,k,G+1} = \begin{cases} \text{mean}_{WL}(S_F) & \text{if } S_F \neq \varnothing \\ M_{F,k,G} & \text{otherwise} \end{cases} \tag{22}$$

An index k ($1 < k < \text{H}$) determines the position in the memory to update. At the beginning of the search k is initialized to 1. k is incremented whenever a new element is inserted into the history. If > H, k is set to 1. In generation G, the k-th element in the memory is updated. In the update formula (21) and (22), when all individuals in generation G fail to generate an individual better than the parent, i.e., $S_{CR} = S_F = \varnothing$, the memory is not updated. Also, the weighted mean $mean_{WA}(S_{CR})$ is computed according to formula (23) by Peng et al. [44]. The weighted Lehmer mean $mean_{WL}(S_F)$ is computed using the formula below, and as with $mean_{WA}(S_{CR})$:

$$mean_{WL}(S_F) = \sum\nolimits_k^{|S_F|} w_k \cdot S_{F,k}^2 / \sum\nolimits_k^{|S_F|} w_k \cdot S_{F,k} \tag{23}$$

$$mean_{WA}(S_{CR}) = \sum\nolimits_{k=1}^{|S_{CR}|} w_k \cdot S_{CR,k} \tag{24}$$

$$w_k = \triangle f_k / \sum\nolimits_{k=1}^{|S_{CR}|} \triangle f_k \tag{25}$$

where $\triangle f_k = \left| f(\vec{u}_{i,G}) - f(\vec{x}_{i,G}) \right|$. In the same paper, they also proposed a restart strategy for JADE [38].

LSHADE put forward a new method of NP setting—LPSR which reduces the population linearly. The population size at generation 1 is N_{init}, and the population at the end of the run is N_{min}. After each generation G, the population size in the next generation, N_{G+1} is computed according to formula (26):

$$N_{G+1} = \text{round}[((N_{min} - N_{init})/MAXNFE) \cdot NFE + N_{init}] \tag{26}$$

If N_{min} is set to the smallest possible value, the evolutionary operators can be applied in the case of iLSHADE, N_{min} = 6 because the mutation strategy "current to pbest/2-rand" showed as formula (17) requires 4 individuals. NFE is the current number of fitness evaluations, and $MAXNFE$ is the maximum number of fitness evaluations. Whenever $N_{G+1} < N_G$, the $(N_G - N_{G+1})$ worst-ranking individuals are deleted from the population. Similarly, the external archive size $|A|$ is set to N_{init} multiplied by a parameter r^{arc}, $|A| = \text{round}(N_{init} \times r_{arc})$. In addition, external archive A is same as population, whenever $|A|_{G+1} < |A|_G$, the $(|A|_G - |A|_{G+1})$ worst-ranking individuals are deleted from the A. The p value for "current to pbest/2-rand" in each generation G is computed as follows:

$$p = (p_{max} - p_{min}) \cdot \text{rand}(0,1) + p_{min} \tag{27}$$

where rand$(0,1)$ returns a uniformly distributed random number in $[0,1]$, p is a random value in $[p_{max}, p_{min}]$. Finally, the pseudo-code of the iLSHADE algorithm is given in Figure 4.

```
G ← 1, FES ← 0 ,Archive A ← ∅;
Initialize population P_g = (x_{1,G},···,x_{NP,G}) randomly
Initialization-IFEN // described in Figure 3
Set M_F[H] = 0.2 , Set other values in M_F to 0.5,
Set M_CR[H] = 0.8 , Set other values in M_CR to 0.8,
k ← 1  // index counter
while the termination criteria are not meet do
   S_CR ← ∅ , S_F ← ∅ ,
   for i = 1 to NP do
      r_i ← select from [1, H] randomly
      if M_{CR,ri} < 0 then CR_{i,g} ← 0
      else CR_{i,g} ← randn_i(M_{CR,r_i}, 0.1)
      end if
      F_{i,g} ← randc_i(M_{F,r_i}, 0.1)
      u_{i,G} ← Mutation-PIRFE with "current to pbest/2-rand" // described in Figure 3
   end for
   for i = 1 to NP do
      if u_{i,G} is better than x_{i,G} then
         x_{i,G+1} ← u_{i,G} ;
         x_{i,G} → A, CR_{i,g} → S_CR , F_{i,g} → S_F
      else
         x_{i,G+1} ← x_{i,G} ;
      end if
      Update-IFEN IFEN_{i,G} // described in Figure 3
   end for
   Shrink A , if necessary
   Update M_{CR,i} and M_{F,i} except i = H
   Apply LPSR strategy using formula (26)// linear population size reduction
   Update p using formula (27)
   G ← G + 1, FES ← FES + NP
end while
```

Figure 4. Pseudocode of iLSHADE.

4. Numerical Experiment

The iLSHADE algorithm was tested in both low and high dimension on a set of 10 benchmark functions demonstrated in Table 1. Table 1 indicates benchmark problems with different structures and characteristics. In the table, "O-V" means the optimum fitness and "O-S" stands for the optimum solution. f1, f2, f3, f5 and f6 are unimodal optimization problems to evaluate the convergence, while f4, f7, f8,f9 and f10 are multimodal optimization problems with a huge number of local optima to test the convergence precision [18].

Table 1. Details of benchmark problems.

Benchmark Function	Name	Domain	O-V	O-S
$f_1 = \sum_{i=1}^{n} x_i^2$	Sphere	$[-100, 100]^n$	0	$\{0, 0, \cdots, 0\}$
$f_2 = \sum_{i=1}^{n} \lvert x_i \rvert + \prod_i^n \lvert x_i \rvert$	Schwefel (2.2)	$[-100, 100]^n$	0	$\{0, 0, \cdots, 0\}$
$f_3 = \sum_{i=1}^{n} \left(\sum_{j=1}^{i} x_j \right)^2$	Schwefel (1.2)	$[-100, 100]^n$	0	$\{0, 0, \cdots, 0\}$
$f_4 = \sum_{i=1}^{n-1} \left[100 (x_{i+1} - x_i^2)^2 + (x_i - 1)^2 \right]$	Rosenbrock	$[-30, 30]^n$	0	$\{0, 0, \cdots, 0\}$
$f_5 = \sum_{i=1}^{n} \left(\lfloor x_j + 0.5 \rfloor \right)^2$	Step	$[-100, 100]^n$	0	$\{0, 0, \cdots, 0\}$
$f_6 = \sum_{i=1}^{n} i x_i^4$	Quartic	$[-1.28, 1.28]^n$	0	$\{0, 0, \cdots, 0\}$
$f_7 = \sum_{i=1}^{n} -x_i \sin \left(\sqrt{\lvert x_i \rvert} \right)$	Schwefel (2.26)	$[-500, 500]^n$	−418.9n	*
$f_8 = \sum_{i=1}^{n} [x_i^2 - 10 \cos(2\pi x_i) + 10]$	Rastrigin	$[-5.12, 5.12]^n$	0	$\{0, 0, \cdots, 0\}$
$f_9 = -20 \exp\left(-0.2 \sqrt{1/n \sum_{i=1}^{n} x_i^2} \right) - \exp(1/n \cos(2\pi x_i)) + 20 + e$	Ackley	$[-32, 32]^n$	0	$\{0, 0, \cdots, 0\}$
$f_{10} = 1 + \sum_{i=1}^{n} \frac{x_i^2}{4000} + \prod_i^n \cos\left(\frac{x_i}{\sqrt{i}} \right)$	Griewank	$[-600, 600]^n$	0	$\{0, 0, \cdots, 0\}$

Note: * (420.9876, 420.9876, . . . , 420.9876).

The iLSHADE is compared to DE and other improved DE like LSHADE, JADE, CoDE and jDE. The number of function evaluations is used to appraise the convergence. These experiments are made on a personal computer, Windows10, Intel(R) Core(TM) i7-5500U CPU@ 2.40GHZ, RAM 8.00 GB. The dimension of benchmark functions is D = 10 in low dimension and 30 in high dimension, and 51 runs of an algorithm were needed for each function. The maximum number of objective function evaluations is D $\times$ 10,000. The optimal values are known for all benchmark functions.

In the experiments, the parameters in LSHADE, JADE, CoDE, jDE and DE were kept unchanged refer to [33,36,38,39,42], and the parameter setting in the iLSHADE is same as LSHADE except the following parameters:

- Using "current to pbest/2-rand" mutation strategy,
- The p value for mutation strategy is computed as $p_G = \text{rand}[p_{\min}, p_{\max}]$, where $p_{\min} = 2/NP$ is set such that when $\vec{x}_{pbest,G}$ is selected, at least 2 individuals are needed, and $p_{\max} = 0.25$.
- Initial population size $N_{init} = 15log(D)\sqrt{D}$, the control parameter of external archive size $r^{arc} = 2$.
- Historical memory size H = 6; set a final pair of parameters $M_F[\text{H}] = 0.2$ and $M_{CR}[\text{H}] = 0.8$, other M_F values are initialized to 0.5 and other M_{CR} are initialized to 0.8.
- PIRFE parameter $LEG = 50$.

The aggregate results of statistical testing ($+, -, \approx$) on 10 functions are shown in Tables 2 and 3. The symbols $+, -, \approx$ indicate that a given algorithm performed significantly better ($+$), significantly worse ($-$), or not significantly different better or worse ($\approx$) compared to iLSHADE using the Wilcoxon rank-sum test [45] (significantly, $p < 0.05$).

Table 2. Experimental results of 10 test functions in low dimension.

f	**LSHADE Mean (Std Dev)**	**JADE Mean (Std Dev)**	**CoDE Mean (Std Dev)**	**jDE Mean (Std Dev)**	**DE Mean (Std Dev)**	**iLSHADE Mean (Std Dev)**
f_1	0.00×10^{0} $(0.00 \times 10^{0}) \approx$	8.61×10^{-36} (7.18×10^{-36}) –	5.93×10^{-38} (7.34×10^{-38}) –	1.15×10^{-38} (1.18×10^{-38}) –	2.73×10^{-46} (1.42×10^{-45}) –	0.00×10^{0} (0.00×10^{0})
f_2	6.49×10^{-49} (4.38×10^{-48}) –	8.75×10^{-20} (4.68×10^{-20}) –	7.88×10^{-22} (6.58×10^{-22}) –	1.38×10^{-22} (1.01×10^{-22}) –	8.03×10^{-25} (1.65×10^{-24}) –	3.36×10^{-64} (2.05×10^{-63})
f_3	1.10×10^{-91} (5.90×10^{-91}) –	2.81×10^{-35} (2.75×10^{-35}) –	2.43×10^{-39} (4.02×10^{-39}) –	1.24×10^{-40} (1.79×10^{-40}) –	4.58×10^{-45} (2.03×10^{-44}) –	0.00×10^{0} (0.00×10^{0})
f_4	0.00×10^{0} $(0.00 \times 10^{0}) \approx$	0.00×10^{0} $(0.00 \times 10^{0}) \approx$	1.47×10^{-05} (1.96×10^{-05}) –	7.39×10^{-03} (9.11×10^{-03}) –	7.82×10^{-02} (5.53×10^{-01}) –	0.00×10^{0} (0.00×10^{0})
f_5	0.00×10^{0} $(0.00 \times 10^{0}) \approx$	0.00×10^{0} $(0.00 \times 10^{0}) \approx$	0.00×10^{0} $(0.00 \times 10^{0}) \approx$	0.00×10^{0} $(0.00 \times 10^{0}) \approx$	0.00×10^{0} $(0.00 \times 10^{0}) \approx$	0.00×10^{0} (0.00×10^{0})
f_6	0.00×10^{0} $(0.00 \times 10^{0}) \approx$	1.40×10^{-71} (3.27×10^{-71}) –	6.47×10^{-71} (1.52×10^{-70}) –	1.40×10^{-72} (3.66×10^{-72}) –	8.39×10^{-88} (4.17×10^{-87}) –	0.00×10^{0} (0.00×10^{0})
f_7	−4189.83 $(2.73 \times 10^{-12}) \approx$	−4189.83 $(2.73 \times 10^{-12}) \approx$	−4189.83 $(2.73 \times 10^{-12}) \approx$	−4189.83 $(2.73 \times 10^{-12}) \approx$	−4189.83 $(2.73 \times 10^{-12}) \approx$	−4189.83 (2.73×10^{-12})
f_8	0.00×10^{0} $(0.00 \times 10^{0}) \approx$	0.00×10^{0} $(0.00 \times 10^{0}) \approx$	0.00×10^{0} $(0.00 \times 10^{0}) \approx$	0.00×10^{0} $(0.00 \times 10^{0}) \approx$	3.06×10^{0} (2.47×10^{0}) –	0.00×10^{0} (0.00×10^{0})
f_9	3.72×10^{-15} $(9.55 \times 10^{-16}) \approx$	3.86×10^{-15} (6.90×10^{-16}) –	4.00×10^{-15} (2.37×10^{-30}) –	3.93×10^{-15} (4.93×10^{-16}) –	3.93×10^{-15} (4.93×10^{-16}) –	3.72×10^{-15} (9.55×10^{-16})
f_{10}	0.00×10^{0} $(0.00 \times 10^{0}) \approx$	2.42×10^{-12} (6.51×10^{-12}) –	0.00×10^{0} $(0.00 \times 10^{0}) \approx$	3.22×10^{-04} (2.28×10^{-03}) –	8.58×10^{-02} (6.17×10^{-02}) –	0.00×10^{0} (0.00×10^{0})
–	2	6	6	7	8	
+	0	0	0	0	0	
≈	8	4	4	3	2	

Table 3. Experimental results of 10 test functions in high dimension.

f	**LSHADE Mean (Std Dev)**	**JADE Mean (Std Dev)**	**CoDE Mean (Std Dev)**	**jDE Mean (Std Dev)**	**DE Mean (Std Dev)**	**iLSHAD EMean (Std Dev)**
f_1	1.12×10^{-90} (6.44×10^{-90}) –	0.00×10^{0} $(0.00 \times 10^{0}) \approx$	9.85×10^{-19} (7.04×10^{-19}) –	4.31×10^{-41} (4.35×10^{-41}) –	9.34×10^{-44} (2.74×10^{-43}) –	0.00×10^{0} (0.00×10^{0})
f_2	2.09×10^{-42} (1.03×10^{-41}) –	4.11×10^{-27} (4.89×10^{-27}) –	4.01×10^{-12} (1.34×10^{-12}) –	3.48×10^{-24} (1.96×10^{-24}) –	1.16×10^{-05} (8.17×10^{-05}) –	4.88×10^{-58} (1.55×10^{-57})
f_3	3.85×10^{-81} (1.74×10^{-80}) –	3.58×10^{-49} (7.53×10^{-49}) –	4.22×10^{-19} (3.41×10^{-19}) –	7.27×10^{-43} (1.05×10^{-42}) –	1.16×10^{-46} (6.85×10^{-46}) –	0.00×10^{0} (0.00×10^{0})
f_4	1.40×10^{-25} (9.70×10^{-25}) –	$1.85 \times 10^{+01}$ $(1.01 \times 10^{+01})$ –	$1.82 \times 10^{+01}$ (3.29×10^{0}) –	$1.15 \times 10^{+01}$ (8.23×10^{0}) –	2.62×10^{0} (2.60×10^{0}) –	0.00×10^{0} (0.00×10^{0})
f_5	0.00×10^{0} $(0.00 \times 10^{0}) \approx$	1.96×10^{-02} (1.39×10^{-01}) –	0.00×10^{0} $(0.00 \times 10^{0}) \approx$	0.00×10^{0} $(0.00 \times 10^{0}) \approx$	4.22×10^{0} (7.61×10^{0}) –	0.00×10^{0} (0.00×10^{0})
f_6	0.00×10^{0} $(0.00 \times 10^{0}) \approx$	0.00×10^{0} $(0.00 \times 10^{0}) \approx$	2.35×10^{-33} (2.93×10^{-33}) –	1.79×10^{-69} (3.86×10^{-69}) –	1.74×10^{-59} (1.10×10^{-58}) –	0.00×10^{0} (0.00×10^{0})
f_7	−12,569.49 $(1.82 \times 10^{-12}) \approx$	−12,567.16 $(1.64 \times 10^{+01})$ –	−12,569.49 $(1.82 \times 10^{-12}) \approx$	−12,569.49 $(1.82 \times 10^{-12}) \approx$	−11552.14 $(3.68 \times 10^{+02})$ –	−12,569.49 (1.88×10^{-05})
f_8	1.74×10^{-16} (6.35×10^{-16}) +	0.00×10^{0} $(0.00 \times 10^{0}) \approx$	8.38×10^{-12} (9.18×10^{-12}) +	4.83×10^{0} (3.86×10^{0}) –	$3.62 \times 10^{+01}$ $(1.44 \times 10^{+01})$ –	3.16×10^{-11} (1.66×10^{-10})
f_9	4.00×10^{-15} $(2.37 \times 10^{-30}) \approx$	4.76×10^{-15} (1.46×10^{-15}) –	2.74×10^{-10} (1.08×10^{-10}) –	5.60×10^{-15} (1.77×10^{-15}) –	2.64×10^{-01} (5.21×10^{-01}) –	4.00×10^{-15} (2.37×10^{-30})
f_{10}	0.00×10^{0} $(0.00 \times 10^{0}) \approx$	1.55×10^{-03} (3.81×10^{-03}) –	3.05×10^{-17} (1.08×10^{-16}) –	0.00×10^{0} $(0.00 \times 10^{0}) \approx$	7.99×10^{-03} (1.46×10^{-02}) –	0.00×10^{0} (0.00×10^{0})
–	4	6	7	7	10	
+	1	1	1	0	0	
≈	5	3	2	3	0	

1 Simulation in low dimension.

Table 2 summarizes the experimental results in low dimension. It shows that iLSHADE eventually converges to optimum for 8 test functions except f_2 and f_9. Moreover, iLSHADE significantly outperforms JADE, CoDE, jDE, DE. Compared to LSHADE, iLSHADE beats LSHADE on two test functions f_2 and f_3. Especially in f_3, only iLSHADE converges to the global optimal value. Overall, iLSHADE performs better than other algorithms on low dimension optimization problems.

2 Simulation in high dimension.

The experimental results in high dimension are summarized in Table 3. iLSHADE eventually converges to optimum for 7 test functions except f_2, f_8 and f_9. iLSHADE performs the best in f_2 and f_9, although it does not converge to the global optimal. While on f_8, JADE performs the best. The proposed algorithm is not good as expected. Above all, iLSHADE has an obvious advantage over other algorithms on high dimension optimization problems.

To sum up, iLSHADE is suitable for both low dimension and high dimension, meaning that the improvement proposed in this paper is effective.

5. Implementation of iLSHADE for LSLCHS

5.1. Solution Structure and Initial Population

To handle constraints and calculate objective function, solution structure for the LSLCHS comprises a group of monthly water levels as the decision variables shown as follows.

$$X = \begin{bmatrix} X^1 \\ X^2 \\ \vdots \\ X^M \end{bmatrix} = \begin{bmatrix} x_1^1, x_2^1, \cdots x_T^1 \\ x_1^2, x_2^2, \cdots x_T^2 \\ \vdots \\ x_1^M, x_2^M, \cdots x_T^M \end{bmatrix} \tag{28}$$

where M is the number of hydropower stations, T (12 month in a year) is the number of intervals. In algorithms relying on heuristic search, initial population is an important issue to convergence speed and population diversity. The iLSHADE has a large initial population size based on LPSR and ensures the diversity of the population by random initialization.

5.2. Constraint Handling

It is multiple and complicate for flow constraint, power generating constraint, amplitude of water level variation and hydraulic connection of cascade. The handling measure currently used for water balance constraint, water level constraint and boundary condition is often corrected to the boundary [18,46]. It has defects such as: (1) The direction of the entering feasible domain is relatively simple and centered on the boundary because of the excessive attention to the rapid into the feasible area; (2) When there are multiple feasible domains, it is easy to ignore small feasible areas;

The ε-constrained method is first proposed by Takahama et al. [47], which relaxes the greed of the feasibility criterion to the constraint conditions. The ε value is set as threshold value in ε-constrained method. In general, constrained optimization maximum problems can be mathematically formulated as follows:

$$\max f\left(\vec{x}\right),\ \vec{x} = (x_1, \ldots, x_D) \tag{29}$$

$$\text{subject to} \begin{cases} g_i\left(\vec{x}\right) \le 0 \ \ i = 1, 2, \cdots, m \\ h_j\left(\vec{x}\right) = 0 \ \ j = 1, 2, \cdots, n \end{cases} \tag{30}$$

where $g_i(\vec{x})$ is inequality constraints, m is the total number of inequality constraint. $h_j(\vec{x})$ is equality constraints and n is the total number of equality constraints. The value of constraint violation $\varphi(\vec{x})$ can be calculated in the following formula (30), and $\varphi(\vec{x})$ of infeasible solutions is bigger than 0.

$$\varphi(\vec{x}) = \sum_{i=0}^{m} \max\left(0, g_i(\vec{x})\right) + \sum_{j=1}^{n}\left(\left|h_j(\vec{x})\right|, 0\right) \tag{31}$$

When the constraint violation values of both solutions are smaller than ε value, the one with better objective function value is selected. Otherwise, the one with smaller constraint violation value is selected. Overall, when any of the following conditions are met, $\vec{x}_i$ is superior to $\vec{x}_j$:

$$\begin{cases} f(\vec{x}_i) < f(\vec{x}_j), \text{ if } g(\vec{x}_i) \le \varepsilon \cap g(\vec{x}_j) \le \varepsilon \\ f(\vec{x}_i) < f(\vec{x}_j), \text{ if } g(\vec{x}_i) = g(\vec{x}_j) \\ g(\vec{x}_i) < g(\vec{x}_j), \text{ otherwise} \end{cases} \tag{32}$$

and the final effectiveness of e-constrained method strongly depends on the control method of ε value. Takahama et al. [47] proposed the following method,

$$\varepsilon(0) = \varphi\left(P_0^{\theta}\right) \tag{33}$$

$$\varepsilon(t) = \begin{cases} \varepsilon(0)(1 - t/T_c)^{cp} & 0 < t < T_c \\ 0 & t > T_c \end{cases} \tag{34}$$

where P_0^{θ} is the top θ-th individual in the initial population, cp is a control parameter. If the number of iterations t is less than a given threshold value T_c, the ε value declines in an exponential way. Otherwise, ε is set to 0 (see formula (33) and (34)).

The ε-constrained method can expand search space, avoid the constraint correction for unfeasible solutions and enable to search infeasible region that is around feasible region. However, the constraints consist of flow, power generating and amplitude of water level. Operating water level in LSLCHS problem is multiple and complex. Their units are not integrated, and the physical quantities corresponding to the same level of different reservoirs are different. To solve the above problems, we proposed ε-constrained in cascade reservoir operation method (ε-CRO) with unify different physical quantities constraint violation. The ε-CRO chooses water to unify different physical quantities constraint violation because all constraints can be converted to outflow constraint and flow accumulated over time as water. The feasible range of outflow $Q_{i,t}$ is expressed in formula (35) and (36).

$$\overline{Q}_{i,t+1} = \min\begin{cases} Q_{i,t+1}^{\max} \\ \left(V(Z_{i,t}) - V(Z_{i,t+1}^{\min})\right)/\Delta t \\ (V(Z_{i,t}) - V(Z_{i,t} - \Delta Z_i))/\Delta t \end{cases} \tag{35}$$

$$\underline{Q}_{i,t+1} = \max\begin{cases} Q_{i,t+1}^{\min} \\ I_{i,t} + \left(V(Z_{i,t}) - V(Z_{i,t+1}^{\max})\right)/\Delta t \\ I_{i,t} + (V(Z_{i,t}) - V(Z_{i,t} + \Delta Z_i))/\Delta t \\ Q_{i,t+1}^{\min\ N} \end{cases} \tag{36}$$

where $Z_{i,t}$ represents the i-th reservoir water level at the -th period, $Z(V)$ is the relationship between water level and storage. $\overline{Q}_{i,t+1}$ and $\underline{Q}_{i,t+1}$ stand for the minimum and maximum outflow under all constraints. $Q_{i,t+1}^{\min N}$ is the outflow for guaranteed output and Δt stands for time horizon. The value of constraint violation in LSLCHS problem $\varphi_{CRO}\left(\overrightarrow{x}\right)$ can be calculated as follows:

$$\varphi_{CRO}\left(\overrightarrow{x}\right) = \sum_{i=1}^{M}\sum_{t=1}^{T}\left[\max(0, Q_{i,t} - \overline{Q}_{i,t+1}) + \max\left(0, \underline{Q}_{i,t+1} - Q_{i,t}\right)\right] \tag{37}$$

$\varepsilon_{CRO}(0)$ and $\varepsilon_{CRO}(t)$ are calculated like in [47] except $\theta = 0.5$, $cp = 1$ and $T_c = 0.5 * MAXNFE$.

6. Case Study

6.1. Description of Case Study

The Jinsha River, the upper stretch of the Yangtze River, is 2290 km long with a 485,000 km^2 basin area flowing through the provinces of Qinghai, Sichuan, and Yunnan in western China (See in Figure 5). Along the river, there are four large hydropower stations with large installed capacity, huge regulating storage and high water head. The total installed capacity of the four large hydropower stations is twice more than the Three Gorges Project (the largest hydropower station in the world). The main parameters of these hydropower stations are listed in Table 4.

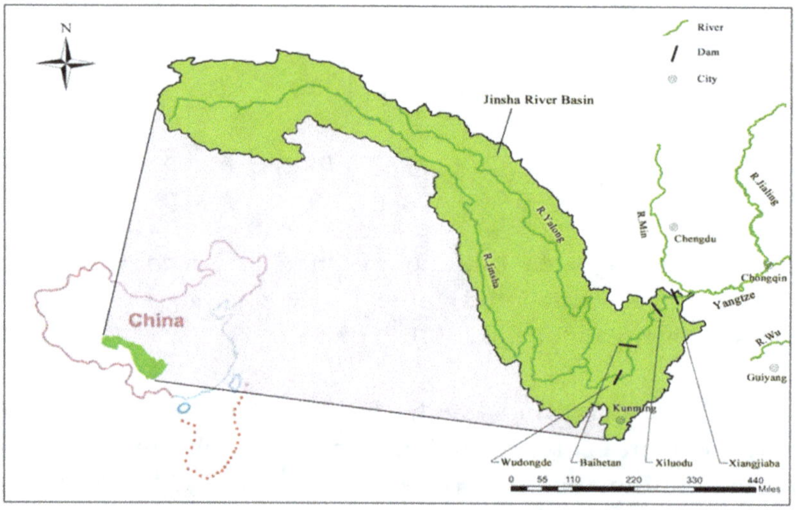

Figure 5. The location of the Jinsha River Basin in China.

Table 4. The main parameters of four large hydropower stations in Jinsha River.

Parameter	Wudongde	Baihetan	Xiluodu	Xiangjiaba
Adjustment ability	Season	Annual	Annual	Season
Regulating storage (billion m^3)	2.60	10.40	6.46	0.90
Hydro plant discharge range (m^3/s)	[49,400, 906]	[49,700, 905]	[43,700, 1500]	[49,800, 1500]
Upriver water level range (m)	[975, 945]	[825, 765]	[600, 540]	[380, 370]
Installed capacity (MW)	12000	16000	13860	6400
Normal water level (m)	975	825	600	380

6.2. Results and Analysis

In the case, the four large hydropower stations are all taken into consideration. According to historical runoff from 1959 to 2014 in the basin, three typical years are chosen to be the inflow conditions: wet year (historical runoff of 1999), normal year (2008) and dry year (1969). Simulation results of iLSHADE are compared to LSHADE, JADE and CoDE in three typical years. The initial water level and terminal water level of all the hydropower stations are set to the normal water level. The schedule period consists of 12 intervals with one month for each interval. In addition, the parameters in all algorithms are the same as those mentioned in Section 4. The maximum evaluation time is set to 40,000.

Table 5 represents that iLSHADE gains the best benefit of power production in three typical years. The convergence process of different algorithms in dry year is shown in Figure 6. Compared to LSHADE, JADE and CoDE on average optimal benefit of 51 independent simulations are illustrated in Table 5, iLSHADE increases the power production by 2.02, 4.04, 2.39 (10^8 KWh) in wet year, 3.03, 7.37, 3.76 (10^8 KWh) in normal year, 3.48, 5.68, 1.96 (10^8 KWh) in dry year. Obviously, the proposed iLSHADE is superior when solving LSLCHS problem by obtaining the maximal benefit of power production efficiently. In particular, the standard deviation of 51 independent simulations in iLSHADE is 0.02 in wet year, 0.01 in normal year, 0.01 in dry year, which shows that the convergence stability of iLSHADE is better than other algorithms. Meanwhile, it can be seen easily from Figure 6a that iLSHADE can avoid premature convergence effectively, at the same evaluation times keep a fast convergence speed compared to LSHADE, CoDE and JADE. Figure 6b depicts that the $\varphi_{CRO}\left(\vec{x}\right)$ of iLSHADE and LSHADE frequent changes and always lower than $\varepsilon_{CRO}(t)$, until evaluation times is greater than T_c, $\varphi_{CRO}\left(\vec{x}\right)$ is limited to 0.

Table 5. Results of 51 independent simulations on generated energy optimization (10^8 KWh).

Method	Wet Year (1999)			Normal Year (2008)			Dry Year (1969)		
	Max	Mean	Std	Max	Mean	Std	Max	Mean	Std
iLSHADE	2425.03	2425.01	0.02	2268.13	2268.11	0.01	1814.36	1814.35	0.01
LSHADE	2423.86	2422.99	0.48	2266.88	2265.08	0.94	1812.64	1810.87	0.99
Diff	1.17	2.02		1.25	3.03		1.72	3.48	
JADE	2424.43	2420.97	1.32	2267.51	2260.74	2.43	1814.00	1808.67	2.234
Diff	0.6	4.04		0.62	7.37		0.36	5.68	
CoDE	2423.39	2422.62	0.30	2265.62	2264.35	0.64	1813.08	1812.39	0.39
Diff	1.64	2.39		2.51	3.76		1.28	1.96	

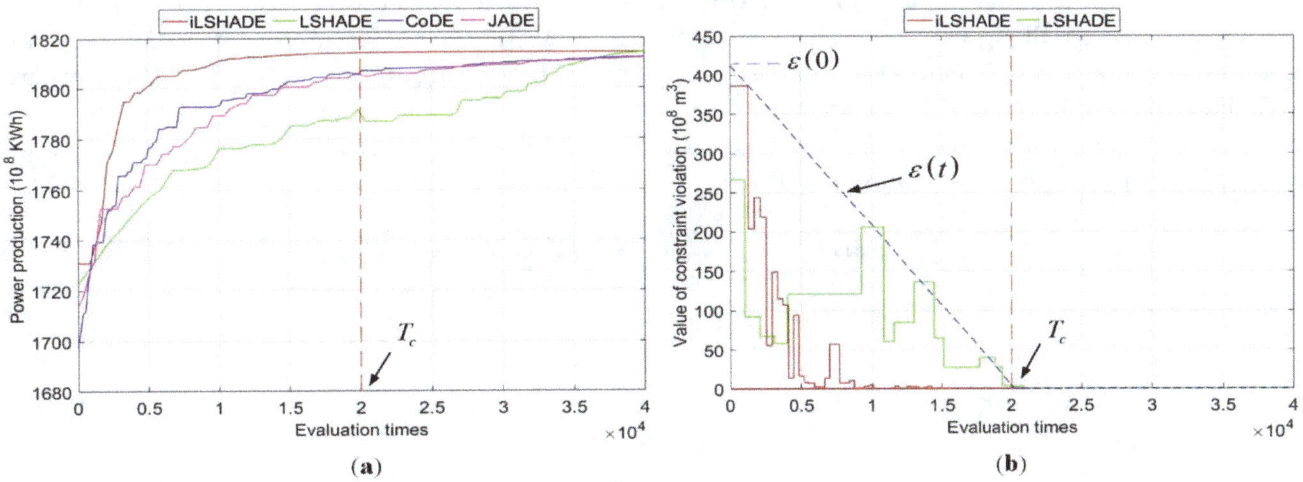

Figure 6. Convergence process of different algorithms.

The monthly reservoir water levels and outflow, as well as the optimal schedule result of Wudongde, Baihetan, Xiluodu and Xiangjiaba result in normal year obtained by iLSHADE are shown in Figure 7. Due to sufficient inflow, the cascade reservoir does not need to release storage capacity to meet the constraints of minimum outflow limit during the dry season from January to March. The inflow of Wudongde is very low in April, so the cascade reservoir needs to release storage capacity to meet the constraint requirements. As one of the upstream reservoirs, Wudongde first lowers the water level. To reduce the water spillage before flood season, Baihetan will lower its water level below the flood control level in advance and impound some water at the last period. Xiluodu and Xiangjiaba lower the water level in the last period. During the impoundment period, Wudongde and Baihetan store water to normal water level before Xiluodu and Xiangjiaba in October. In this way, the water in upstream reservoirs can utilize the downstream high hydraulic head to generate more electric power.

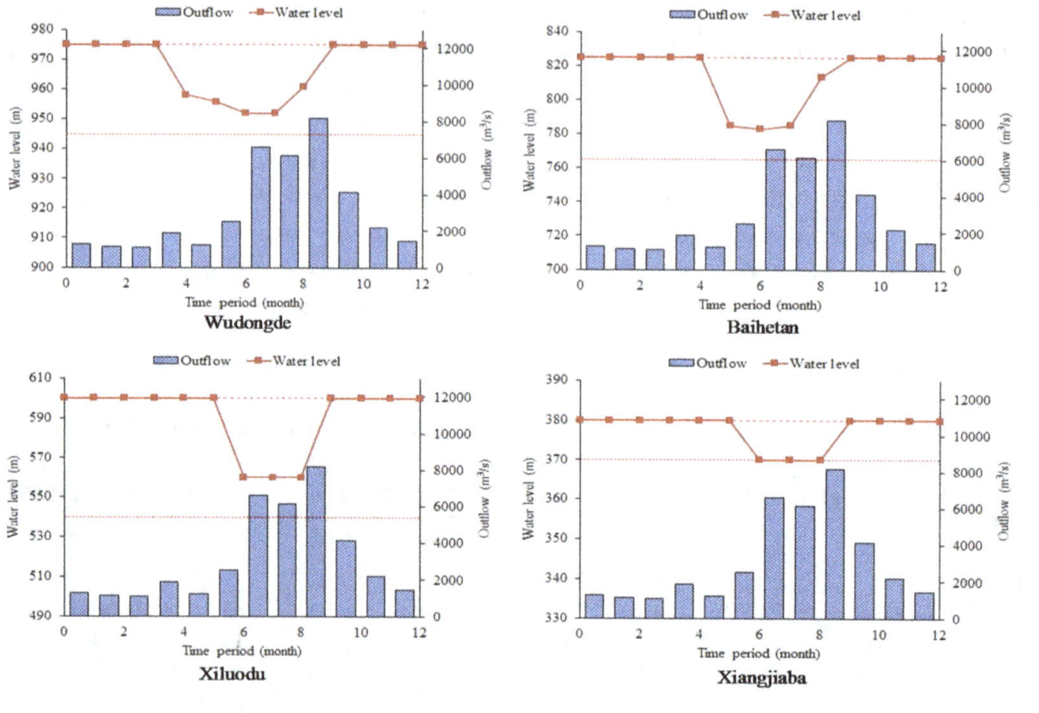

Figure 7. Optimal results in normal year by month.

Furthermore, the historical runoff data from 1959 to 2014 are selected for long sequence calculation. The parameters in all algorithms are the same as those mentioned in Section 6.2 and the maximum evaluation time is set to 40,000. The average adding annual power production that iLSHADE compares to LSHADE, JADE and CoDE is presented in Figure 8. It can be seen clearly from Figure 8 that iLSHADE is superior compared to other algorithm in solving the LSLCHS problem with different types of historical runoff from 1959 to 2014.

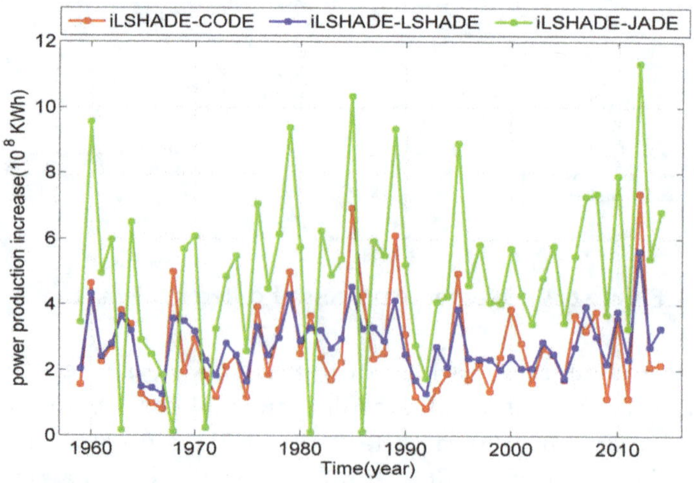

Figure 8. Historical runoff data from 1959 to 2014 for the annual power production increase that iLSHADE compares to LSHADE, JADE and CoDE.

According to the above analysis, proposed iLSHADE is superior when solving LSLCHS problem in different types of runoff by obtaining the maximize benefit of electric power production. All these experiment results fully demonstrate that iLSHADE is a competitive method to solve the LSLCHS problem.

7. Conclusions

An iLSHADE algorithm with new mutation strategy "current to pbest/2-rand" and PIRFE strategy has been developed in this paper to solve the LSLCHS problem. The significant modifications are mainly focused on preventing premature convergence and accelerating convergence. To verify the performance of iLSHADE, numerical simulation in both low and high dimension on a set of 10 benchmark functions has been done. Compared with other improved differential evolution algorithms, iLSHADE obtains better performance with all ten functions in the low dimension and nine functions in the high dimension. This indicates that the proposed new mutation strategy "current to pbest/2-rand" and PIRFE strategy in iLSHADE enhance the performance of original algorithm LSHADE effectively. Then iLSHADE is applied to solve LSLCHS problem for four large hydropower stations in Jinsha River. Compared to LSHADE, JADE and CoDE on average optimal benefit, iLSHADE increases the power production by 3.03, 7.37, 3.76 (10^8 KWh) in a normal year. In particular, the standard deviation of 51 independent simulations in iLSHADE is far lower than other algorithms. Moreover, according to its successful simulation performance with the historical runoff data from 1959 to 2014, iLSHADE can obtain better schedule results with lager generation benefits and better convergence property compared to LSHADE, JADE and CoDE. Above all, iLSHADE is a valid and reliable tool in solving the LSLCHS problem. Future research should consider the iLSHADE algorithm combined with other methods when solving multi objective scheduling problems in LSLCHS problem.

Acknowledgments: This work is supported by the National Natural Science Foundation Key Project of China (No. 91547208), the National Natural Science Foundation of China (No. 51579107), the National Key R&D Program of China (No. 2016YFC0402205) and the Foundation of Huazhong University of Science and Technology (No. 2017KFYXJJ209). Special thanks are given to the anonymous reviewers and editors for their constructive comments.

Author Contributions: Xiaohao Wen and Zhongzheng He designed and conducted the experiments. Xiaohao Wen wrote the draft of the paper. Zhongzheng He prepared the figures for this paper. Jianzhong Zhou proposed the main structure of this study. Chao Wang provided useful advice and made some corrections. All authors read and approved the final manuscript.

References

1. Lior, N. Sustainable energy development: The present (2011) situation and possible paths to the future. *Energy* **2012**, *43*, 174–191. [CrossRef]
2. Wu, X.-Y.; Cheng, C.-T.; Shen, J.-J.; Luo, B.; Liao, S.-L.; Li, G. A multi-objective short term hydropower scheduling model for peak shaving. *Int. J. Electr. Power Energy Syst.* **2015**, *68*, 278–293. [CrossRef]
3. Madani, K. Hydropower licensing and climate change: Insights from cooperative game theory. *Adv. Water Resour.* **2011**, *34*, 174–183. [CrossRef]
4. Xu, J.; Ni, T.; Zheng, B. Hydropower development trends from a technological paradigm perspective. *Energy Convers. Manag.* **2015**, *90*, 195–206. [CrossRef]
5. Chang, X.; Liu, X.; Zhou, W. Hydropower in china at present and its further development. *Energy* **2010**, *35*, 4400–4406. [CrossRef]
6. Tasdemiroglu, E. Development of small hydropower in Türkiye. *Energy* **1993**, *18*, 699–702. [CrossRef]
7. Zhou, Y.; Guo, S.; Xu, C.-Y.; Liu, P.; Qin, H. Deriving joint optimal refill rules for cascade reservoirs with multi-objective evaluation. *J. Hydrol.* **2015**, *524*, 166–181. [CrossRef]
8. Jiang, Z.; Li, A.; Ji, C.; Qin, H.; Yu, S.; Li, Y. Research and application of key technologies in drawing energy storage operation chart by discriminant coefficient method. *Energy* **2016**, *114*, 774–786. [CrossRef]
9. Chen, L.; Singh, V.P. Entropy-based derivation of generalized distributions for hydrometeorological frequency analysis. *J. Hydrol.* **2017**. [CrossRef]

10. Chen, L.; Singh, V.P.; Huang, K. Bayesian technique for the selection of probability distributions for frequency analyses of hydrometeorological extremes. *Entropy* **2018**, *20*, 117. [CrossRef]
11. Chen, L.; Singh, V.P.; Guo, S.; Zhou, J.; Zhang, J.; Liu, P. An objective method for partitioning the entire flood season into multiple sub-seasons. *J. Hydrol.* **2015**, *528*, 621–630. [CrossRef]
12. Chen, L.; Singh, V.P.; Lu, W.; Zhang, J.; Zhou, J.; Guo, S. Streamflow forecast uncertainty evolution and its effect on real-time reservoir operation. *J. Hydrol.* **2016**, *540*, 712–726. [CrossRef]
13. Djebou, D.C.S. Spectrum of climate change and streamflow alteration at a watershed scale. *Environ. Earth Sci.* **2017**, *76*, 653. [CrossRef]
14. Sohoulande Djebou, D.C.; Singh, V.P. Entropy-based index for spatiotemporal analysis of streamflow, precipitation, and land-cover. *J. Hydrol. Eng.* **2016**, *21*, 05016024. [CrossRef]
15. Xu, B.; Zhong, P.-A.; Wu, Y.; Fu, F.; Chen, Y.; Zhao, Y. A multiobjective stochastic programming model for hydropower hedging operations under inexact information. *Water Resour. Manag.* **2017**, *31*, 4649–4667. [CrossRef]
16. Mahmoud, M.; Dutton, K.; Denman, M. Dynamical modelling and simulation of a cascaded reserevoirs hydropower plant. *Electr. Power Syst. Res.* **2004**, *70*, 129–139. [CrossRef]
17. Cârdu, M.; Bara, T. Romanian achievement in hydro-power plants. *Energy Convers. Manag.* **1998**, *39*, 1193–1201. [CrossRef]
18. Wang, C.; Zhou, J.; Lu, P.; Yuan, L. Long-term scheduling of large cascade hydropower stations in Jinsha river, China. *Energy Convers. Manag.* **2015**, *90*, 476–487. [CrossRef]
19. Windsor, J.S. Optimization model for the operation of flood control systems. *Water Resour. Res.* **1973**, *9*, 1219–1226. [CrossRef]
20. Yoo, J.-H. Maximization of hydropower generation through the application of a linear programming model. *J. Hydrol.* **2009**, *376*, 182–187. [CrossRef]
21. Brandao, J.L.B. Performance of the equivalent reservoir modelling technique for multi-reservoir hydropower systems. *Water Resour. Manag.* **2010**, *24*, 3101–3114. [CrossRef]
22. Catalão, J.P.D.S.; Pousinho, H.M.I.; Mendes, V.M.F. Scheduling of head-dependent cascaded hydro systems: Mixed-integer quadratic programming approach. *Energy Convers. Manag.* **2010**, *51*, 524–530. [CrossRef]
23. Jiang, Z.; Qin, H.; Ji, C.; Feng, Z.; Zhou, J. Two dimension reduction methods for multi-dimensional dynamic programming and its application in cascade reservoirs operation optimization. *Water* **2017**, *9*, 634. [CrossRef]
24. Lee, J.H.; Labadie, J.W. Stochastic optimization of multireservoir systems via reinforcement learning. *Water Resour. Res.* **2007**, *43*. [CrossRef]
25. Nanda, J.; Bijwe, P.R. Optimal hydrothermal scheduling with cascaded plants using progressive optimality algorithm. *IEEE Trans. Power Appar. Syst.* **1981**, *PAS-100*, 2093–2099. [CrossRef]
26. Jiang, Z.; Ji, C.; Qin, H.; Feng, Z. Multi-stage progressive optimality algorithm and its application in energy storage operation chart optimization of cascade reservoirs. *Energy* **2018**, *148*, 309–323. [CrossRef]
27. Jiang, C.; Bompard, E. A self-adaptive chaotic particle swarm algorithm for short term hydroelectric system scheduling in deregulated environment. *Energy Convers. Manag.* **2005**, *46*, 2689–2696.
28. Cheng, C.T.; Liao, S.L.; Tang, Z.T.; Zhao, M.Y. Comparison of particle swarm optimization and dynamic programming for large scale hydro unit load dispatch. *Energy Convers. Manag.* **2009**, *50*, 3007–3014. [CrossRef]
29. Hınçal, O.; Altan-Sakarya, A.B.; Ger, A.M. Optimization of multireservoir systems by genetic algorithm. *Water Resour. Manag.* **2011**, *25*, 1465–1487. [CrossRef]
30. Yuan, X.; Yuan, Y. Application of cultural algorithm to generation scheduling of hydrothermal systems. *Energy Convers. Manag.* **2006**, *47*, 2192–2201. [CrossRef]
31. Wang, W.; Li, C.; Liao, X.; Qin, H. Study on unit commitment problem considering pumped storage and renewable energy via a novel binary artificial sheep algorithm. *Appl. Energy* **2017**, *187*, 612–626. [CrossRef]
32. Jalali, M.R.; Afshar, A.; Mariño, M.A. Multi-colony ant algorithm for continuous multi-reservoir operation optimization problem. *Water Resour. Manag.* **2007**, *21*, 1429–1447. [CrossRef]
33. Storn, R.; Price, K. Differential evolution—A simple and efficient heuristic for global optimization over continuous spaces. *J. Glob. Optim.* **1997**, *11*, 341–359. [CrossRef]
34. Lu, Y.; Zhou, J.; Qin, H.; Li, Y.; Zhang, Y. An adaptive hybrid differential evolution algorithm for dynamic economic dispatch with valve-point effects. *Expert Syst. Appl. Int. J.* **2010**, *37*, 4842–4849. [CrossRef]
35. Yuan, X.; Cao, B.; Yang, B.; Yuan, Y. Hydrothermal scheduling using chaotic hybrid differential evolution. *Energy Convers. Manag.* **2008**, *49*, 3627–3633. [CrossRef]

36. Brest, J.; Greiner, S.; Boskovic, B.; Mernik, M.; Zumer, V. Self-adapting control parameters in differential evolution: A comparative study on numerical benchmark problems. *IEEE Trans. Evolut. Comput.* **2006**, *10*, 646–657. [CrossRef]
37. Qin, A.K.; Huang, V.L.; Suganthan, P.N. Differential evolution algorithm with strategy adaptation for global numerical optimization. *IEEE Trans. Evolut. Comput.* **2009**, *13*, 398–417. [CrossRef]
38. Zhang, J.; Sanderson, A.C. Jade: Adaptive differential evolution with optional external archive. *IEEE Trans. Evolut. Comput.* **2009**, *13*, 945–958. [CrossRef]
39. Wang, Y.; Cai, Z.; Zhang, Q. Differential evolution with composite trial vector generation strategies and control parameters. *IEEE Trans. Evolut. Comput.* **2011**, *15*, 55–66. [CrossRef]
40. Mallipeddi, R.; Suganthan, P.N.; Pan, Q.K.; Tasgetiren, M.F. Differential evolution algorithm with ensemble of parameters and mutation strategies. *Appl. Soft Comput.* **2011**, *11*, 1679–1696. [CrossRef]
41. Tanabe, R.; Fukunaga, A. Success-History Based Parameter Adaptation for Differential Evolution. In Proceedings of the 2013 IEEE Congress on Evolutionary Computation, Cancun, Mexico, 20–23 June 2013; pp. 71–78.
42. Tanabe, R.; Fukunaga, A.S. Improving the search performance of shade using linear population size reduction. In Proceedings of the 2014 IEEE Congress on Evolutionary Computation, Beijing, China, 6–11 July 2014; pp. 1658–1665.
43. Liang, J.J.; Qu, B.Y.; Suganthan, P.N. *Problem Definitions and Evaluation Criteria for the CEC 2014 Special Session and Competition on Single Objective Real-Parameter Numerical Optimization*; Computational Intelligence Laboratory, Zhengzhou University, Zhengzhou China and Technical Report; Nanyang Technological University: Singapore, 2013.
44. Peng, F.; Tang, K.; Chen, G.; Yao, X. Multi-start jade with knowledge transfer for numerical optimization. In Proceedings of the Eleventh Conference on Congress on Evolutionary Computation, Trondheim, Norway, 18–21 May 2009; pp. 1889–1895.
45. Cortina-Borja, M. *Handbook of Parametric and Nonparametric Statistical Procedures*, 5th ed.; Sheskin, J., Ed.; CRC Press: Boca Raton, FL, USA, 2012; p. 382.
46. Liao, X.; Zhou, J.; Zhang, R.; Zhang, Y. An adaptive artificial bee colony algorithm for long-term economic dispatch in cascaded hydropower systems. *Int. J. Electr. Power Energy Syst.* **2012**, *43*, 1340–1345. [CrossRef]
47. Takahama, T.; Sakai, S.; Iwane, N. Constrained optimization by the ε constrained hybrid algorithm of particle swarm optimization and genetic algorithm. In Proceedings of the Australasian Joint Conference on Artificial Intelligence, Sydney, Australia, 5–9 December 2005; pp. 389–400.

7

Watershed Sediment and its Effect on Storage Capacity: Case Study of Dokan Dam Reservoir

Mohammad Ezz-Aldeen [1], Rebwar Hassan [1], Ammar Ali [2], Nadhir Al-Ansari [1,*] and Sven Knutsson [1]

[1] Department of Civil, Environmental and Natural Resources Engineering, Lulea University of Technology, 971 87 Lulea, Sweden; mohezz@ltu.se (M.E.-A.); rebwar.hassan@ltu.se (R.H.); Sven.Knutsson@ltu.se (S.K.)

[2] Department of Water Resources Engineering, Baghdad University, Baghdad 10071, Iraq; ammali_75@yahoo.com

* Correspondence: nadhir.alansari@ltu.se

Abstract: Dokan is a multipurpose dam located on the Lesser Zab River in the Iraq/Kurdistan region. The dam has operated since 1959, and it drains an area of 11,690 km^2. All reservoirs in the world suffer from sediment deposition. It is one of the main problems for reservoir life sustainability. Sustainable reservoir sediment-management practices enable the reservoir to function for a longer period of time by reducing reservoir sedimentation. This study aims to assess the annual runoff and sediment loads of the Dokan Dam watershed using the soil and water assessment tool (SWAT) model to evaluate the relative contributions in comparison with the total values delivered from both watershed and Lesser Zab River and to identify the basins with a high sediment load per unit area. These help in the process of developing a plan and strategy to manage sediment inflow and deposition. The SUFI-2 program was applied for a model calibrated based on the available field measurements of the adjacent Derbendekhan Dam watershed, which has similar geological formations, characteristics and weather. For the calibration period (1961–1968), the considered statistical criteria of determination coefficients and Nash–Sutcliffe model efficiency were 0.75 and 0.64 for runoff while the coefficients were 0.65 and 0.63 for sediment load, respectively. The regionalization technique for parameter transformation from Derbendekhan to Dokan watershed was applied. Furthermore, the model was validated based on transformed parameters and the available observed flow at the Dokan watershed for the period (1961–1964); they gave reasonable results for the determination coefficients and Nash–Sutcliffe model efficiency, which were 0.68 and 0.64, respectively. The results of SWAT project simulation for Dokan watershed for the period (1959–2014) indicated that the average annual runoff volume which entered the reservoir was about 2100 million cubic meters (MCM). The total sediment delivered to the reservoir was about 72 MCM over the 56 years of dam life, which is equivalent to 10% of the reservoir dead storage. Two regression formulas were presented to correlate the annual runoff volume and sediment load with annual rain depth for the studied area. In addition, a spatial distribution of average annual sediment load was constructed to identify the sub basin of the high contribution of sediment load.

Keywords: Dokan Dam; runoff; sediment load; SWAT

1. Introduction

Most of the dams, storage schemes, and different hydraulic structures around the world suffer from sedimentation problems. For dams and reservoirs, this effect is mainly concerned with the design capacity and operation schedule. The main source of reservoir sediments is the main river flow in addition to the runoff water, carrying the sediment load from watersheds and valleys surrounding the reservoir.

After a long period of dam operation, it is usually necessary to evaluate the current storage capacity of the reservoir relative to that in the design stage. The runoff and sediment load delivered to the reservoir could be estimated based on measured values of continuous river flow. Schleiss et al. [1] highlights and discusses the main matters concerning reservoir sedimentation. The reservoir sedimentation problem should be considered from the early stages of planning design and operation. In addition, the sedimentation process can create problems downstream from the dam which should also be considered in the planning and design stages.

Physically based models are usually used in cases where runoff and sediment load data records are not available. These models are of two types, which are referred to as single storm models and continuous simulation models. The Areal Non-Point Source Watershed Environment Response Simulation (ANSWERS) is developed by Beasley et al. [2]; and the European Soil Erosion Model (EUROSEM) was improved by Morgan et al. [3]; all of those models mentioned are examples of the former models. Examples of the latter are the spatially distributed erosion and sediment yield component, chemicals, runoff and erosion from agricultural management systems CREAMS (Science and Education Administration; Department of Agriculture, Washington, WA, USA) which are proposed by Knisel [4]; the SHESED model (the hydrologic and sediment transportation model of hydrological modeling system (SHE)) which is proposed by Wicks and Bathurst [5]; and the soil and water assessment model (SWAT) that was developed by Arnold et al. [6]. The SWAT model is the most commonly used model and, for this reason, a number of researchers have modified this model for different purposes (see [7,8]). Durao et al. [9] estimates the transported nutrient load in the Ardila River watershed in Spain by applying the SWAT model in order to identify the contribution of this load to the whole watershed. The model is applied to simulate long-period data; the real daily precipitation data is considered for the period 1930–2000. The considered flow data for model calibration and validation extend from 1950 to 2000, and nutrient data stretch from 1981 to 1999. The results indicate that the main source of diffusion prolusion comes from the main tributaries of Spain. Wang et al. [10] tests the possible conservation practices within a rangeland watershed using the agricultural policy/environmental extender (APEX). The model is calibrated and validated for both flow and sediment yield for the Cowhouse Creek watershed in north Texas. The analysis of the scenario extends from 1951 to 2008. It shows that a significant reduction reaches to 58.8% of overland sediment losses from the area covered by a range brush to range grass. This reduction is due to the replacement of shrub species with herbaceous species within the subareas. Samaras and Koutitas [11] evaluates the effect of land-use changes in a watershed on coastal erosion for a selected area in north Greece. They apply both the SWAT model and a shoreline evolution model, a shoreline evolution model, for this purpose. The simulation is applied before and after land use change using three formulas of sediment transportation. The result indicates that a reduction in crop/land use cover from 23.3% to 5.1% leads to a reduction in both watershed sediment yield and sediment discharge at the outlet by (56.4%) and (26.4 to 12.8%), respectively. This study can be considered as a suitable tool and guide for future work in the same field. Samaras and Koutitas [12] studies the effect of climate change on sediment transport and morphology. The study is applied to a selected sandy coast area and its watershed in North Greece. Both SWAT models are implemented for the watershed and PELNCON-M is implemented for the coastal area to achieve the study aims. Two scenarios are employed; the first one is considered to be an extreme rise in the precipitation depth on the watershed, and the second one is considered to be an extreme rise in waves in the coastal area. Results of the first scenario shows a significant effect on erosion, sediment transport, sediment yield and discharge at the watershed outlet, while the second scenario indicates a lower effect on the coastline variation. Arnold [13] developed the SWAT + CUP model (SWAT Calibration and Uncertainty Programs, Swiss Federal Institute of Aquatic Science and Technology, Zurich, Switzerland), which provides a semi-automatic tool for decision-making for the SWAT model by applying both manual and automated calibration and incorporating both sensitivity and uncertainty analysis. A number of previous studies [14–17] were applied using the SWAT model to estimate runoff, sediment yield and/or other soluble materials for ungagged watersheds based on neighboring or similar property watersheds. Another technique is used for flow and erosion, and sediment transport is the distributed mode. Juez et al. [18] simulates the hydro-sedimentary response of the Western Mediterranean catchment to a representative rainfall storm. The simulation combines the distributed flow surface model with the empirical model for infiltration, the Soil Conservation Services Model (SCS), and the erosion model, which is the Hillslope Erosion Model (HEM), considering water depth and flow as a 2D model. The present model is a tool for analyzing the hydro-sedimentary process at a temporal and special scale.

Most countries in the Middle East suffer from water shortage problems, where the annual allocation per capita does not exceed 500 m^3 [19]. For this reason, water is essential to life, socioeconomic development, and political stability in this region. Future prospects are negative; therefore, this problem is expected to be more chronic and severe in future [20]. Iraq used to be considered to be a relatively rich country in terms of its water resources, until the mid-1970s, because of the presence of the Tigris and Euphrates Rivers [21]. Due to regional and internal problems in Iraq, the estimation of the overall water required is about 75–81 billion cubic meters (BCM) [22], while the available quantity is 59–75 BCM and will drop to 17.61 BCM in 2025 [23]. In view of this situation, it is very important to know the actual storage capacity of the reservoirs—which are unknown now—so that prudent water resources planning can be done. The sedimentation rate of several reservoirs was recently investigated in Mosul and Dohuk. This is the third reservoir to be dealt with in Iraq. The bed of the Dokan Dam reservoir (located in the northeast of Iraq) is surveyed by Hassan et al. [24], and this studied the bed sediment using 32-bed samples distributed spatially over the reservoir. The results indicate that the bed sediments of the reservoir are composed of silt (48%), clay (23%), gravel (15%) and sand (14%).

All reservoirs in the world suffer from sediment deposition. This is one of the main problems for reservoir life sustainability. Sustainable reservoir sediment-management practices enable continued reservoir functioning for a longer period by reducing reservoir sedimentation. Iraq suffers from water shortage problems, especially after the construction of a series of storage reservoirs in source countries (Turkey, Syria and Iran), so the evaluation of the actual live storage capacity of dams is important for the prudent management of the operation schedule. The aim of this study is to assess the annual runoff and sediment loads of the Dokan Dam watershed (ungagged area) using a SWAT model set-up based on the parameter transformation technique of the modeling-gauged Derbendekhan watershed to learn the hydrological behavior of the area and to assess its contribution to the total values pouring into the reservoir. Moreover, the set-up model helps us to find the spatial distribution of erosion and annual sediment yield for the sub basins. This will help us to find the sub basins with a high sediment yield and evaluate effective factors for them. These assessments help in the process of developing a plan and a strategy for managing the sediment inflow and deposition.

2. Study Area

2.1. Location and Topography

The considered study area is the watershed of the Dokan Dam reservoir, situated in the northeast of Iraq (Figure 1). Dokan dam is a concrete arch dam located in the Lower Zab River, about 65 km southeast of Sulaimaniyah city and 295 km north of Baghdad, the capital city of Iraq. The dam height is about 116 m at maximum river depth, having a total storage capacity of 6.87 $\times$ 10^9 m^3 (6.14 $\times$ 10^9 m^3 live storage and 0.73 $\times$ 10^9 m^3 dead storage) at normal operation level of 511 m.a.s.l. [22]. The dam has been built to serve irrigation, power generation, water supply and flood control needs. Due to the limited observed data of flow at the Dokan Dam watershed and the unavailability of sediments load data, the second watershed considered for this research is the Derbendekhan Dam watershed; it is the nearest watershed to the study area. The properties of the Dokan and Derbendikhan watersheds are shown in Table 1. The digital elevation model (DEM) with 30 m resolution is considered to identify the watershed boundary, classification of overland and channel flow, slopes and other properties.

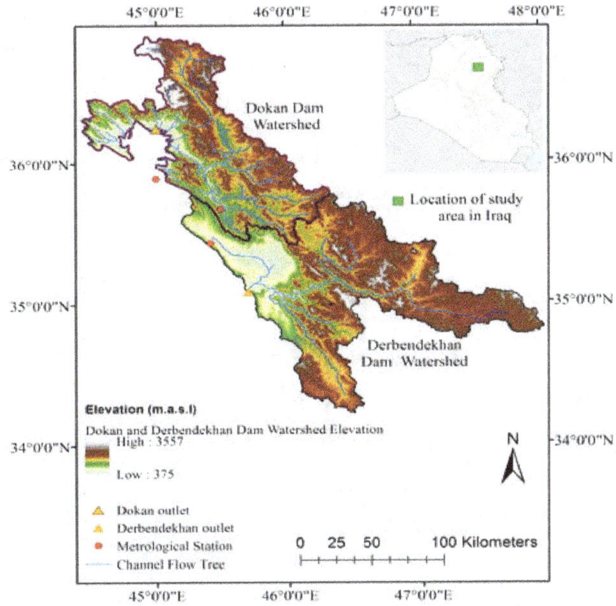

Figure 1. Topographic map of the watershed areas of the Dokan and Derbendekhan Dams and their locations in Iraq.

Table 1. Properties of the Dokan and Derbendekhan watersheds.

Watershed	Area (km^2)	Max. Elevation (m.a.s.l)	Min. Elevation (m.a.s.l)	Average Slope (%)	Maximum Annual Rain (mm)	Minimum Annual Rain (mm)
Dokan Dam	11,690	3557	489	26.5	1125	182
Derbendekhan Dam	15,280	3332	375	23.3	970	174

2.2. Soil Type and Land Use

The exposed rocks at the Dokan and Derbendekhan watershed areas are mainly limestone and minor exposures of dolomitic limestone, dolomite, and Quaternary alluvial deposits [25,26]. Based on the Reconnaissance Soil Map of the three Northern Governorates, Iraq [26] and the Food and Agriculture Organization of the United Nations (FAO) soil map [27], both watersheds are located on a common extended type of soil classification. Samples for different soil classes were taken depending on the soil map of the study area. A map of soil types is prepared for this study as a shape file for each watershed to be used in the SWAT model (see below for model details). The soil samples analysis includes grain size distribution in different types, organic matter content and hydraulic conductivity. The analysis of soil samples indicates that the area generally consists of four major soil types. Most of the area (85.6%) is covered by gravelly sandy mud; 6.9% is gravelly mud; and 7.5% of two types of muddy gravel (the main differences between the two types are the percent of gravel, which is 74% and 56% for types 1 and 2, respectively). Figure 2 shows the shape file of soil type considered in the SWAT model for the Dokan watershed.

The land use map for the years (1976–1979) [28] and available satellite image (NASA's Landsat GeoCover, 2007, with a spatial resolution of 14.25 m) indicates that the winter plants and pastures represent the main part of the land use map of the studied area. This depends on rainfall as a main source of irrigation water. The other parts are forests, vegetables and urban areas (villages). The land use change is limited (see Table 2). This is mainly due to the geological nature of both watersheds. In addition, the topography of the area does not enhance any changes. It is noteworthy to mention that rain is the main source of irrigation. For these reasons, the land cover did not change widely through the study period since the operation of the dam from the year 1959 to the year 2014. The Dam and the studied watershed are located away from the main cities, so changes in the urban and rural areas are limited. Table 2 shows the percentage of different land use cover for the two periods of the available land use map.

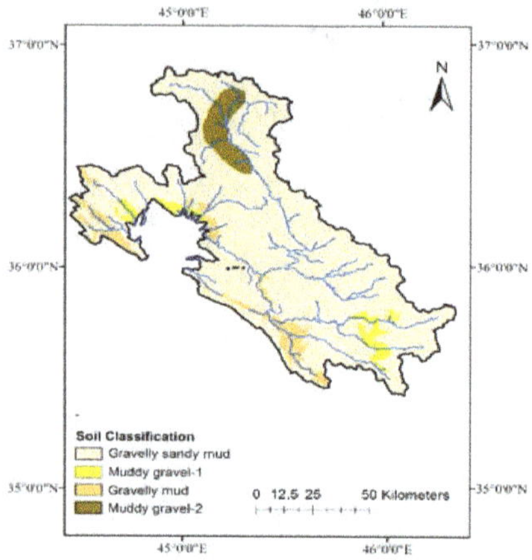

Figure 2. Soil type classification of the Dokan Reservoir watershed.

Table 2. Percentages of different land use types for two years of the studied period.

Year	Winter Plant and Pasture	Forest	Vegetables	Urban Area (Village)
1976–1979	77.3	22.0	0.5	0.2
2007	82.7	15.6	1.6	0.1

Due to the small difference between the percentage of land cover for the two available years, a map of land use for the study area is prepared as a shape file (Figure 3). The area consists of four types of land use/cover. Winter plants (pasture) and forests of different types of trees cover the main part of the study area, while the remaining small area is planted with vegetables near the reservoir boundary and/or in urban areas (villages).

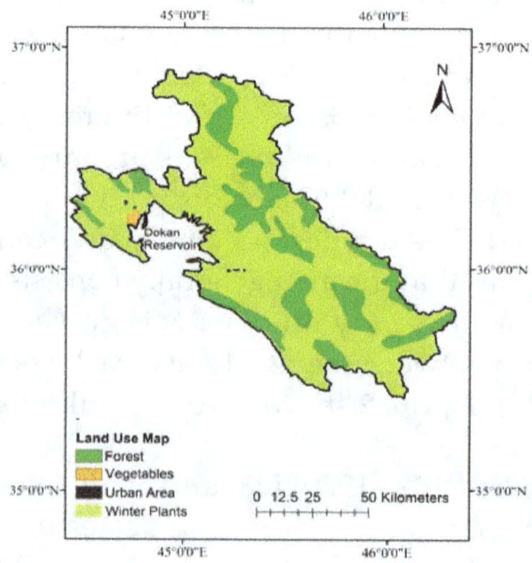

Figure 3. Land use and land cover classification of the Dokan Reservoir watershed.

3. Applied Model

The soil and water assessment tool (SWAT) is a physically based continuous simulation model for short or long times that can be applied to large river basins and complex watersheds. It was developed by the US Department of Agriculture, Agricultural Research Service [6]. The model is an efficient tool to estimate the flow and sediment load in addition to different chemical and nutrient materials. The model divides the watershed into sub basins based on their DEM date and hydrological response units (HRU); each unit has the same soil type, land use and land slope.

The hydrological simulation is based on the topographical terrain, soil type, land use and hydrological data of daily precipitation, maximum and minimum temperature, wind speed, relative humidity and solar radiation. The flow can be estimated based on a water balance equation; this equation is simulated in the SWAT model by different modular: the land phase and routing phase [14]. For the land phase, the soil water balance calculation is based on the following form [29]:

$$SW_t = SW_0 + \sum_{i=1}^{t}(R_{day} - Q_{sur} - E_a - w_{seep} - Q_{qw}) \quad (1)$$

where,

SW_t: Water content of the soil (mm);
SW_0: Initial water content (mm);
R_{day}: Depth of precipitation (mm);
Q_{sur}: Equivalent depth of surface runoff (mm);
E_a: Evapotranspiration depth (mm);
w_{seep}: Depth of water seepage out of considered surface profile (mm);
Q_{qw}: Equivalent depth of return flow (mm).

The Penman–Monteith method is considered for potential evapotranspiration estimation. The required input data to estimate the potential evapotranspiration (PET) using the Penman–Monteith method are daily solar radiation, air temperature, relative humidity and wind speed. The formula of this method considers three effective factors for evapotranspiration, which are the required energy to sustain evaporation, the required strength to remove the water vapor and the aerodynamic in addition to resistance of the surface. The Penman–Monteith method is in the following form [29]:

$$\lambda E = \frac{\Delta(H_{net} - G) + \rho_{air} \cdot c_p \cdot [e_z^o - e_z]/r_a}{\Delta + \gamma \cdot \left(1 + \frac{r_c}{r_a}\right)} \quad (2)$$

where,

λE: Latent heat flux density (MJ/m^2/day);
E: Evaporation rate (mm/day);
Δ: Saturation vapor pressure-temperature slope (de/dt) (kPa/Co);
H_{net}: Net radiation (MJ/m^2/day);
G: Density of heat flux to the ground (MJ/m^2/day);
ρ_{air}: Density of the air (kg/m^3);
c_p: Specific heat at constant pressure (MJ/kg/day);
e_z^o: Saturated vapor pressure of air at height z, (kPa);
e_z: Water vapor pressure of air at height z (kPa);
λ: Psychrometric constant (kPa/Co);
r_c: Resistance of plant canopy (s/m);
r_a: Diffusion resistance of the air layer (s/m).

Also, the different parameters of land management are recognized based on soil type, land use and land cover. The soil water content and soil infiltration can be estimated by two methods based on the available data, either by the Green–Ampt infiltration equation or curve number methods. The Green–Ampt equation requires rainfall data of a sub daily interval, which is not available in Iraqi weather stations, so the curve number method is utilized throughout this work using the following form [29]:

$$Q_{surf} = \frac{\left(R_{day} - 0.2S\right)^2}{\left(R_{day} + 0.8S\right)} \tag{3}$$

where,

Q_{surf}: Equivalent depth of surface runoff (mm);
R_{day}: Rainfall depth of the considered day (mm);
S: Retention parameter (mm).

The value of S can be estimated by the following equation [29]:

$$S = 25.4\left(\frac{1000}{CN} - 10\right) \tag{4}$$

where, CN is the curve number of that considered day.

The second process includes the estimation of soil erosion from the overland due to rainfall detachment and surface runoff in addition to channel erosion and deposition. The sediments, routing in both the overland and channel flow, are estimated based on rainfall data, soil properties, land use/land cover and topography. The maps of soil type and land use are required with the digital elevation model (DEM) data to identify the topography of the watershed and to classify it into overland and channel sediment flow. The modified universal soil loss equation (MUSLE) is considered in the following form [29]:

$$sed = 11.8 \times \left(Q_{surf} \cdot q_{peak} \cdot are_{hru}\right)^{0.56} K_{USLE} \cdot C_{USLE} \cdot P_{USLE} \cdot LS_{USLE} \cdot CFRG \tag{5}$$

where,

sed: Yield of sediment for the considered storm or day (ton.);
Q_{surf}: Volume of surface runoff (mm/ha);
q_{peak}: Greatest surface runoff rate (m^3/s);
are_{hru}: Hydrologic response unit area (ha);
K_{USLE}: Soil erodibility factor of Universal Soil Loss Equation (USLE);
C_{USLE}: Cover and management factor of USLE;
P_{USLE}: Soli practice factor of USLE;
LS_{USLE}: Topographic factor of USLE;
$CFRG$: Factor of coarse fragment.

4. Model Calibration and Validation

4.1. Runoff and Sediment Load Calibration

Although the mathematical and conceptual models are considered widely in hydrological studies to simulate different events, such as runoff flow, sediment and both suspended and dissolved material transport, they still require calibration with measured values to ensure the accuracy of the model outputs.

Two types of dataset are prepared to be applied to SWAT model. The metrological (climate) data include daily precipitation, maximum and minimum temperature, wind speed, relative humidity and solar radiation; hydrometric data are also present. The second dataset is the topography data, which includes the DEM map.

In view of the limited available measurements of flow and unavailable sediment records of the Dokan Dam watershed, the available data of the Derbendekhan Dam watershed (the adjacent watershed to Dokan, Figure 1), were considered for both flow and sediment model calibration. Due to the similarity in the geological formation, soil type, land use, topographical and watershed characteristics (Table 1), the parameters of the calibrated watershed can be transformed to an ungauged watershed model [17]. To calibrate the results of the SWAT model for both runoff and sediment, the SWAT-CUP software is applied. It is an efficient tool to adjust different parameters of the SWAT model to obtain optimal local results and create an uncertainty analysis of SWAT model parameters to provide an easy and quick method of calibration and standardized calibration [30]. The considered software for the model is the Sequential Uncertainly Fitting version 2 (SUFI-2, Swiss Federal Institute of Aquatic Science and Technology, Zurich, Switzerland). In this program, all the uncertainty parameters can be used in the model calibration, including uncertainly in driving variables, parameters of the conceptual model and considered data [30]. Different statistical criteria can be considered in the model objective function to evaluate model performance, such as the determination coefficient, Nash–Sutcliff model efficiency, root mean square error and Chi-square. The determination coefficient is considered to be effective criteria to obtain the optimal values of flow and sediment concentration between the observed and measured data.

The SWAT project for the Derbendekhan watershed is set-up including the required DEM data, soil type land use, as shown in Figures 1–3, respectively, and meteorological data based on the nearest stations to the area as shown in Figure 1. The monthly average flow rate data at Derbendekhan station are considered for model calibration. To obtain an enhanced calibration of the model and for more understanding of the model's performance, the monthly recorded flow data are separated into base flow and surface runoff [14]. The recursive digital filter technique [31] is used to obtain a monthly separation based on the original daily separation technique. The separated monthly runoff from the total flow as measured values is applied in SUFI-2 to calibrate the model parameters. The statistical criterion of the determination coefficient is used as an objective function criterion. Besides this, the Nash–Sutcliffe model is employed to evaluate the model performance. For monthly runoff flow, the highest obtained values are 0.75 and 0.64 for the coefficient of determination and Nash–Sutcliffe model, respectively. The model uncertainty was measured using two factors: the P-factor reflects the percentage of measured data bracketed by 95% prediction uncertainty (95PPU). This means that one minus the P-factor represents the presence of poorly simulated values. The R-factor is another measurement of model uncertainly equal to the average thickness of the 95PPU band divided by the standard deviation of measured data. For the runoff calibration period, the P-factor is 0.78 and the R-factor is 1.27. Figure 4 shows the observed and simulated runoff at the Derbendekhan watershed outlet and the uncertainty band (95PPU) for the period (1961–1968).

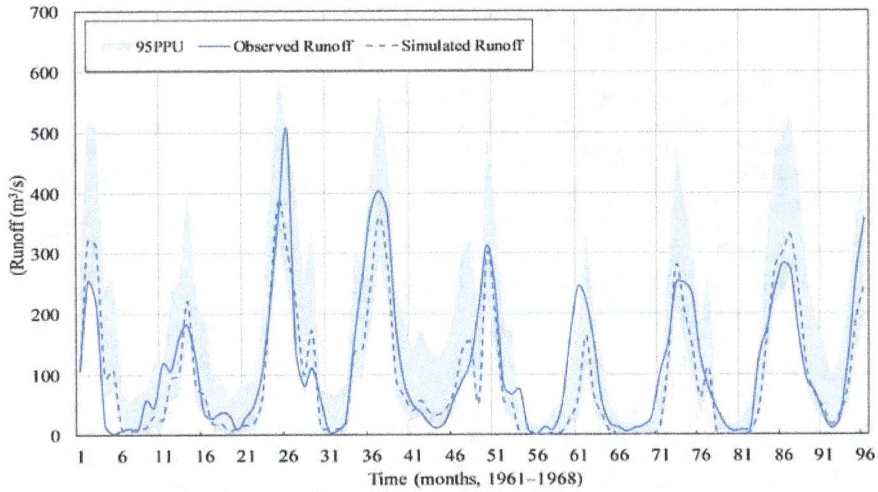

Figure 4. Monthly observed and simulated runoff at the Derbendekhan outlet and 95PPU for the period 1961–1968.

The available measured sediment concentration data of the Diyala River at Derbendekhan station, as presented by Assad [32], were utilized to calibrate the SWAT model parameters for the same period of runoff flow calibration. Figure 5 shows the measured and optimal simulated values of sediment load concentration at Derbendekhan outlet. The same statistical criteria are implemented, and the optimal resultant values are 0.65 and 0.63 for determination coefficient and Nash–Sutcliffe model efficiency, respectively, while the P-factor and R-factor equal to 0.68 and 2.27 respectively.

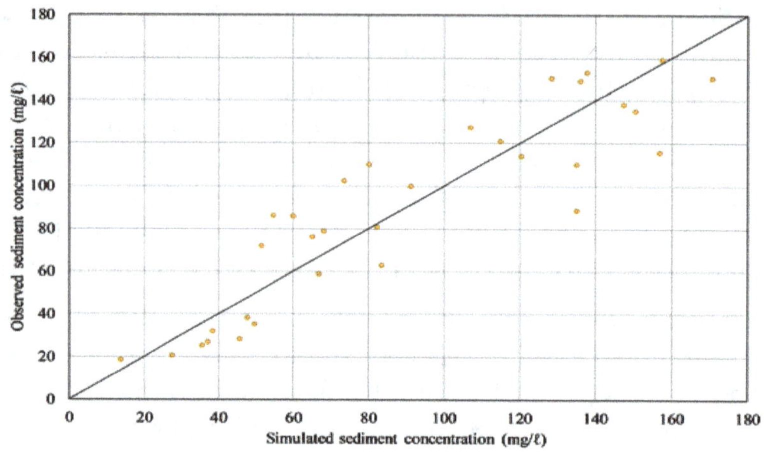

Figure 5. Observed and simulated sediment concentration for period (1961–1968) at the Derbendekhan watershed.

The most effective parameters are selected for the runoff and sediment model calibration as proposed by [28] in addition to other parameters. The resultant best-fitted values (optimal values) of the parameters and the considered range are shown in Table 3. Parameters listed in Table 3 are also mentioned in other previous literature [14,33].

Table 3. The range, optimal (fitted) values and sensitivity analysis of considered parameters.

Parameter	Description	Min_Value	Max_Value	Fitted_Value
V_GWQMN.gw	Threshold depth in shallow aquifer (mm).	0	2	1.698
V_CH_COV1.rte	Channel erodibility factor.	0.05	0.6	0.38715
R_USLE_K(..).sol	USLE, equation soil erodibility (k) factor.	−0.8	0.8	−0.1168
R_LAT_SED.hru	Sediment conc. In lateral and ground flow.	0	5000	2525
R_SPCON.bsn	Linear par. for calculating max. amount of sediment that can be re-entrained during channel sediment routing.	0.0001	0.01	0.008743
V_SPEXP.bsn	Exponential Par. for calculating max. amount of sediment that can be re-entrained during channel sediment routing.	1	2	1.549
V_GW_REVAP.gw	Groundwater "revap" coefficient.	0.02	0.2	0.07166
V_ALPHA_BF.gw	Base flow alpha factor (days).	0	1	0.555
V__CH_COV2.rte	Channel cover factor.	0.001	1	0.281719
V_GW_DELAY.gw	Groundwater delay (days).	0.02	0.2	0.07166
R_SOL_BD(..).sol	Moist bulk density.	−0.5	1	−0.3755
R_USLE_C(..)plant.dat	Min value of USLE-C factor for land cover/plant.	−0.5	0.5	0.151000
R_CH_N2.rte	Manning's "n" value for the main channel.	−0.5	0.5	−0.279
R_USLE_P.mgt	USLE, support practice parameter.	0	1	0.941
R_SOL_K(..).sol	Saturated hydraulic conductivity.	−0.5	0.5	0.177
R_CN2.mgt	SCS runoff curve number.	−0.2	0.2	0.198
R_SOL_AWC(..).sol	Available water content of the soil layer.	−0.5	1	0.5005

Note: **R**: Relative, **V**: Replace. (..) for different soil or plant type.

4.2. *Runoff Validation*

A SWAT model project is set-up for the Dokan Dam watershed, the main part of the study area. The limited recorded data of the monthly flow rate at the outlet of the Dokan watershed and the absence of any sediment load measurements leads to utilizing the regionalization technique to transfer the effective parameters from the adjacent gauged (Derbendekhan) watershed. Due to its similarity in geological formation, soil type, land use, watershed characteristics and weather data, the effective hydrological parameters obtained from the Derbendekhan (gauged) watershed can be transformed to the Dokan (ungauged) watershed. The process of parameter transformation is called regionalization. There are a number of presented methods for the regionalization of the watershed hydrological parameters: Kokkonen [34] applies the regression approach, while Parajka et al. [35] employs kringing and a similarity approach and Heuvelmans et al. [36] investigates the application of artificial neural nets and other methods. Since the Derbendekhan watershed is adjacent to the Dokan watershed (Figure 1), and the physical, topographical properties and rainfall are similar (Table 1) along with the geological formation, land use/land cover, and soil type, both watersheds have similar flow and sediment parameters. In this case, the effective parameters can be transformed from a donor watershed to an ungauged watershed. The fitted values of the Derbendekhan parameters calibrated by the SUFI-2 program are transferred to the Dokan watershed SWAT project. The SUFI-2 program is implemented for the calibration, uncertainly analysis and regionalization of the considered parameters of the SWAT model for the Debendekhan watershed for runoff and sediment load of concentration. Here, the SUFI-2 algorithm is used for the calibration, validation and measurement of the uncertainty for input data, model and sensitive parameters. The degree of uncertainty is measured by two values: P-factor and R-factor. The percent of measured values bracketed by 95% prediction uncertainly represent (95PPU), which is the P-factor while the ratio of 95PPU thickness divided by standard deviation of measured values is equal to the R-factor. When the simulated values are exactly the measured ones, the value of P-factor equals 1 and the value of R-factor equals zero [37].

Based on the transformed parameters, the simulated runoff flows are compared with measured values at the Dokan watershed outlet after the separation of the base flow for the period 1961–1964 (Figure 6) to evaluate the effectiveness of the transformation of the hydrological parameters. For the validation period of runoff, the obtained values of the determination coefficient and Nash–Sutcliffe model efficiency are 0.68 and 0.64, respectively, indicating that the transformation process is successful. The P-factor and R-factor for the model uncertainty also indicate a reasonable model performance: the P-factor is 0.71 and R-factor is 0.97.

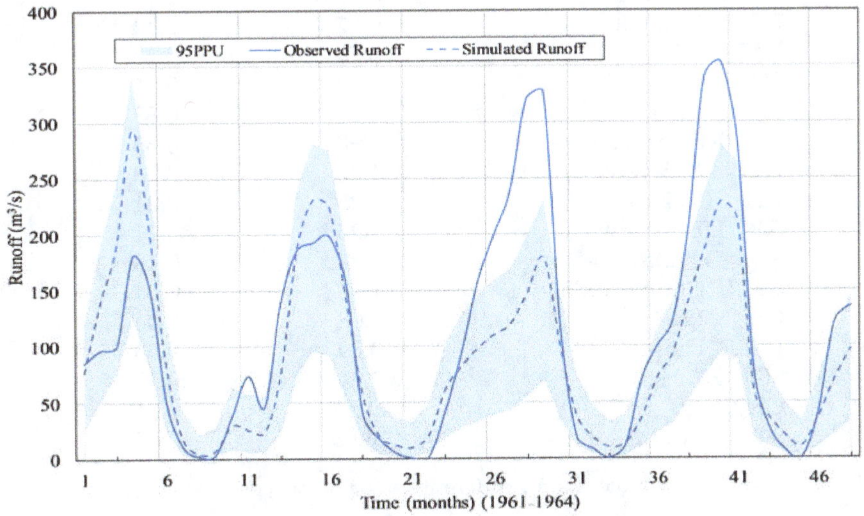

Figure 6. Monthly observed and simulated runoff at the Dokan outlet and 95PPU for the period 1962–1965.

5. Results and Discussion

After Durbendekhan SWAT project calibration for runoff and sediment data, the best-fitted values of the hydrological watershed parameters are transformed by the regionalization technique to the

Dokan SWAT project, the main project of the study. The sensitivity analysis is also studied for the most effective parameters on both the runoff and sediment load. The sensitivities are accomplished to identify the effective parameters on runoff and sediment load values for the watershed. The parameter sensitivity is estimated in the SUFI-2 model based on the multiple regression system presented by Abbaspour [38] to evaluate the effect of the considered parameter value (b_i) on the objective function (g); its sensitivity is in the following form:

$$g = \alpha + \sum_{i=1}^{m} \beta_i b_i \tag{6}$$

This formula calculates the average changes in the objective function due to the change in a given parameter while other parameters are changing. The comparative significance and sensitivity of each parameter are estimated based on the statistical criteria of the *t*-stat and *p*-value. The *t*-stat value is obtained from the coefficient of a parameter in the multiple regression analysis divided by its standard error. If the coefficient value is large in comparison to the standard errors, this mean that the parameter is sensitive. The *p*-value can be obtained by comparing the *t*-stat value with the student's distribution table [37]. The *p*-value of each term test is the null hypothesis, in which the coefficient is not affected. If the *p*-value is less than 0.05, it indicates that the null hypothesis can be rejected. The *t*-stat and *p*-values of different effective parameters are shown in Table 4. The parameters are arranged from low to high sensitivity, i.e., from low *t*-stat value or high *p*-value. The result of the test indicates that the soil water content, soil curve number at normal conditions (CN2) and the soil saturated hydraulic conductivity are the most effective parameters while threshold depth in a shallow aquifer, the channel erodibility factor and soil erodibility (k) factor in USLE have the lowest effect on runoff and sediment load simulation.

Table 4. Effective parameters arranged from low to high sensitivity based on *t*-stat and *p*-value.

Parameter	Absolute *t*-Stat	*p*-Value
GWQMN.gw	0.61	0.54
CH_COV1.rte	0.62	0.54
USLE_K(..).sol	0.77	0.44
LAT_SED.hru	0.78	0.43
SPCON.bsn	0.92	0.36
SPEXP.bsn	0.97	0.33
GW_REVAP.gw	0.98	0.33
ALPHA_BF.gw	1.20	0.23
CH_COV2.rte	1.25	0.21
GW_DELAY.gw	1.25	0.21
SOL_BD(..).sol	1.34	0.18
USLE_C(..)plant.dat	1.96	0.05
CH_N2.rte	2.08	0.04
USLE_P.mgt	3.69	0.03
SOL_K(..).sol	9.34	0.02
CN2.mgt	11.57	0.01
SOL_AWC(..).sol	16.05	0.00

Note: (..) for different soil or plant type.

The model is applied for the study period to estimate the runoff and sediment load reaching the Dokan Reservoir. The considered years of simulation have begun since the operation of the dam in 1959 to the year of the bathometry survey (2014) carried out by [39]. The resultant annual runoff volume that enters the Dukan Reservoir from the HRUs ranges from 300 to 4600 MCM (Figure 7), depending on rainfall intensity, depth and distribution through the rainy season. The runoff average volume from the watershed represents 35% of the live storage capacity of the dam, indicating that watershed runoff makes a significant contribution to reservoir inflow.

The SWAT model is an efficient tool to estimate the runoff hydrograph, but, in some hydrological studies, such as scheduling reservoir operations to supply the demand rate, the assessment of water resource income only is required. A regression formula is determined based upon the input–output data of SWAT model of the study area. This is a simple and quick tool to correlate the annual runoff depth with annual rainfall with good correlation (R^2 = 0.9) results without the need for more detailed input data as required in SWAT projects. The relationship used is in the following form:

$$Run_{Ann} = 0.075 \times R_{Ann}^{1.21} - 1.92,\ R^2 = 0.90\ (for\ R_{Ann} > 16\ \text{mm}) \tag{7}$$

where

Run_{Ann}: Annual runoff depth mm);
R_{Ann}: Annual rainfall depth (mm).

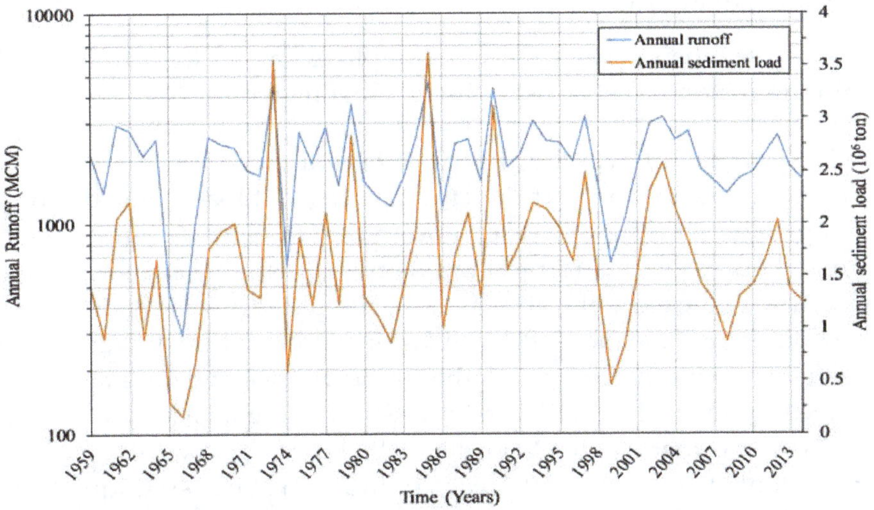

Figure 7. Annual runoff volume and sediment load delivered to the Dokan Reservoir for the period 1959–2014.

The formula is suitable when the annual runoff depth is greater than 15 mm to avoid negative runoff; this value is already much lower than the minimum historical recorded value.

The sediment load delivered to the Dokan reservoir is also estimated based on MUSLE programmed into the SWAT model for each single storm. The results are presented here annually. The average annual sediment load concentration is 650 mg/ℓ. This concentration can be considered relatively low in comparison with other locations or measurements in the region as proposed by [40,41] as well as the worldwide rate [42]. This is due to the nature of the rocks of the area and the effect of plant cover, such as winter pasture and plants and some forest trees throughout the region which reduce the detachment of soil particles transported with runoff flow. The estimated annual sediment load delivered to the Dokan reservoir from the watershed ranges from 3.6 to 0.16 $\times$ 10^6 ton for studied period, Figure 7. The average annual value is 1.63 $\times$ 10^6 ton.

The sediment trap efficiency of the reservoir is estimated based on the method presented by Garg V. and Jothiprakash V. [43]. This depends on reservoir storage capacity and annual inflow. The trap efficiency of the Dokan reservoir changes through the study period from 1 to 0.985.

Based on the results obtained from the simulation model, the estimated sediment load volume deposited in the reservoir for the considered period is about 10% of the dead storage capacity. This value is for the watershed only, which is considered a reasonable value and does not affect the designed project life. However, the Lesser Zab load should be also considered to evaluate the total amount of sediment load delivered and deposited within the reservoir. The total amount of sediment deposited in the Dokan reservoir for the period (1959–2014) is 209 MCM [38]. This means that the sediment load delivered into the reservoir from the watershed based on simulated results is about 34% of the total sediments deposited within the reservoir, which mean that the watershed sediment contribution is an effective value.

Due to the huge amount of data required, including topographical, metrological and hydrological data, different maps of soil, and land use to estimate the sediment load based on the applied model, a simple regression formula is used. It is based on simulated values to correlate the sediment load per unit area of the Dokan watershed with annual runoff depth in the following form:

$$Sed_{Ann} = 0.056 \cdot R_{Ann}^{1.195} - 7.33,\ \mathrm{R}^2 = 0.97\ (for\ R_{Ann} > 60\mathrm{mm}) \quad (8)$$

where

Sed_{Ann}: Annual sediment load per unit area, (ton/km^2);
R_{Ann}: Annual rainfall depth (mm).

The formula is suitable when the annual runoff depth is greater than 60 mm; this value is already lower than the minimum historical recorded value in Dokan area.

A special distribution map of average annual sediment yield per unit area of sub basins is also prepared (Figure 8a). It can be noticed that the annual sediment load contribution ranges from 13 to 950 ton/km^2 approximately. The rate of erosion and sediment yield depends on a number of factors: topography, soil type, land cover and rainfall intensity. Comparing the sub basins of different soil types and land uses, in both the annual sediment yield map (Figure 8a) and the sub basin slope map (Figure 8b), it can be noticed that the most effective factor for the sediment yield is the land slope rather than other factors. This can be clearly noticed in that some basins have the same soil type and land use but a higher slope gives higher sediment yield. Sub basins having an average slope between 25 to 45% represent the area of high sediment yields from 400 to 950 t/km^2; however, when the sub basin slope is less than 20%, the sediment yield per unit area is reduced to about 40 t/km^2.

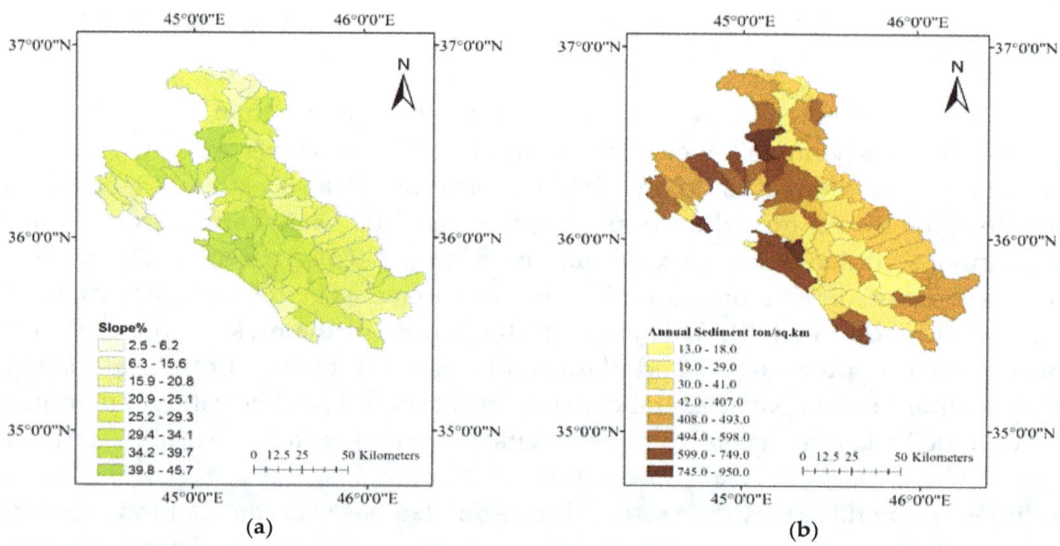

Figure 8. (**a**) Spatial distribution of average annual sediment load for the Dokan watershed sub basins; (**b**) average slope of sub basins.

This map is a tool that can enable decision-makers to apply a suitable method to reduce the erosion load, especially from high erosion rate areas. Depending on the selected area, the treatment may include practicing strip planting, terracing, or contour forming to reduce the effect of slope on surface runoff flow velocity, erosion and sediment transport capacity.

6. Conclusions

The soil and water assessment tool (SWAT) model is applied to assess the runoff and sediment delivered from the Dokan Dam watershed. Due to the limited recorded flow data and the absence of sediment measurements data at a station near the inlet of Dokan Reservoir, the model is calibrated for both runoff and sediment load for the Derbendekhan watershed adjacent to the Dokan watershed. The regionalization technique is employed to transfer the calibrated parameters of the SWAT project from a gauged (Derbendekhan) to an ungauged (Dokan) watershed. The resultant monthly runoff flow for the Dokan SWAT project is based on transformed parameters which were compared with measured values to evaluate the regionalization technique and model performance. The determination coefficient (R^2) and Nash–Sutcliffe model efficiency (Eff.) are 0.68 and 0.64, respectively, indicating a reasonable model performance with this technique. The average watershed contribution for annual runoff represents 35% of the dam life storage; this percentage is considered effective in the dam operation schedule. The total sediment load delivered to the Dokan reservoir from the watershed for the studied period is about 72 MCM. This load forms about 10% of the dead storage capacity of the reservoir. Generally, the total sediment load delivered and deposited in the reservoir for the period of dam operation is considered acceptable within the allowed limits. The map of special distribution of annual sediment load yield per unit area of each sub basins is presented; the average slopes map reflects a good agreement with the map of annual sediment load yield in comparison to other effective factors to be considered. This indicates that the land slope is the most effective factor on erosion and sediment transport. This can be used for soil conservation treatment to reduce the erosion rate.

Author Contributions: M.E.-A. and R.H. did the field, methodology and modelling, A.A. helped in the modelling, N.A.-A. and S.K. did the supervision.

Acknowledgments: The authors would like to express their thanks to John McManus (University of St. Andrews, St. Andrews, UK) and Ian Foster (Northampton University, Northampton, UK) for reading the manuscript and for their fruitful discussions and suggestions, and to Lulea University of Technology for some of the financial support for this research.

References

1. Schleiss, A.J.; Franca, M.J.; Juez, C.; De Cesare, G. Reservoir Sedimentation—Vision Paper. *J. Hydraul. Res.* **2016**, *54*, 595–614. [CrossRef]
2. Beasley, D.B.; Huggins, L.F.; Monke, E.J. ANSWERS: A model for watershed planning. *Trans. ASAE* **1980**, *23*, 938–944. [CrossRef]
3. Morgan, R.P.C.; Quinton, J.N.; Smith, R.E.; Govers, G.; Poesen, J.W.; Auerswald, K.; Chisci, G.; Torri, D.; Styczen, M.E.; Folly, A.J. The European soil erosion model (EUROSEM): A dynamic approach for predicting sediment transport from fields and small catchments. *Earth Surface Process. Landf.* **1998**, *23*, 527–544. [CrossRef]
4. Knisel, W.G. *CREAMS: A Field-Scale Model for Chemical, Runoff and Erosion from Agricultural Management Systems*; USDA Conservation Research Report; Conservation Model Development: Washington, WA, USA, 1980.
5. Wicks, J.M.; Bathurst, J.C. SHESED: A physically based, distributed erosion and sediment yield component for the SHE hydrological modelling system. *J. Hydrol.* **1996**, *175*, 213–238. [CrossRef]
6. Arnold, J.G.; Srinisvan, R.; Muttiah, R.S. Large area hydrologic modeling and assessment. Part I: Model development. *Am. Water Resour. Assoc.* **1998**, *34*, 73–89. [CrossRef]
7. Easton, Z.M.; Fukal, D.R.; White, E.D.; Collick, A.S.; Ashagre, B.B.; McCartney, M.; Awulachew, S.B.; Ahmed, A.A.; Steenhuis, T.S. A multi basin SWAT model analysis of runoff and sedimentation in the Blue Nile, Ethiopia. *Hydrol. Earth Syst. Sci.* **2010**, *14*, 827–1841. [CrossRef]
8. Swami, V.A.; Kulkarni, S.S. Simulation of runoff and sediment yield for a Kaneri Watershed Using SWAT Model. *J. Geosci. Environ. Protect.* **2016**, *4*, 1–15. [CrossRef]
9. Durao, A.; Leitao, P.C.; Brito, D.; Fernandes, R.; Neves, R.; Morais, M.M. Estimation of Transported Pollutant Load in Ardila Catchment Using the SWAT Model. In Proceedings of the International SWAT Conference, Toledo, Spain, 15–17 June 2011.

10. Wang, X.; Amonett, C.; Williams, J.R.; Fox, W.E.; Tu, C. Assessing Impacts of Rangeland Conservation Practices Prior to Implementation: A Simulation Case Study using APEX. In Proceedings of the International SWAT Conference, Toledo, Spain, 15–17 June 2011.
11. Samaras, A.G.; Koutitas, C.G. The impact of watershed management on coastal morphology: A case study using an integrated approach and numerical modeling. *Geomorphology* **2014**, *211*, 52–63. [CrossRef]
12. Samaras, A.G.; Koutitas, C.G. Modeling the impact of climate change on sediment transport and morphology in coupled watershed-coast systems: A case study using an integrated approach. *Int. J. Sediment Res.* **2014**, *29*, 304–315. [CrossRef]
13. Arnold, J.G.; Youssef, M.A.; Yen, H.; White, A.Y.; Sheshukov, A.M.; Sadeghi, D.N.; Moriasi, D.N.; Steiner, J.L.; Amatya, D.M.; Haney, E.B.; et al. Hydrological processes and model representation: Impact of soft data on calibration. *Trans. ASABE* **2015**, *58*, 1637–1660.
14. Emam, A.R.; Kappas, M.; Linh, N.H.K.; Renchin, T. Hydrological Modeling and Runoff Mitigation in an Ungauged Basin of Central Vietnam Using SWAT Model. *Hydrology* **2017**, *4*, 1–17.
15. Emam, A.R.; Kappas, M.; Linh, N.H.K.; Renchin, T. Hydrological Modeling in an Ungauged Basin of Central Vietnam. *Hydrol. Earth Syst. Sci. Discuss.* **2016**, 1–33. [CrossRef]
16. Prabhanjan, A.; Rao, E.P.; Eldho, T.I. Application of SWAT Model and Geospatial Techniques for Sediment-Yield Modeling in Ungauged Watersheds. *J. Hydrol. Eng.* **2014**, *20*, C6014005. [CrossRef]
17. B'ardossy, A. Calibration of hydrological model parameters for ungauged catchments. *Hydrol. Earth Syst. Sci.* **2007**, *11*, 703–710. [CrossRef]
18. Juez, C.; Tena, A.; Fernández-Pato, J.; Batalla, R.J.; Garcia-Navarro, P. Application of A Distributed 2d Overland Flow Model for Rainfall/Runoff and Erosion Simulation in A Mediterranean Watershed. *Cuadernos de Investigación Geográfica* **2018**. (In Spanish) [CrossRef]
19. Al-Ansari, N. Hydropolitics of the Tigris and Euphrates Basins. *Engineering* **2016**, *8*, 140–172. [CrossRef]
20. Bazzaz, F. Global climatic changes and its consequences for water availability in the Arab world. In *Water in the Arab Word: Perspectives and Prognoses*; Roger, R., Lydon, P., Eds.; Harvard University: Cambridge, MA, USA, 1993; pp. 243–252.
21. Al-Ansari, N.; Knutsson, S. Toward Prudent management of Water Resources in Iraq. *Adv. Sci. Eng. Res.* **2011**, *1*, 53–67.
22. World Bank. *Iraq. Country Water Resources, Assistance Strategy: Addressing Major Threats to People's Livelihoods*; Water, Environment, Social and Rural Development Department: Washington, DC, USA, 2006.
23. UNESCO. *Iraq's Water in the International Press*; UNESCO Office for Iraq: Erbil, Iraq, 2010.
24. Hassan, R.; Al-Ansari, N.; Ali, S.A.; Ali, A.A.; Abdullah, T.; Knutsson, S. Dukan Dam Reservoir Bed Sediment, Kurdistan Region, Iraq. *Engineering* **2016**, *8*, 582–596. [CrossRef]
25. Karim, K.H.; Al-Hakari, S.H.S.; Kharajiany, S.O.A.; Khanaqa, A.P. Surface Analysis and Critical Review of the Darbandikhan (Khanaqin) Fault, Kurdistan Region, Northeast Iraq. *Kurdistan Acad.* **2016**, *12*, 61–75.
26. Berding, F. *Reconnaissance Soil Map of the Three Northern Governorates*; FAO Erbil Coordination Office: Erbil, Iraq, 2001.
27. Food and Agriculture Organization of the United Nations. *The Digital Soil Map of the World and Derived Soil Properties*; Version 3.6; FAO: Rome, Italy, 2003.
28. Ministry of Agricultural and Irrigation, Republic of Iraq. *Map of Soil and Land Use*; Ministry of Agricultural and Irrigation: Baghdad, Iraq, 1990.
29. Neitsch, S.L.; Arnold, J.G.; Kiniry, J.R.; Williams, J.R. *Soil and Water Assessment Tool*; Theoretical Documentation Version 2009; Texas Water Resources Institute: College Station, TX, USA, 2011.
30. Abbaspour, K.C.; Vejdani, M.; Haghighat, S. SWAT-CUP Calibration and Uncertainty Programs for SWAT. In Proceedings of the International Congress on Modelling and Simulation (MODSIM 2007), Melbourne, Australia, 10 December 2007.
31. Smakhtin, V.U. Estimating continuous monthly baseflow time series and their possible applications in the context of the ecological reserve. *Water SA* **2001**, *27*, 213–218. [CrossRef]
32. Assad, M.N. Sediments and Sediment Discharge in the River Diyala. Master's Thesis, Baghdad University, Baghdad, Iraq, 1978.
33. Salimi, T.E.; Nohegar, A.; Malekian, A.; Hosseini, M.; Holisaz, A. Runoff simulation using SWAT model and SUFI-2 algorithm. *Casp. J. Environ. Sci.* **2016**, *14*, 69–80.
34. Kokkonen, T.; Jakeman, A.; Young, P.; Koivusalo, H. Predicting daily flows in ungauged catchments:

Model regionalization from catchment descriptors at the Coweeta Hydrologic Laboratory, North Carolina. *Hydrol. Process.* **2003**, *11*, 2219–2238. [CrossRef]

35. Parajka, J.; Merz, R.; Bloschl, G. A comparison of regionalisation methods for catchment model parameters. *Hydrol. Earth Syst. Sci.* **2005**, *9*, 157–171. [CrossRef]
36. Heuvelmans, G.; Muys, B.; Feyen, J. Regionalisation of the parameters of a hydrological model: Comparison of linear regression models with artificial neural nets. *J. Hydrol.* **2006**, *319*, 245–265. [CrossRef]
37. Tang, F.F.; Xu, H.S.; Xu, Z.Z. Model calibration and uncertainty analysis for runoff in the Chao River Basin using sequential uncertainty fitting. *Procedia Environ. Sci.* **2012**, *13*, 1760–1770. [CrossRef]
38. Abbaspour, K.C. *SWAT-CUP: SWAT Calibration and Uncertainty Programs—A User Manual*; Swiss Federal Institute of Aquatic Science and Technology, Eawag: Dubendorf, Switzerland, 2015.
39. Hassan, R.; Al-Ansari1, N.; Ali, A.; Ali, S.; Knutsson, S. Bathymetry and siltation rate for Dokan Reservoir, Iraq. *J. Lakes Reserv. Res. Manag.* **2017**, *20*, 1–11. [CrossRef]
40. Walling, D.E. *The Impact of Global Change on Erosion and Sediment Transport by Rivers: Current Progress and Future Challenges*; United Nations Educational, Scientific and Cultural Organization: Paris, France, 2009.
41. Basson, G. *Sedimentation and Sustainable Use of Reservoirs and River Systems*; International Commission on Large Dams: Paris, France, 2009.
42. Mahmood, R. *Reservoir Sedimentation: Impact, Extent, Mitigation*; World Bank: Washington, DC, USA, 1987.
43. Garg, V.; Jothiprakash, V. Trap Efficiency Estimation of a Large Reservoir. *J. Hydraul. Eng.* **2008**, *14*, 88–101. [CrossRef]

8

Real-Time Flood Control by Tree-Based Model Predictive Control Including Forecast Uncertainty

Gökçen Uysal [1,2,*], Rodolfo Alvarado-Montero [1,3], Dirk Schwanenberg [1,4] and Aynur Şensoy [2]

1 Institute of Hydraulic Engineering and Water Resources Management, University of Duisburg-Essen, 45141 Essen, Germany; Rodolfo.AlvaradoMontero@deltares.nl (R.A.-M.); Dirk.Schwanenberg@kisters.de (D.S.)

2 Department of Civil Engineering, Anadolu University, 26555 Eskişehir, Turkey; asensoy@anadolu.edu.tr

3 Deltares, Operational Water Management, Deltares, Rotterdamseweg 185, 26 MH Delft, The Netherlands

4 KISTERS AG, Business Unit, Pascalstraße, 52076 Aachen, Germany

* Correspondence: gokcenuysal@anadolu.edu.tr

Abstract: Optimal control of reservoirs is a challenging task due to conflicting objectives, complex system structure, and uncertainties in the system. Real time control decisions suffer from streamflow forecast uncertainty. This study aims to use Probabilistic Streamflow Forecasts (PSFs) having a lead-time up to 48 h as input for the recurrent reservoir operation problem. A related technique for decision making is multi-stage stochastic optimization using scenario trees, referred to as Tree-based Model Predictive Control (TB-MPC). Deterministic Streamflow Forecasts (DSFs) are provided by applying random perturbations on perfect data. PSFs are synthetically generated from DSFs by a new approach which explicitly presents dynamic uncertainty evolution. We assessed different variables in the generation of stochasticity and compared the results using different scenarios. The developed real-time hourly flood control was applied to a test case which had limited reservoir storage and restricted downstream condition. According to hindcasting closed-loop experiment results, TB-MPC outperforms the deterministic counterpart in terms of decreased downstream flood risk according to different independent forecast scenarios. TB-MPC was also tested considering different number of tree branches, forecast horizons, and different inflow conditions. We conclude that using synthetic PSFs in TB-MPC can provide more robust solutions against forecast uncertainty by resolution of uncertainty in trees.

Keywords: reservoir operation; multi-stage stochastic optimization; TB-MPC; flood control; real-time control

1. Introduction

Reservoirs are one of the main components of integrated water resources systems. Their control poses a challenging problem, since it must cope with different conflicting objectives as water supply, hydropower, and flood control [1–5]. Operating with guide curves is a common practice in typical reservoir operation [6–8]. These operating strategies however, rely on long term records that can filter and underestimate extreme events such as drought and flood conditions [9]. Flood events strain reservoir operators to refill and keep maximum water levels. In response to short-term fluctuations, the operators need to anticipate actions and release the excess amount of water in order to have sufficient flood storage volume in the reservoir [10]. However, the storage after the event must be recovered in order to satisfy water supply during the following dry season [11]. Therefore, flood control and water conservation require careful planned strategies for short-term operation. On the other hand, system optimization on this time scale becomes more complex due to the high-dimensionality, non-linearity of the system, and dynamic structure of the control process.

In general, a reservoir operation is assumed to be a dynamic system in which the future states are a function of the current states. Throughout optimization of sequential decision processes, Dynamic Programming (DP) developed by Bellman [12] is an important milestone. The main problem can be divided into sub-problems, separately solved successively over each stage to get an overall optimum solution. Later, researchers contributed on this technique in different aspects such as using sampling stochastic DP, incremental DP, stochastic dual DP [13–15]. Though the solution is well suited for highly nonlinear, nonconvex problems, the main difficulty arises with the so-called "curse of dimensionality". This means more (exponentially growing) computational time with increasing dimensions of states, decisions, and disturbances [12,16].

Operation of the reservoir system is an optimal control problem, thus an alternative solution is proposed by adapting Model Predictive Control (MPC, also known as Receding Horizon Control) [17]. The concept has been tested in different application areas such as rivers, reservoirs, and irrigation canals [18–20] in water resources management. The system is optimized for a forecast horizon by solving an open-loop control problem simultaneously in every time step, then only the first time step control value of the computed sequence is applied and the rest are discarded. At the next step, system states are updated and the process is repeated again. This feedback mechanism is called closed-loop optimization. In the real-world, the reservoir system is operated according to forecast based control decisions, but updated to the realization of the disturbance (inflows). Thus, in the study closed-loop experiments are preferred due to the feedback mechanism to mimic the real-world reservoir operation decisions with respect to open-loop experiments.

MPC requires prediction of disturbance in the real-time control of a water structure, but streamflow forecasts always bring along forecast uncertainty. In real-time forecasting applications, forecasts can be biased and tend to over- or underestimate the actual streamflow [21] and it is hard to fully avoid this [10]. In practice, there are several sources of uncertainty that reduce the accuracy of peak flow estimation. Essentially, these are the hydrological forecast model structure [22], model external forcing e.g., precipitation [23,24], model parameters [25], and initial conditions [26].

Although it is not possible to eliminate all types of uncertainties, one solution might be the consideration of forecast uncertainty using ensemble forecasts in short-term operation. Labadie [16] sorted the hindrances of reservoir system operations, among them he emphasized the need to incorporate uncertainty and risk issues in optimization models. Recently, different researchers showed how probabilistic flood forecasts are more robust and effective than the traditional deterministic ones; correspondingly systems are improved by models that quantify forecast uncertainty [27–29].

In short-term management, Numerical Weather Prediction (NWP) based Ensemble Prediction Systems (EPS) can provide probabilistic streamflow forecasts. The idea behind it is to represent future states of the atmosphere by perturbing the initial conditions (uncertainty). Uncertainty becomes much larger when managing small basins and small rivers [30]. In the case of non-availability or inadequacy of the numerical weather forecast data, synthetic streamflow generation is a common practice for reservoir simulation and optimization studies especially in hypothetical studies [28]. There are various approaches to produce synthetic data e.g., the Thomas-Fiering model [28], Fractional Gaussian noise model [31], artificial neural network [32] or hybrid models [33]. Among them, only a limited number of studies [28,31] demonstrate the forecast uncertainty evolution process such that future periods have larger uncertainty and that there is a correlation between uncertainties.

While MPC seeks optimum trajectories based on a single disturbance, i.e., the inflow into a reservoir, its stochastic counterparts try to find an optimum solution for the entire ensemble and incorporate forecast uncertainty. However, this might yield an infeasible solution due to the burden of multiple disturbance in combination with hard constraints. An innovative technique to deliberate ensemble forecasts with an adaptive control is proposed as Tree-Based MPC (TB-MPC) [34]. This approach reduces the number of ensemble members by its tree generation algorithms using all trajectories and then proper problem formulation is set by Multi-Stage Stochastic Programming. The method is relatively new in reservoir operation [34–37], especially closed-loop hindcasting experiments and its assessment is quite rare [24] in the literature.

This paper demonstrates a real-time flood control in consideration of streamflow forecast uncertainty especially for limited storage multi-purpose reservoirs using synthetic ensemble inflows and the mass-conservative TB-MPC method. While we focus on the trade-off between flood risks and water conservation benefits, its implementations into the test case follow: (1) development of a novel Probabilistic Streamflow Forecasts (PSFs) method to produce ensemble inflows, (2) the configuration of an hourly (deterministic and stochastic) predictive reservoir optimization model, (3) testing of different PSF scenarios in closed-loop hindcasting mode in comparison with deterministic counterparts (perfect data and Deterministic Streamflow Forecasts, DSF) and (4) the assessment of TB-MPC model features considering different number of tree branches, different inflow conditions, and different forecast horizons in the area of interest.

In this study, the methodology is applied to the Yuvacik dam reservoir, fed by a catchment area of 258 km^2 and located in Turkey, owing to its challenging operation due to downstream flow constraints. The reservoir serves as a main water supply for Kocaeli province. First, a hydro-meteorological rule based decision support system is developed for daily and hourly operation [38]. Later, it is shown that underestimated daily inflow forecasts during a flood event operation in critical reservoir level may result in underestimated spillage releases when using deterministic MPC [39]. Therefore, the case establishes a precedent for similar relatively small reservoirs with multi-purpose operational characteristics. At this point, this paper complements deterministic methods by PSF integrated TB-MPC including forecast uncertainty.

2. Materials and Methods

Hindcasting experiments are the representation of a real-time system by an iterative process: (i) MPC anticipates required spillages with the optimization model for a given time length (so-called forecast horizon). (ii) The reservoir simulation model updates with actual inflows and anticipated spillages from step (i) to find updated forebay elevations. (iii) This process is repeated every hour (so-called receding horizon) until the whole hindcasting period (96 h in this study) has been completed. The general methodology is outlined in Figure 1. We apply closed-loop hindcasting experiments by the following three modes:

1. Perfect Hindcast Experiments: The flood hydrograph corresponding to a 100 years return period (Q_{100}) of the basin is utilized as perfect forecast data in deterministic MPC. This represents the best solution since it exhibits the optimized releases and forebay elevations under perfect knowledge of the future inflows, and it is evaluated as reference case for comparative analyses with forecast based models.

2. Deterministic Hindcast Experiments: This represents the skill of single value DSF evaluation by deterministic MPC. Randomly perturbed inflows in each receding horizon are employed in deterministic MPC mode. The random perturbation having a forecast horizon up to 48 h is updated in each hour.

3. Probabilistic Hindcast Experiments: This represents the skill of ensemble PSF evaluation by multi-stage stochastic TB-MPC. Synthetically generated ensemble inflows in each receding horizon are employed in stochastic MPC mode (Exp-A). This provides forecast uncertainty consideration in the application. Moreover, features of TB-MPC are investigated with selection of different forecast horizons, tree branching numbers, and different inflow conditions (Exp-B).

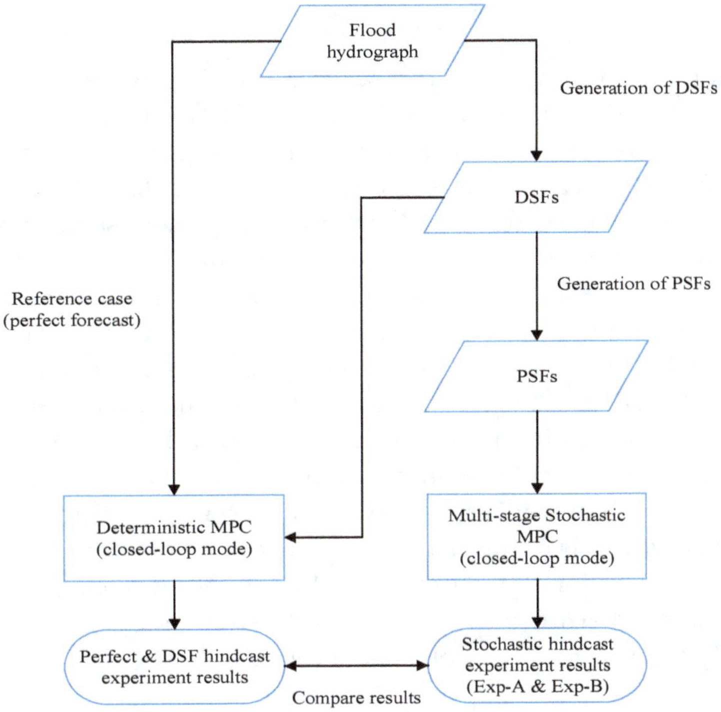

Figure 1. The general framework of the experiments. DSFs stands for Deterministic Streamflow Forecasts. PSFs stands for Probabilistic Streamflow Forecasts. MPC stands for Model Predictive Control.

2.1. *Deterministic and Probabilistic Synthetic Streamflow Generation*

Since forecast inflows include uncertainties, they can be represented by the relative inflow forecasting error [10]. In this paper, we refer to them as DSF scenarios which are perturbed from original flood hydrographs. While the DSF sequence is a single value, the PSFs are generated by applying a dispersion around the DSF with a specified probability distribution (Figure 2). They are eventually repeated in each time step (depending on selected receding horizon) to mimic a dynamic real-time system.

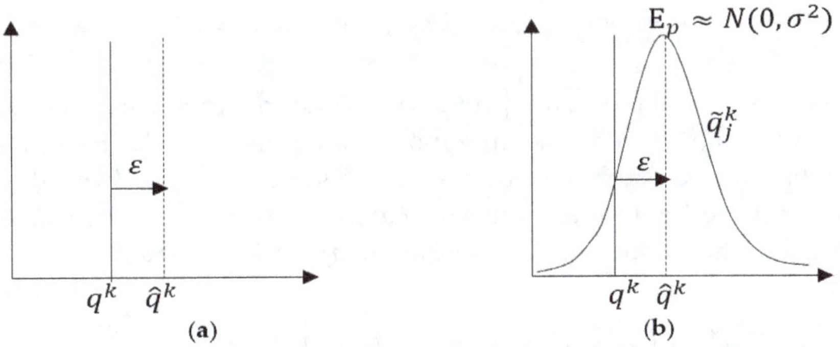

Figure 2. Schematic of single time-step streamflow forecast uncertainty: (**a**) DSF schematization; (**b**) PSF schematization. q^k stands for observed inflow. ε stands for relative inflow forecasting error. $\hat{q}^k$ stands for DSF member. $\tilde{q}_j^k$ stands for j^{th} PSF member. k stands for time index.

DSF is represented by random perturbation using relative inflow forecasting error (ε).

$$\varepsilon^k = \frac{\hat{q}^k - q^k}{q^k} \tag{1}$$

$$DSF: \hat{q}^k = q^k\left(1+\varepsilon^k\right) \tag{2}$$

where ε, $\hat{q}$, q stands for relative inflow forecasting error, forecasted inflows, observed inflows, respectively over a forecast horizon represented by $k = 1, \ldots, N$ time instants.

Even though a range of relative inflow forecasting errors can be detected in the literature [40], acceptable ε during a flood forecasting is within a 0.15–0.30 interval in real-world hydrological forecasting studies [4,10,41]. These studies cover different project scales and climatology. In this study ε values are randomly selected from uniform distribution and range between -0.3 and 0.3 to generate individual DSF values from observed inflows. Also, ε values change for each k time instant.

We produce a set of synthetized members (ensembles, PSF) that are spread around single valued DSF. Conventionally, uncertainty is defined by relative standard error which is expected to be independent (also referred as white noise), usually Gaussian with zero-mean. Therefore, the error pdf is assumed to be $E_p \approx N(0,\sigma^2)$ for all time steps [21,28,42,43].

We introduced a forecast evolution process which is similar to that of Zhao et al. [28] with increasing forecast uncertainty through the forecast horizon. Our implementation of correlation between forecast errors however, is defined by the algorithm below. Synthetic PSFs are generated as an empirical conditional distribution of the perturbed DSF. In other words, they have certain mean (μ) from DSF and relative standard error ($\hat{\sigma} = \sigma * \mu$), considered as an uncertainty level.

We consider the following notation of probabilistic scenario in an ensemble forecast (PSF) matrix as:

$$PSF = \begin{bmatrix} \tilde{q}_1^1 & \tilde{q}_2^1 & \cdots & \tilde{q}_j^1 & \cdots & \tilde{q}_M^1 \\ \tilde{q}_1^2 & \tilde{q}_2^2 & \cdots & \tilde{q}_j^2 & \cdots & \tilde{q}_M^2 \\ \cdots & \cdots & \cdots & \cdots & \cdots & \cdots \\ \tilde{q}_1^k & \tilde{q}_2^k & \cdots & \tilde{q}_j^k & \cdots & \tilde{q}_M^k \\ \tilde{q}_1^N & \tilde{q}_2^N & \cdots & \tilde{q}_j^N & \cdots & \tilde{q}_M^N \end{bmatrix} \tag{3}$$

where N denotes the length of forecast data (so-called forecast lead-time), M denotes the number of ensemble members, $\tilde{q}_j^k$ denotes an arbitrary ensemble member in matrix, and k, j correspond to lead-time and ensemble member indexes, respectively

The procedure to generate synthetic PSFs is accomplished mainly in two successive steps:

1. In the initial time step, PSF ($\tilde{q}_j^1$) members are generated as:

$$\tilde{q}_j^k = N \sim \left(\hat{q}^k, \hat{q}^k * \hat{\sigma}^k\right),\ k = 1\ \&\ j = 1,2,\ldots,M \tag{4}$$

2. The PSF should be correlated with previous members, therefore, the differences between successive DSF values (referred to k time instants for the same DSF sequence) are calculated, then normally distributed errors having mean and standard error from the previous time step are generated and added to the differences. Maximum function is added in order to eliminate negative values, and the remaining PSF members ($\tilde{q}$) are formulated as:

$$\begin{gathered} \tilde{q}_j^k = \left(\hat{q}^k - \hat{q}^{k-1}\right) + \max\left(N \sim \left(\hat{q}^{k-1}, \hat{q}^{k-1} * \hat{\sigma}^k\right), 0\right) \\ k = 2,\ldots,N\ \&\ j = 1,2,\ldots,M \end{gathered} \tag{5}$$

According to intuition, a longer forecast lead-time results in a less reliable forecast. The level of forecast uncertainty (in terms of relative standard error) is assumed to be the same for all members but increasing towards the forecast horizon:

$$\hat{\sigma}^{k-1} \le \hat{\sigma}^k \le \hat{\sigma}^{k+1} \tag{6}$$

Uncertainty in the precipitation is typically much greater (even the largest source of uncertainty from input data) compared to the other meteorological variables [44]. Since higher precipitation leads to higher discharge data, total uncertainty of the forecast chain is expected to be higher during rainfall events, and less during no rain condition. Therefore, this condition is considered in the application by updated relative standard errors depending on DSF values as:

$$\hat{\sigma}^k = \left\{ \begin{array}{lll} \hat{\sigma}^k = \hat{\sigma}^{k-1} & if & \hat{q}^k \leq \hat{q}^{k-1} \\ \hat{\sigma}^k > \hat{\sigma}^{k-1} & if & \hat{q}^k > \hat{q}^{k-1} \end{array} \right\} \tag{7}$$

In the application, a cumulative forecast uncertainty (relative standard error of the final time step, $\hat{\sigma}^N$) is selected beforehand and incremental standard error values are calculated for each time instant. This is a posteriori information based on trial-and-error by the assessment of forecast uncertainty.

2.2. Deterministic Model Predictive Control (MPC)

Contrary to feedback controls where control actions are determined with the current state of the system, MPC takes the future state of the system into account. MPC is commonly implemented by having several components in a receding horizon strategy. These components are a system description (model), a set of disturbances (water inflow into the system, water extractions etc.), objective and physical or operational constraints to represent physical operational management objectives and constraints of the system. Deterministic MPC considers a discrete time-dynamic system according to

$$x^k = f\left(x^{k-1}, x^k, u^k, d^k\right) \tag{8}$$

$$y^k = g\left(x^k, u^k, d^k\right) \tag{9}$$

where x, y, u, d are, respectively, the state, dependent variable (output), control, and disturbance vectors, $f(.)$ and $g(.)$ are linear or non-linear, time-variant vector functions representing an arbitrary water resources model [45]. If being applied in MPC, Equations (8) and (9) are used for predicting future trajectories of the state x and dependent variable y over a finite time horizon represented by $k = 1, \ldots, N$ time instants, in order to determine the optimal set of control variables u by an optimization algorithm. Under the assumption of knowing the realization of the disturbance d^k over the time horizon N, for example the inflows into the reservoir system $\left\{d^k\right\}_1^N$, the simultaneous (or collocated) MPC [45] becomes:

$$\min_{u, x \in \{0, \ldots, T\}} \sum_{k=1}^{N-1} J\left(x^k, u^k, d^k\right) + E\left(x^N, u^N, d^k\right) \tag{10}$$

$$s.t. : h\left(x^k, y^k, u^k, d^k\right) \leq 0, k = 1, \ldots, N \tag{11}$$

$$x^k - f\left(x^{k-1}, x^k, u^k, d^k\right) = 0 \tag{12}$$

where $J()$ is a cost function associated with each state transition, $E()$ is an additional cost function related to the final state condition, and $h()$ are hard constraints on control variables and states, respectively. It receives a single disturbance vector, so that it might be either an observed or deterministic forecast. In a simultaneous or collocated mode, the related process model (Equation (8)) becomes an equality constraint of the optimization problem at dedicated time steps in Equation (12) [46]. The optimization problem is solved using an efficient gradient-based optimizer such as the nonlinear optimizers IPOPT [47] with MA27 solver of the HSL library in combination with adjoint modeling. Adjoint modeling is essential for the efficient computation of the derivative of the objective function with respect to the controlled variables ($dJ = d\left(x^k, u^k, d^k\right)/du^k$). The model itself is implemented in RTC-Tools [45].

2.3. Multi-Stage Stochastic Tree-Based MPC (TB-MPC)

In this part, the deterministic model is extended to a multi-stage stochastic optimization model. The stochastic nature of inflows is reflected by a probabilistic forecasts ensemble, d_j^k, where j denotes the ensemble index ($j = 1, \ldots, M$) and k denotes the time instant ($k = 1, \ldots, N$). Hence, Equation (10) is rearranged for J and E, then it becomes the probability-weighted sum of the objective function terms of the individual ensemble. The formula is shown below as:

$$\underset{u,x \in \{0,\ldots,T\}}{min} \sum_{j=1}^{M} p_j \sum_{k=1}^{N-1} J(x_j^k, u_j^k, d_j^k) + E\left(x_j^N, u_j^N, d_j^N\right) \quad (13)$$

where p_j stands for the probability of the ensemble member (equal for each one), M stands for the number of the ensembles.

The key factor of the approach depends on the control variable (u_j^k) definition for setting up of a stochastic optimization problem due to the limitation of the usable storage. A simple way would be to find the control trajectory which is optimal for the whole ensemble (on average, worst case, chance constrained, etc.). In case of an average definition, it might give feasible solutions on consideration of the ensemble having small variations. However, if the operation of the reservoir system is highly constrained due to a limitation of storage, the solution of the optimization can be dominated by these constraints or it becomes infeasible.

To avoid this problem, multi-stage stochastic optimization is proposed and successfully applied using scenario trees for disturbance, states, and control trajectories [34,48]. Thus, the approach becomes adaptive, the control trajectory does take into account the resolution of uncertainty. This means uncertainty is gradually resolved by showing tree branches at specific time instants along the forecast horizon. While scenario reduction is employed in various ways and represented by a tree nodal partition matrix $P(j,k) \in \mathbb{Z}^{MxN}$, a recently proposed novel mass conservative approach which keeps the probability-weighted sum of a quantity of the original ensemble is used in this study. Details are available in [24,37]. We present the first implementation of the TB-MPC in application to a synthetically generated ensemble streamflow forecast for a limited storage multi-purpose reservoir under flood case.

3. A Real World Test Case and Model Set-Up

3.1. Study Area

The proposed methodology was tested in the Yuvacik Dam reservoir, with a 258 km^2 catchment area, located in the east part of Marmara Region, Turkey (Figure 3). The earth-filled dam was constructed in 1999 for the water supply of Kocaeli city. An annual 142 hm^3 of drinking, domestic and industrial water for 1.5 million inhabitants should be supplied. The relatively limited reservoir has an active storage capacity of approximately 51.2 hm^3 at maximum operating level of 169.3 m whereas 169.8 m is the maximum water level. The spillway is controlled by four radial gates. It should be noted that while a volume of 36.60 hm^3 is stored between minimum operation level of 112.50 m and spillway crest elevation of 159.95 m, a volume of 14.60 hm^3 is kept above spillway crest elevation and behind the radial gates. On the other hand, a Dam Management System (DMS) which has been developed as a part of a Supervisory Control and Data Acquisition (SCADA) system by the reservoir operators provides data collection and transmission from automatic gauges.

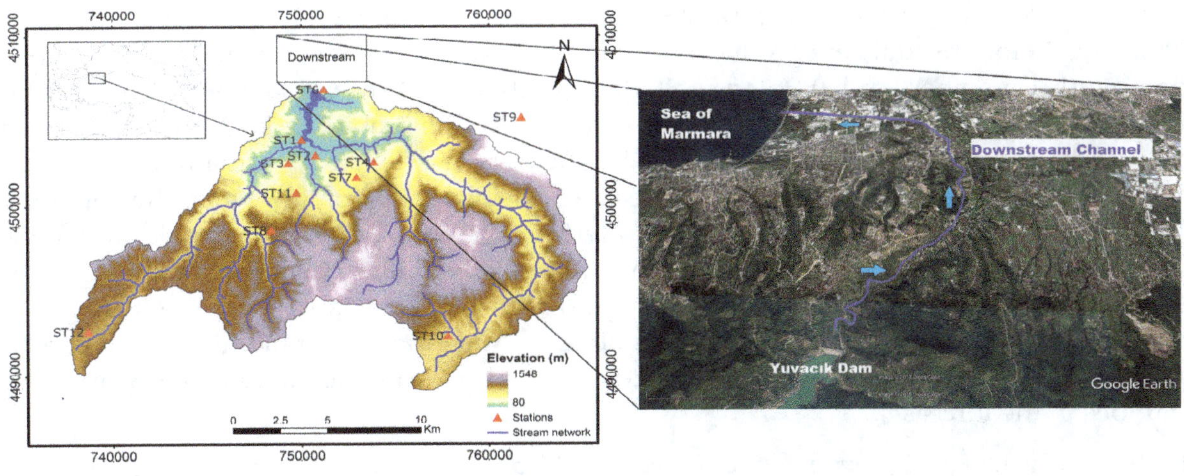

Figure 3. Yuvacık Basin DEM, station network and downstream region.

3.2. MPC Inputs

The maximum amount of water to be released during daily operation is set as 100 m^3/s by the regional water authority taking the drainage discharge conditions of the downstream canal and lateral flows into consideration. The main reason for that is a 12 km long downstream reach that passes from a narrow valley near a rural district and flows into the Marmara Sea after a sharp curvature by a manmade channel next to industrial and urban areas. The dam is built to protect the downstream region against extreme flood events, and the maximum spillway release is set to 200 m^3/s during hourly flood control. This value is the maximum allowable flood limit without severe damage in the downstream. The operation of the dam is multi-purpose subject to two main objectives: (i) water supply and (ii) flood control.

Since available data are insufficient to precisely assess extreme events [49], design flood hydrographs are generally derived based on maximum instantaneous records at the dam site, selected probability distributions using several parameters [50,51], and considered to be representative of extreme conditions expected in the future. In the Yuvacik catchment, the annual peak flow values over 19 years were statistically analyzed and it was detected that the series is in line with the Gumbel dispersion function according to the DSI (State Hydraulic Works, the main governmental water organization in Turkey) report [52]. The peak flow flood values are derived for different return periods (Table 1).

Table 1. Flood hydrograph peak values.

Return Periods (Years)	Project Value (m^3/s)
5	208
10	297
25	410
50	506
100	597

The study adopted a 24 hourly flood hydrograph having an expected occurrence of 100 years that is utilized as a perfect streamflow forecast (Q_{100} flood hydrograph) in the main test application. The flood peak-occurrence time from the beginning is 6 h (i.e., the response time of the catchment to the rainfall event) and the peak flow corresponds to 597 m^3/s. The total flood volume equals to 17.1 hm^3. Being an extreme case, the peak flow of the selected flood hydrograph is three times greater than the downstream channel capacity (200 m^3/s). The hindcasting experiments were conducted in an arbitrary year during a critical operation period (May) when the initial forebay elevation was high. The whole closed-loop hindcasting period covers 96 h, from [1-May-2012 00:00:00] to [4-May-2012 23:00:00]. It is assumed that a 24-h long flood hydrograph occurs between [3-May-2012 00:00:00] and [3-May-2012 23:00:00]. Hourly forecast data are produced for 48 h lead-time. This means that in each 1 h, 48-h long DSF data (with 1 member) and 48-h long PSF data (with M = 50 ensemble members) are generated throughout the whole hindcasting period. Given the lack of probabilistic hydrological forecasts for this case study, we decided to recreate the stochasticity by assuming a normally distributed noise around the deterministic forecast. This in fact becomes an innovation on its own in the paper, since an objective approach has not been reported in the literature before. Although it was desired to increase the number of PSF members in order to cover much more possibilities, this ended up with the reverse situation in the optimization model due to the high-dimensional data space. Different ensemble sizes were also tried in the control model, and it was observed that increasing member number estimates higher uncertainty range, 50 members are considered to provide enough spread to capture the major uncertainties in the forecasts.

In this study, DSFs with error range between −0.3 and 0.3 and PSFs with final relative standard error ($\hat{\sigma}^{48} = 0.2$) are depicted with three different randomized scenarios (Sce-Q100a, Sce-Q100b and Sce-Q100c). Here, a scenario corresponds to randomly generated and independent forecasting data sets (both DSFs and PSFs) to be employed in the same hindcasting period. In each scenario, randomly perturbed DSFs from perfect data are produced in each hour for a given lead-time and an error interval. Then, similarly (in each scenario and in each hour), ensemble PSFs are produced from the mother DSFs with a given uncertainty. Since the generated streamflows are produced in each hour, the receding horizon is always set to one hour in all deterministic and stochastic MPC hindcasting experiments. In terms of operational point of view, the control trajectory changes at each time step and is not a fixed one for the complete event (it changes depending on the forecast and its uncertainty, as well as the forecast horizon).

As explanatory example from two scenarios (Sce-Q100a and Sce-Q100b) is shown in Figure 4 for a lead-time of 48 h. Fifty members are generated for PSFs, but only two members (min and max values) are given in the figure to show the uncertainty ranges (referred to uncertainty band). In the figure, [T0] corresponds to "time zero" namely the starting time of the optimization. Two time instants [T0: 01-May-2012 12:00:00] (Figure 4a) and [T0: 01-May-2012 13:00:00] (Figure 4b) are exampled here, but the process is dynamically repeated in each one hour throughout the hindcasting period in closed-loop experiments for each scenario. Finally, each scenario is composed of all generated 96 forecast sequences (both in DSFs and their PSFs) having a 48 h lead-time in the hindcasting period.

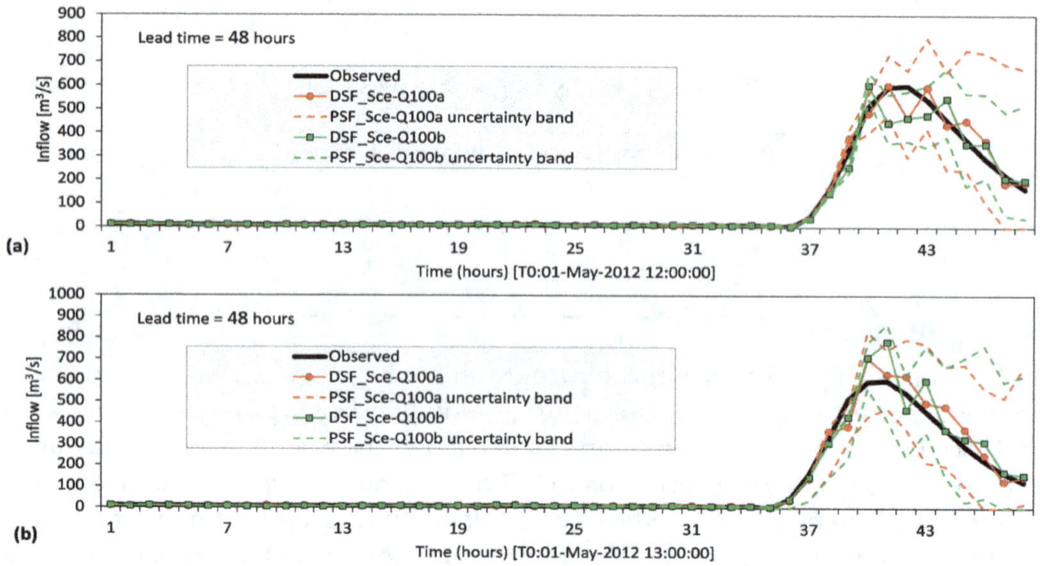

Figure 4. Graphical representation of the scenarios (Sce-Q100a and Sce-Q100b) for: (**a**) T0: 01-May-2012 12:00:00; (**b**) T0: 01-May-2012 13:00:00.

Different inflow conditions such as Q_{25} and Q_{50} (from Table 1) are also tested under Section 4.3.3. Characteristics of the PSF scenarios are presented with the performance assessment by using a mean Continuous Ranked Probability Score (CRPS) which generalizes the Mean Absolute Error (MAE) in the case of probabilistic forecasts (Supplementary Figure S1). Mean CRPS summarizes the quality of a probability forecast into a number by comparing the integrated square difference between the cumulative distribution function of forecasts and observations [24]. According to that, mean CRPS increases with forecast lead-time, while each scenario shows a different performance. The scenario number is not critical in the study because the focal idea is to develop an objective approach for stochasticity of the flows, to use them in stochastic optimization set-up and to compare the results with a deterministic equivalent. Thus, scenarios can be deliberated as different source based forecast data sets.

An example of generated forecast data sets (Sce-Q100a) is given for different time steps shown by T0 in Figure 5. Similarly, [T0] corresponds to "time zero" namely the starting time of the optimization in the figure. The red dashed line shows DSFs which are derived from a perfect forecast by 30% random noise. Synthesized PSF data are shown in gray ink. While generating PSF, updated forecast data have "different starting time" and "different duration to the peak flow" which affect the uncertainty band in the end. For example, the starting time of the 01-May-2012 20:00:00 (fourth panel) and 02-May-2012 00:00:00 (fifth panel) have less duration to peak flow, thus they have higher dispersion at peak flow compared to the last panel and subsequently uncertainty is kept high in the falling limb in the fourth and fifth panels. However moving forward in time, the duration to peak flow at T0 decreases, so that the relative standard error cannot increase too much due to the flow condition. That is the reason why the last panel has less spread in the end. It is directly related to the approach used to generate the ensembles.

Contrary imperfect data (DSFs and PSFs), perfect forecast (shown in Figure 5 as a black line) is a single chain and utilized in the initial MPC hindcasting experiment. This test allows the performance of the developed real-time reservoir system control model to be assessed without forecast uncertainty; as a result a sufficiently long forecast horizon is decided before starting imperfect forecast tests. During TB-MPC hindcasting experiments, PSF data are converted to trees depending on the selected branch numbers. Hereby, previously generated forecast data (from Sce-Q100a) PSFs with 50 members (Figure 5) are transformed into optimization tree inflows with 16 branches (members) in Figure 6. Thereby, the number of generated PSF data are limited by this reduction, but still capture the major uncertainty, and the tree is used in the optimization algorithm.

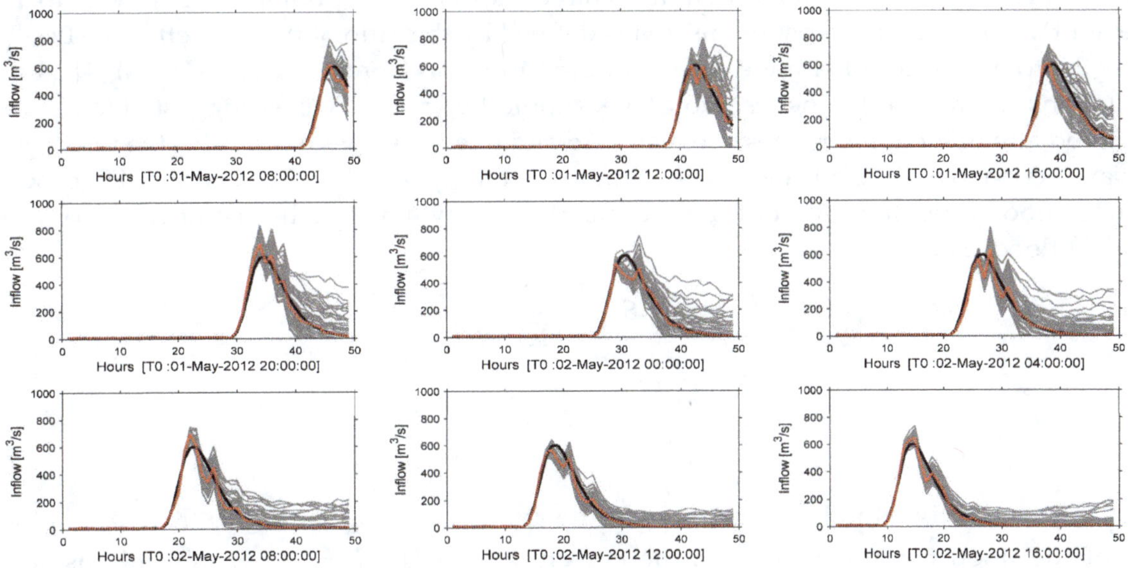

Figure 5. Perfect forecast, DSF vs. PSF for different receding horizons (Sce-Q100a).

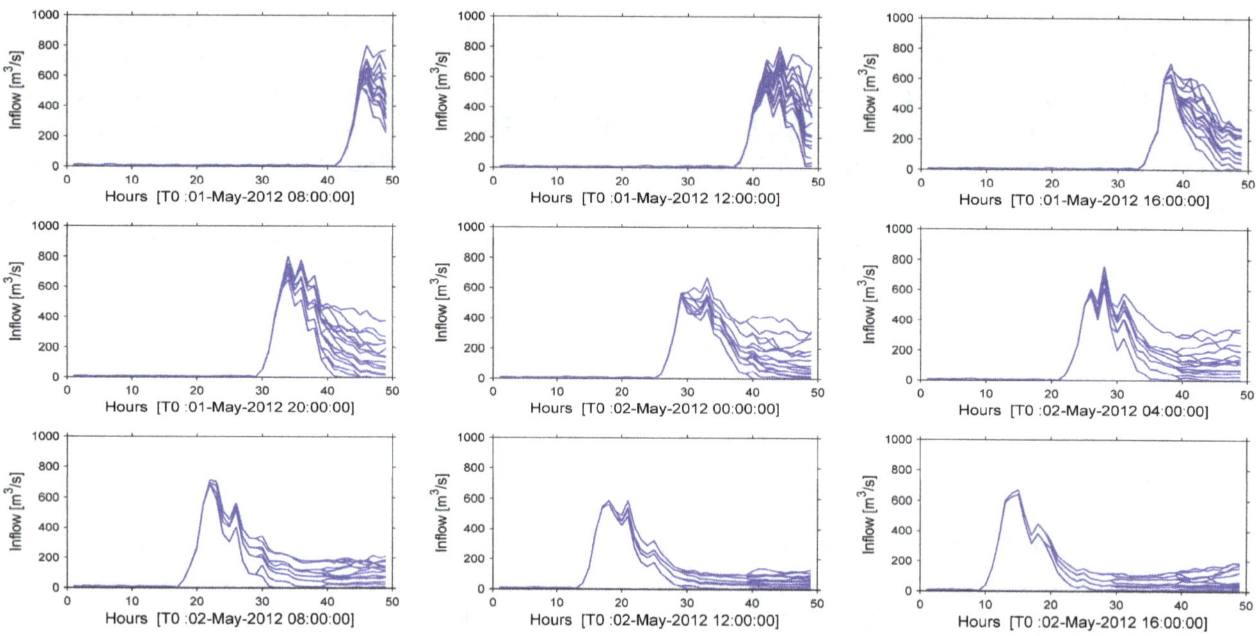

Figure 6. PSF ensemble members (50 members) transformed into optimization trees (16 branches).

Assessment of an effective reservoir operation in the short term is done by quantifying the minimization of downstream damage, the dam safety and maximization of the forebay elevation at the end of the flood event [53]. Similarly, four objective criteria are used in this study to test the experiments. These are:

- Forebay elevation at the end of a flood event: Forebay elevation should be same at the end of a flood event in order to provide long term water supply targets,
- Flooding threshold value: Spillway discharges should be less than channel capacity, thus this is considered as the flooding threshold value and the maximum discharge at the dam outlet is checked,
- Total flood volume at the downstream area: The cumulative volume of the released flood water (only above the maximum flood limit of 200 m^3/s) should be zero for the best flood management,
- Flood storage index (FSI): It is essential to have enough flood pool in the reservoir to attenuate the hourly extremes. To measure this, FSI is defined by the ratio of the total effective flood storage over the total volume of storage corresponding to Flood Control Levels (FCLs) [4] as shown in Equations (14) and (15) . The reservoir level should be kept at FCL as suggested to reserve space for flood attenuation. FSI ranges between zero and one. While zero indicates the reservoir level is always kept above FCL, one indicates operation is totally based on FCL. Higher FSI ensures more reliable flood operation (under forecast uncertainty) by having a high empty reservoir volume against flood risk.

$$FSI_f = \frac{\sum_{k=1}^{N} v_f^k}{\sum_{k=1}^{N} v_{FCL}^k} \tag{14}$$

$$v_f^k = \begin{cases} v_{act}^k \; if \; v^k \leq v_{FCL}^k \\ v_{FCL}^k \; if \; v^k > v_{FCL}^k \end{cases} \quad for \; k = 1, 2, \ldots N \tag{15}$$

where v_f^k is the effective flood storage, v_{FCL}^k is the storage corresponding to FCL, v_{act}^k is the actual volume of the flood control pool; v^k is the current storage by $k = 1, \ldots, N$ time instants. If v^k is below FCL, it is equal to the volume of the flood control pool; otherwise, it is the actual available storage.

While the "flooding threshold value" gives the maximum instantaneous damage in the downstream, "total flood volume at downstream area" reflects the cumulative total penalty from the target set point of operational constraint within a hindcast period. It is also important to note that the perfect data based model results neglect uncertainty completely and serve as reference. Any deviation from these results objectively presents the inefficiency of the DSF and PSF based model results.

3.3. Model Set-Up

The reservoir model is based on the continuity equation for the reservoir. The water balance equation for a single reservoir can be arranged as:

$$s^k = s^{k-1} + \Delta t\left(Q_I^k - Q_S^k - Q_{WS}^k\right) \tag{16}$$

where, s is the storage volume (m^3), Δt is the time difference between k^{th} and $(k-1)^{th}$ time steps, Q_I, Q_S , and Q_{WS} are the reservoir inflows, spillway flow and water supply (m^3/s), respectively. Also, forebay elevation, fb could be computed by:

$$fb^k = f_{ls}\left(s^k\right) \tag{17}$$

where f_{sl} is a piecewise-linear level-storage relation.

In order to satisfy physical conditions capacity curves, discharge limits, and boundary elevations are introduced in constraints.

3.4. Physical and Operational Constraints

The constraints that form the equalities and inequalities are the continuity equation and the physical boundaries. The continuity equation ensures a consistent mass balance definition, accordingly residuum (r^k) is introduced as an equality constraint and it must be zero as formulated below:

$$r^k = s^k - s^{k-1} - \Delta t\left(Q_I^k - Q_S^k - Q_{WS}^k\right) = 0 \tag{18}$$

The system's physical limits should be met and they are defined as hard constraints. First, forebay level physical constraints are defined as:

$$fb_{min} \le fb^k \le fb_{max} \tag{19}$$

where fb_{min} is the minimum reservoir elevation and fb_{max} is the maximum reservoir elevation. Beyond the operational targets, the spillway discharge flow should be within physical limits depending on its capacity curve and is defined as the constraint below:

$$Qs_{min} \le Qs^k \le Qs_{max}^k \tag{20}$$

$$Qs_{max}^k = f_{sdc}\left(fb^k\right) \tag{21}$$

where Qs_{min} is the minimum spillway flow and is set to zero and Qs_{max} is the maximum allowable spillway flow and it is determined by the spillway discharge curve (f_{sdc}) i.e., zero at spillway crest elevation and variable above elevations depending on the current forebay elevation (as a function of fb).

3.5. Objective Function

An objective function determines the target of the operation by its terms. These terms are sometimes called soft constraints and an overall optimum of all terms is targeted. Optimization of a multi-purpose reservoir is aimed to mitigate flooding while maximizing water benefit or hydropower assets as its own objective function. Controlled variables are the forebay elevation and spillway discharges. The term sensitivity is set by the weights and the power of the equation. These weights are case-specific and depend on the significance of the soft constraints. For example, in this study higher penalizations are assigned for w_3 and w_4 in order to prevent downstream flooding and provide the long term water supply target, respectively. The weights are determined by a trial-and-error approach.

A reservoir having a limited capacity should include the terms below for hourly management:

- Differences between optimized forebay elevation and maximum operating elevation are minimized in Equation (22),
- Spillway discharges are minimized in Equation (23),
- Spillway releases above a specified discharge (200 m^3/s) are constrained in Equation (24). This has a high weight in order to prevent damage in the downstream,
- The case of a set point for forebay elevation is given i.e., a variable guide curve (it is the same for each hour in a day) for long term targets [39], this term stands to minimize deviations from it in Equation (25),
- Spillway discharges between two consecutive time steps are constrained by consideration of the mechanical gate operation efficiency (against wear and tear) in Equation (26).

$$J1(fb) = w_1 \sum_{k=1}^{N} \left(f_{max} - fb^k \right) \tag{22}$$

$$J2(Qs) = w_2 \sum_{k=1}^{N} \left(Qs^k \right) \tag{23}$$

$$J3(Qs) = w_3 \sum_{k=1}^{N} \max\left(Qs^k - Q_{set}, 0 \right)^2 \tag{24}$$

$$J4(fb) = w_4 \sum_{k=1}^{N} \max\left(fb^k - fb_{set}, 0 \right)^2 \tag{25}$$

$$J5(Qs) = w_5 \sum_{k=1}^{N} \left(Qs^{k+1} - Qs^k \right)^2 \tag{26}$$

where, $w_j (j = 1, \ldots, 5)$ stands for weights associated with each term ($w_j > 0$), Q_{set} stands for the maximum channel capacity without downstream region flooding, fb_{set} stands for the time-dependent variable guide curve, and N stands for forecast horizon.

Finally, closed-loop optimization minimizes the main objective function (J), which equals the summation of different objectives, in Equation (27) for optimum forebay elevations and spillway discharges agreement with pre-defined constraints in Equations (18)–(21). Optimizations are conducted in each hour (due to the selection of a one hour receding horizon in this study) for N forecast horizon throughout the entire hindcasting period. Only the first value of the computed control sequence (spillway discharge) is applied to the system and the rest is discarded. In the next time step, the state (forebay elevation) is updated and the optimization is repeated with updated forecast data.

$$\begin{matrix} \min J(fb, Qs) \\ k \in \{0, \ldots, T\} \end{matrix} = J1 + J2 + J3 + J4 + J5 \tag{27}$$

4. Numerical Experiments and Results

4.1. Deterministic MPC Hindcasts Using Perfect Forecasts

It is assumed that selecting a sufficiently long forecast horizon in the closed-loop mode provides an approximately actual infinite horizon solution [54]. In the first trial, hindcasting experiments were employed using perfect forecasts with respect to different forecast horizons. The tests were conducted for 6, 12, 18, 24, 36, and 48 h to estimate a sufficiently long forecast horizon, and named as PER6, PER12, PER18, PER24, PER36, and PER48, respectively. The results of the experiments are presented in terms of spillway discharges and forebay elevation in Figure 7. Short forecast horizons (such as 6 h and 12 h) produce delayed releases which in return increase peakflow at the dam outlet. This is an important indicator of flood risk assessment since higher discharges create a larger extension of damage in the downstream area. Therefore, MPC needs a longer forecast horizon to handle the problem even with perfect inflow forecasts. Forecast horizons of 18 h or longer provide downstream safety by reaching reasonable maximum discharges under 200 m^3/s limit at the outlet. Longer forecast horizons than 18 h e.g., 24, 36, and 48 h, result in a similar response. Therefore, the experiment results of PER24, PER36, and PER48 overlap in the figure. According to the results, one can note that the mitigation of a major flood even with maximum operating levels and 200 m^3/s downstream channel constraint can be achieved under perfect future knowledge of flood inflows at least 18 h beforehand.

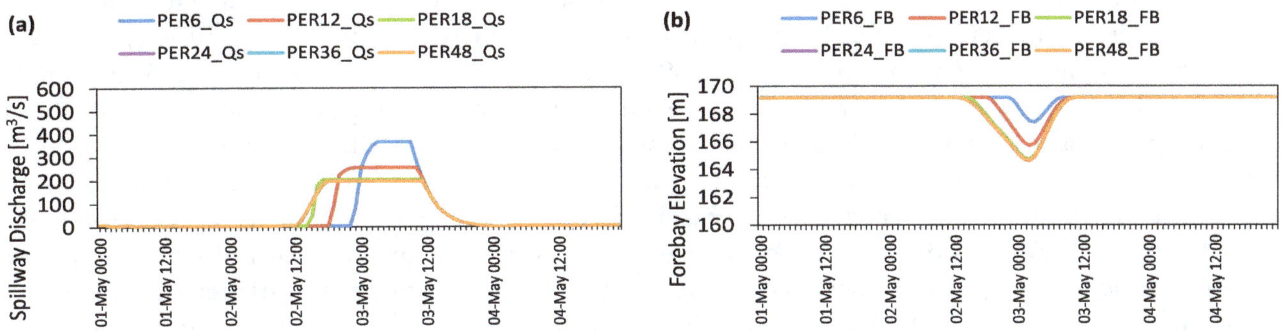

Figure 7. Comparison of closed-loop MPC forecast horizon performance using perfect streamflow forecasts: (**a**) Spillway discharge (m^3/s); (**b**) forebay elevation (m).

4.2. Deterministic MPC Hindcasts Using DSFs

DSFs are produced as single forecast trajectories by adding random perturbations on observations, thus they under/over-estimate the actual flood inflows. This situation becomes more significant for the refilling season in spring when the initial reservoir forebay elevation is at a critical level. DSF data based MPC hindcasting experiment results using Sce-Q100a data are presented in Figure 8. Likewise, longer forecast horizons than 18 h e.g., 24, 36, and 48 h result in a similar response, thus the experiment results of DSF24, DSF36, and DSF overlap in the figure. According to the results, spillway discharges are greater than the upper limit of 200 m^3/s and create flooding in the downstream. This is considered as lower reliability compared to perfect data based experiments, and mainly attributed to the forecast disturbance which introduces 30% bias to the control strategy. Longer forecast horizons (such as 18, 24, 36, 48 h) perform better and releases are shifted to earlier time steps. However, it is not possible to mitigate the flood event even for forecast horizons longer than 18 h due to the given bias in the inflows and the lack of uncertainty in the system optimization. Compared to perfect forecasts based MPC, the variations in spillages are higher due to updated information for each receding horizon.

4.3. Multi-Stage Stochastic TB-MPC Hindcasts Using PSFs

In this part, hindcasting experiments are conducted by means of PSF data and multi-stage stochastic optimization. TB-MPC uses scenario trees for disturbances (inflows), states (forebay elevation) and control trajectories (spillway discharges) to form a related stochastic model. Definition of the multiple stages at branching points using binary trees makes the process a multi-stage stochastic optimization in TB-MPC. Hence, the results become more adaptive to the resolution of uncertainty and have better expected performances.

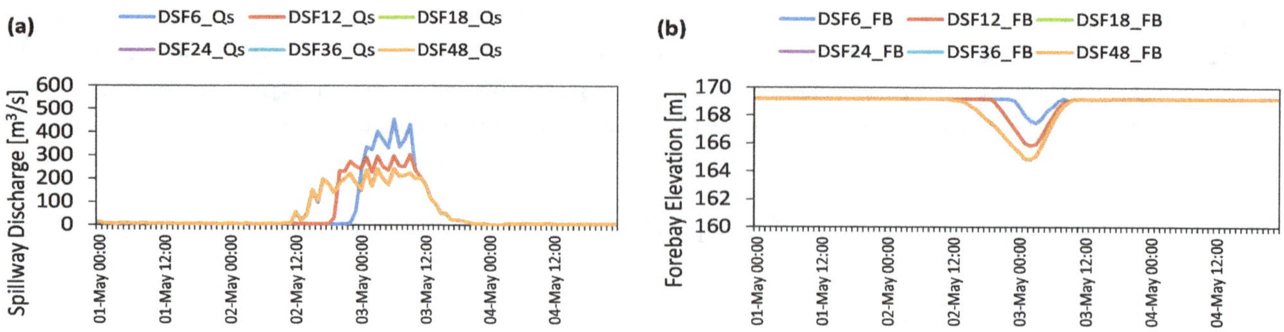

Figure 8. Comparison of closed-loop MPC forecast horizon performances with DSF (Sce-Q100a): (**a**) Spillway discharge (m^3/s); (**b**) forebay elevation (m).

MPC solves an open-loop optimization at every time instant along a forecast horizon and applies only to the first control value to the system. Therefore, an example from a selected single case [T0 = 2-May-2012 14:00:00] of open-loop optimization results is presented with spillway discharge and forebay elevation in smooth trees (Figure 9). One can note that the use of binary trees splits the stochastic variables into two trajectories at each specific branching point. The timing of specific branching points is automatically defined due to the tree-reduction algorithm (check Section 2.3) when the branch number and optimization forecast horizon have been fixed. In this study, 16 tree branches were chosen based on the performance comparison (check Section 4.3.1) for the forecast horizon of 48 h (check Section 4.3.2), thus branching points are fixed every 10 h by the tree reduction algorithm. The root value is an optimal discharge for the entire control sequence considering uncertainty resolution through the forecast horizon. Only single discharge data from a rooted tree of the open-loop optimization is used in the simulation, and the process is shifted to the next time step. At the following time instant, the optimization is reformulated with the updated initial reservoir level and updated forecast data. After applying this procedure through the whole hindcast period (96 h), closed-loop TB-MPC hindcasting experiment results (one single output) are obtained. These results are discussed under the subsections of Section 4.3.

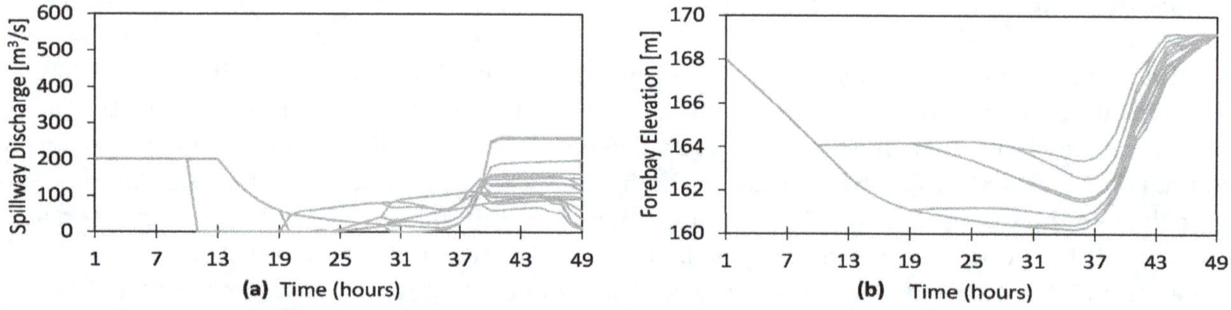

Figure 9. Open-loop optimization results of multi-stage stochastic optimization (from Sce-Q100a): (**a**) Spillway discharge trees (m^3/s); (**b**) Forebay elevation trees (m).

4.3.1. TB-MPC Hindcasts Considering a Different Number of Tree Branches

The number of tree branches used in TB-MPC is user-defined and there is no direct method to select the most appropriate one. In this part, performances of TB-MPC hindcasting experiments using Sce-Q100a data are conducted by selecting a different number of tree branches in each trial and the results are compared with each other. According to TB-MPC methodology, total PSF members can be reduced to 2^X branches, e.g., 1, 2, 4, 8, . . . etc. [24]. Therefore, in this study 50 PSF ensembles were reduced to six different branches (1, 2, 4, 8, 16 and 32) and tested in a hindcast test. The experiments were done using Sce-Q100a PSF as input forecast data and comparison of the optimization results from different tree branches numbers is given in Figure 10 in terms of spillway flows and forebay elevation. This experiment shows the effects of the resolution of tree and correspondingly capturing forecast uncertainty in stochastic optimization. If the forecast horizon is set to 48 h, we can get optimum results after 16 branches as shown (Figure 10a) in terms of the spillway discharge. Since the results for 16–32 trees are exactly the same, they overlap with each other in the same figure. Although higher resolution overestimates the inflows which increases the pre-releases, it is still able to restore the forebay elevation target at the end of the flood event (Figure 10b).

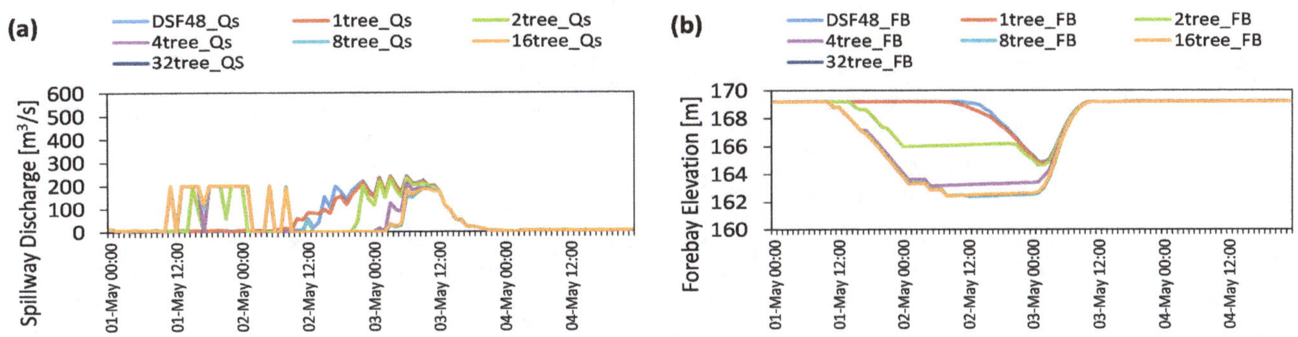

Figure 10. Comparison of closed-loop MPC with different tree reduction branches for 48 h forecast horizon (Sce-Q100a): (**a**) Spillway discharge (m^3/s); (**b**) forebay elevation (m).

The computation time is given for different tree reduction branches in the hindcasting experiments in comparison with deterministic MPC with DSF (Table 2). Note that this is the total time for the entire hindcast period (96 h). The time spent in deterministic MPC is three times less than TB-MPC with one tree branch which can be considered as the ensemble mean. This is mainly attributed to the tree reduction process before the optimization of the model. However, the computation time increases as long as higher tree branches are used in the model. Therefore, the tree number is fixed to 16 branches for the following experiments of the study. These results are also valid for the remaining scenarios.

Table 2. Computation times in MPC hindcasting experiments. MPC stands for Model Predictive Control. DSF stands for Deterministic Streamflow Forecast. TB-MPC stands for Tree-based MPC.

Hindcasting Experiment	Total CPU Time (s)
MPC with DSF	151
TB-MPC with 1 tree branch	491
TB-MPC with 2 tree branches	551
TB-MPC with 4 tree branches	633
TB-MPC with 8 tree branches	677
TB-MPC with 16 tree branches	867
TB-MPC with 32 tree branches	1354

4.3.2. TB-MPC Hindcasts Considering Different Forecast Horizons

Figure 11 presents a comparative visualization of different hindcasting experiments with perfect data and DSF under deterministic configuration and PSF under multi-stage stochastic TB-MPC for 18, 24, 36, and 48 h forecast horizons. Notice that deterministic methods have similar results for a longer forecast horizon whereas TB-MPC start earlier pre-releases. This is mainly attributed to consideration of uncertainty resolution through the forecast horizon in the decision mechanism. When the selected forecast horizon increases in TB-MPC experiments, the uncertainty band in the PSF spread increases and thereby much more water is evacuated from the spillway with the more conservative policy for flood control. It is remarkable to note that even when a longer forecast horizon is selected, forebay elevations at the end of the flood event are always equal to the initial reservoir elevation. These results are also valid for the remaining scenarios.

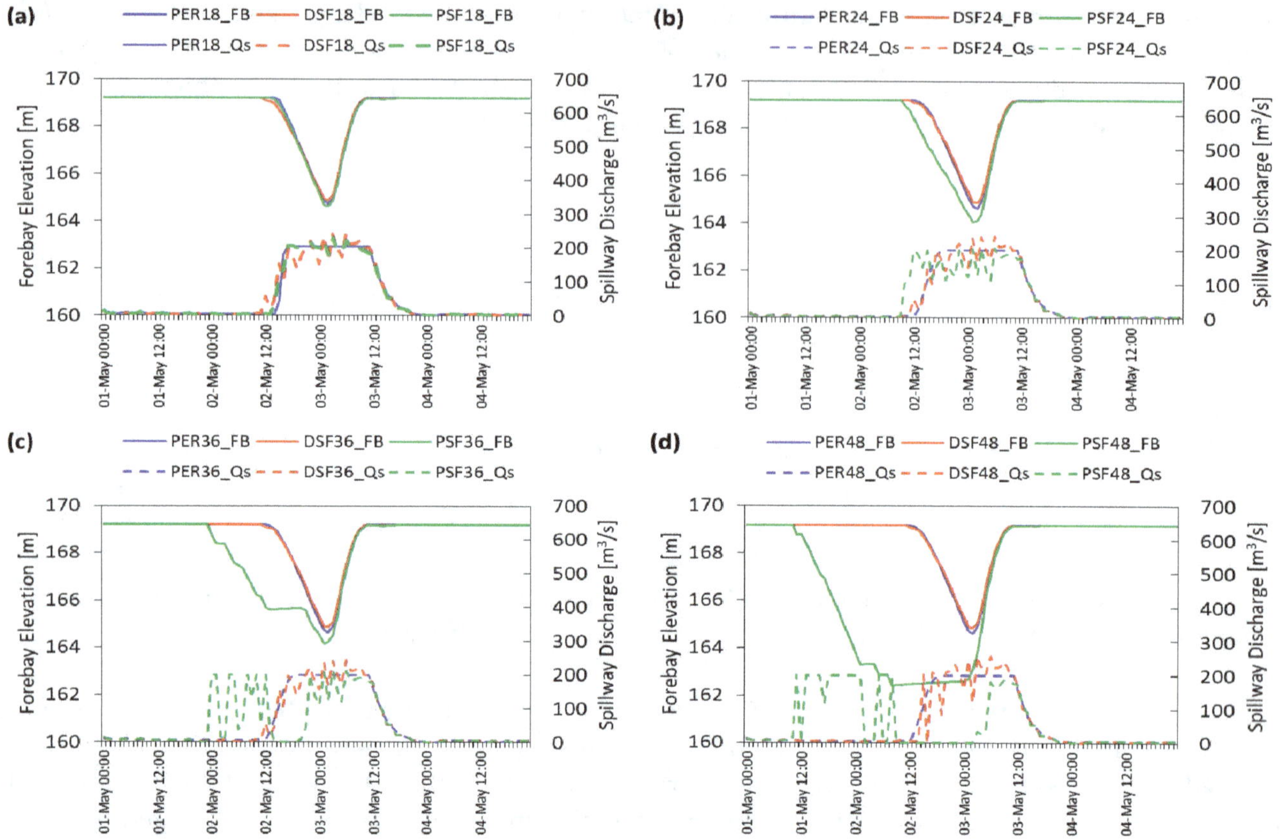

Figure 11. Comparison of deterministic (perfect and DSF) and stochastic (PSF) closed-loop MPC results with different forecast horizons (Sce-Q100a): (**a**) 18 h; (**b**) 24 h; (**c**) 36 h; (**d**) 48 h.

4.3.3. Assessment of the Approach for Different Inflow Conditions and Scenarios

The hindcasting experiments having forecast horizons of 48 h were enriched with different inflow conditions to check the robustness of the developed methodology. To that end, hydrographs characterized by return periods lower than 100 years such as 25 and 50 years, which have peak flows greater than the downstream channel capacity, were also used in the closed-loop hindcasting experiments. All experiments were evaluated by forebay elevation at the end of the flood event, the penalization of the flooding threshold value by the peak flow observed at the Yuvacık outlet, the total flood volume at the downstream area, and FSI. Similar to the previous set-up, flood hydrographs were independently utilized in each hindcasting test as perfect data, DSF and PSF data sets were generated. The tests were also elaborately investigated with three different independent scenarios (a,b,c) for each flow condition. The complete results belonging to the hindcasting experiments are given in Supplementary Figures S2 and S3 for Q_{25} and Q_{50} scenarios, respectively. On the other hand, deterministic (perfect and DSF) and stochastic (PSF) closed-loop MPC results from Sce-Q100a, Sce-Q100b and Sce-Q100c are shown in Figure 12. It is notable that the stochastic set-up always provides pre-releases and takes precautions against flood event. The deterministic model only takes actions over several hours which is similar to the perfect data based reference model, but generates much more spillage above the flood threshold compared to stochastic TM-MPC.

A brief summary of all inflow conditions for the three scenarios are given (Table 3) in terms of peak flow values. Since perfect data based models can give the desired maximum spillway of

200 m^3/s for each condition, they are not shown in the table. According to results, there is always an improvement in spillway discharges for different flows conditions of Q_{25}, Q_{50} and Q_{100}.

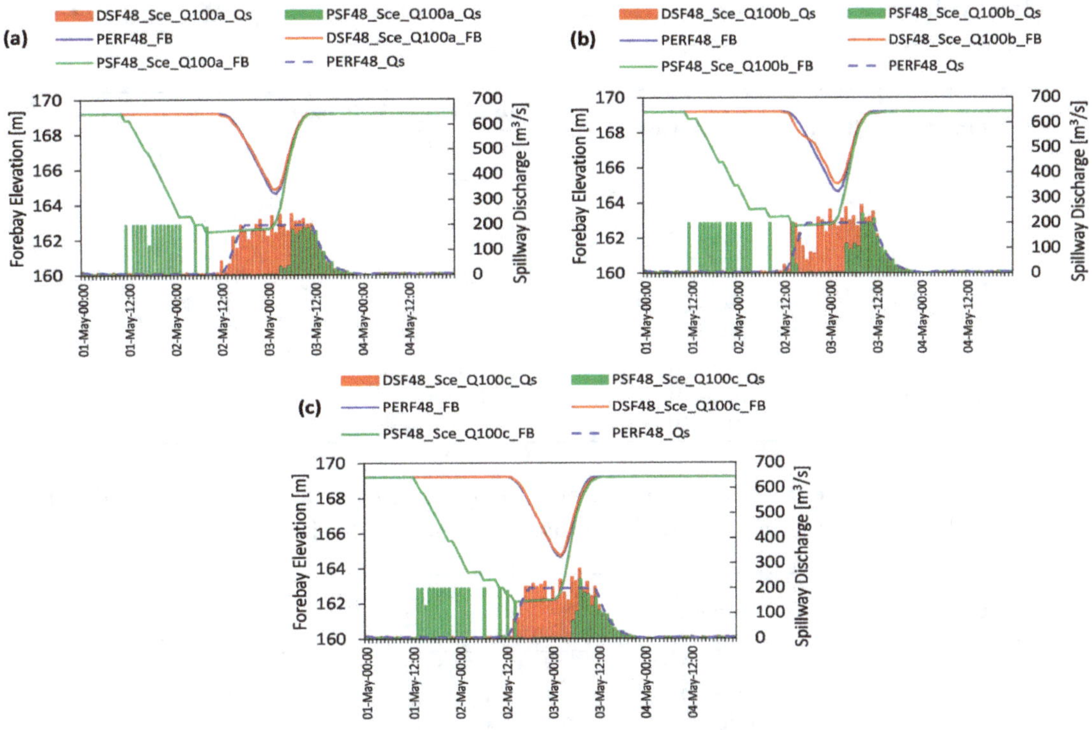

Figure 12. Comparison of deterministic (perfect and DSF) and stochastic (PSF) closed-loop MPC results with different forecast scenarios for Q_{100}: (**a**) Sce-Q100a; (**b**) Q100-Sceb; (**c**) Q100-Scec.

Table 3. Peakflow assessment of deterministic and stochastic closed-loop MPC results for different inflow conditions with forecast horizons of 48 h.

Flood Hydrograph	Scenarios	Peakflow at Yuvacik Outlet (m^3/s)	
		Deterministic MPC	Stochastic MPC
Q_{25}	Sce-Q25a	243	231
	Sce-Q25b	255	243
	Sce-Q25c	248	243
Q_{50}	Sce-Q50a	241	211
	Sce-Q50b	245	200
	Sce-Q50c	246	200
Q_{100}	Sce-Q100a	242	200
	Sce-Q100b	269	235
	Sce-Q100c	278	233

Table 4 presents the total flood volumes at the downstream area calculated in the hindcasting period. Compared to flooding threshold assessment, this indicator provides more insight into spillway operation. The best management (zero volume) is possible by perfect data using a forecast horizon of 48 h. Almost in all inflow cases and scenarios, the total flood volume decreases in the stochastic mode. This change highlights the added value of the stochastic optimization in comparison to the deterministic one especially for different forecast data sets (scenarios) and different flow conditions. FSI values were also calculated for both DSF and PSF scenarios and results are presented in Table 5, respectively. It is important to note that each FSI is calculated according to its own FCL i.e., FCL Q_{100} is used for Q_{100} operation assessment whereas FCL Q_{50} is used for the Q_{50} based flood case. According to this, TB-MPC stands as the more confident by higher FSI but also still can provide a high reservoir level at the end of the event without compromising water supply targets. For an uncertain future, a higher FSI is more reliable and preferable with lower risk for water supply as well.

Table 4. Flood volume assessment of deterministic and stochastic closed-loop MPC for different inflow conditions with forecast horizon of 48 h.

Flood Condition	Scenarios	Total Flood Volume (1×10^6 m^3)	
		Deterministic MPC	Stochastic MPC
Q_{25}	Sce-Q25a	0.507	0.302
	Sce-Q25b	0.549	0.254
	Sce-Q25c	0.438	0.271
Q_{50}	Sce-Q50a	0.666	0.062
	Sce-Q50b	0.471	0.004
	Sce-Q50c	0.331	0.004
Q_{100}	Sce-Q100a	0.690	0.004
	Sce-Q100b	1.256	0.184
	Sce-Q100c	1.018	0.127

Table 5. FSI value assessment of deterministic and stochastic closed-loop MPC according to Flood Control Levels (FCLs) for different inflow conditions with forecast horizon of 48 h.

Flood Condition	Scenarios	Flood Storage Index (FSI)	
		Deterministic MPC	Stochastic MPC
Q_{25}	Sce-Q25a	0.652	0.800
	Sce-Q25b	0.659	0.990
	Sce-Q25c	0.659	0.796
Q_{50}	Sce-Q50a	0.566	0.723
	Sce-Q50b	0.598	0.770
	Sce-Q50c	0.606	0.758
Q_{100}	Sce-Q100a	0.457	0.650
	Sce-Q100b	0.463	0.645
	Sce-Q100c	0.456	0.645

According to results, PSF based multi-stage stochastic MPC optimization has the advantage of including forecast uncertainty in the optimization set-up. Also, the uncertainty in the flows can be represented in the synthesized PSFs by developing the proposed error generation method. There is always performance improvement in the TB-MPC model for flood flows of Q_{25}, Q_{50}, Q_{100} with a forecast horizon up to 48 h with respect to (i) maximum discharge at the dam outlet (compared to flooding threshold value), (ii) total flood volume at the downstream area and (iii) FSI, while keeping the forebay elevation at the desired level for water supply at the end of the flood event. This shows the added value of the approach and provides reasonable outputs compared to the deterministic counterpart. The developed framework also indicates robust solutions against forecast uncertainty along with a different independent hindcasting experiment assessment.

5. Conclusions and Outlook

This study shows the added value of stochastic optimization using a synthetic probabilistic forecast generation and mass-conservative scenario tree reduction technique. TB-MPC provides multi-stage stochastic optimization in comparison to its deterministic counterpart. According to hindcasting experiments, our main conclusions are:

- Forecast uncertainty is indispensable especially for flood management. It is critical for those cases in which wrong or poor decisions may result with loss of life and property. At this point, considering uncertainty provides better management in terms of flood metrics without discarding water supply purposes.

- Independent closed-loop hindcasting experiment scenarios demonstrate the robustness of the system developed against biased information (disturbances).
- Probabilistic data represent forecast uncertainty in comparison to deterministic equivalents. In this study, a new synthetic streamflow generation method is proposed to represent forecast uncertainty for reservoir optimization.
- The synthetic PSF generation model that considers the dynamic evolution of uncertainties is valuable if hydrological model outputs driven by a rainfall and temperature forecast ensemble are not available. This method is very advantageous from the operational standpoint, since it does not require complex computations and is easy to implement while considering conditional (flow dependent) increasing uncertainty through time. It is simple to formulate, comprehend, and easy to repeat.
- Besides the ensemble generation, tree reduction parameters should be carefully investigated in the problem definition phase. In the case of selecting a lower branch and forecast horizon than required, TB-MPC results may converge to deterministic MPC results.
- The system was also tested against different inflow conditions which have greater flood value than the downstream channel capacity. According to the results, the method provides reliable results against different high flood conditions in the hindcasting experiments.

In future work, a hydrological model will be used, thus synthesized PSF might be derived from perturbed hydrological model forcings such as precipitation and temperature. A comparison of these results can contribute to improve synthetic ensemble generation and its consideration in TB-MPC.

Supplementary Materials
Figure S1: mean CRPS of ensemble scenarios: (a) Sce-Q25; (b) Sce-Q50; (c) Sce-Q100, Figure S2: Comparison of deterministic (perfect and DSF) and stochastic (PSF) closed-loop MPC results with different forecast scenarios for Q_{25}: (a) Sce-Q25a; (b) Sce-Q25b; (c) Sce-Q25c, Figure S3: Comparison of deterministic (perfect and DSF) and stochastic (PSF) closed-loop MPC results with different forecast scenarios for Q_{50}: (a) Sce-Q50a; (b) Sce-Q50b; (c) Sce-Q50c.

Acknowledgments: The first author would like to thank The Scientific and Technological Research Council of Turkey (TUBITAK) for the scholarship (2214A program). This study is supported by Anadolu University Scientific Research Projects Commission (under the grant No: 1506F502 and No: 1705F189). Graphs were prepared by Daniel's XL Toolbox (www.xltoolbox.net) and MATLAB 2012a (License number: 991708).

Author Contributions: Aynur Şensoy set the problem and background. Gökcen Uysal organized the main structure of the paper. Gökcen Uysal, Rodolfo Alvarado-Montero, and Dirk Schwanenberg designed the methodology. Gökcen Uysal conducted the experiments, and prepared the draft paper. Finally, Dirk Schwanenberg, Rodolfo Alvarado-Montero, and Aynur Şensoy provided useful advice and made some corrections. All authors read and approved the final manuscript.

References

1. Ahmad, S.; Simonovic, S.P. System dynamics modeling of reservoir operations for flood management. *J. Comput. Civ. Eng.* **2000**, *14*, 190–198. [CrossRef]
2. Plate, E.J. Flood risk and flood management. *J. Hydrol.* **2002**, *267*, 2–11. [CrossRef]
3. Wei, C.C.; Hsu, N.S. Optimal tree-based release rules for real-time flood control operations on a multipurpose multireservoir system. *J. Hydrol.* **2009**, *365*, 213–224. [CrossRef]
4. Şensoy, A.; Uysal, G.; Şorman, A.A. Developing a decision support framework for real-time flood management using integrated models. *J. Flood Risk Manag.* **2016**. [CrossRef]
5. Cheng, W.M.; Huang, C.L.; Hsu, N.S.; Wei, C.C. Risk analysis of reservoir operations considering short-term flood control and long-term water supply: A case study for the Da-Han Creek Basin in Taiwan. *Water* **2017**, *9*, 424. [CrossRef]
6. Oliveira, R.; Loucks, D.P. Operating rules for multireservoir systems. *Water Resour. Res.* **1997**, *33*, 839–852. [CrossRef]
7. Chang, F.J.; Chen, L.; Chang, L.C. Optimizing the reservoir operating rule curves by genetic algorithms. *Hydrol. Process.* **2005**, *19*, 2277–2289. [CrossRef]

8. Liu, P.; Guo, S.; Xu, X.; Chen, J. Derivation of aggregation-based joint operating rule curves for cascade hydropower reservoirs. *Water Resour. Manag.* **2011**, *25*, 3177–3200. [CrossRef]
9. Rani, D.; Moreira, M.M. Simulation–optimization modeling: A survey and potential application in reservoir systems operation. *Water Resour. Manag.* **2010**, *24*, 1107–1138. [CrossRef]
10. Li, X.; Guo, S.; Liu, P.; Chen, G. Dynamic control of flood limited water level for reservoir operation by considering inflow uncertainty. *J. Hydrol.* **2010**, *391*, 124–132. [CrossRef]
11. Wan, W.; Zhao, J.; Lund, J.R.; Zhao, T.; Lei, X.; Wang, H. Optimal hedging rule for reservoir refill. *J. Water Resour. Plan. Manag. ASCE* **2016**, *142*, 04016051. [CrossRef]
12. Bellman, R. The theory of dynamic programming. *Bull. Am. Math. Soc.* **1954**, *60*, 503–516. [CrossRef]
13. Faber, B.A.; Stedinger, J.R. Reservoir optimization using sampling SDP with ensemble streamflow prediction (ESP) forecasts. *J. Hydrol.* **2001**, *249*, 113–133. [CrossRef]
14. Yurtal, R.; Seckin, G.; Ardicliogl u, G. Hydropower optimization for the lower Seyhan system in Turkey using dynamic programming. *Water Int.* **2005**, *30*, 522–529. [CrossRef]
15. Raso, L.; Malaterre, P.O.; Bader, J.C. Effective streamflow process modeling for optimal reservoir operation using stochastic dual dynamic programming. *J. Water Resour. Plann. Manage. ASCE* **2017**, *143*, 04017003. [CrossRef]
16. Labadie, J.W. Optimal operation of multireservoir systems: State-of-the-art review. *J. Water Resour. Plan. Manag. ASCE* **2004**, *130*, 93–111. [CrossRef]
17. Morari, M.; Lee, J.H. Model predictive control: Past, present and future. *Comput. Chem. Eng.* **1999**, *23*, 667–682. [CrossRef]
18. Blanco, T.B.; Willems, P.; Chiang, P.K.; Haverbeke, N.; Berlamont, J.; De Moor, B. Flood regulation using nonlinear model predictive control. *Control Eng. Pract.* **2010**, *18*, 1147–1157. [CrossRef]
19. Van Overloop, P.J.; Horváth, K.; Aydin, B.E. Model predictive control based on an integrator resonance model applied to an open water channel. *Control Eng. Pract.* **2014**, *27*, 54–60. [CrossRef]
20. Horváth, K.; Galvis, E.; Valentín, M.G.; Rodellar, J. New offset-free method for model predictive control of open channels. *Control Eng. Pract.* **2015**, *41*, 13–25. [CrossRef]
21. Zhao, T.; Zhao, J.; Yang, D.; Wang, H. Generalized martingale model of the uncertainty evolution of streamflow forecasts. *Adv. Water Resour.* **2013**, *57*, 41–51. [CrossRef]
22. Chen, J.; Brissette, F.P.; Leconte, R. Uncertainty of downscaling method in quantifying the impact of climate change on hydrology. *J. Hydrol.* **2011**, *401*, 190–202. [CrossRef]
23. Roulin, E.; Vannitsem, S. Post-processing of medium-range probabilistic hydrological forecasting: Impact of forcing, initial conditions and model errors. *Hydrol. Process.* **2015**, *29*, 1434–1449. [CrossRef]
24. Fan, F.M.; Schwanenberg, D.; Alvarado, R.; dos Reis, A.A.; Collischonn, W.; Naumman, S. Performance of deterministic and probabilistic hydrological forecasts for the short-term optimization of a tropical hydropower reservoir. *Water Resour. Manag.* **2016**, *30*, 3609–3625. [CrossRef]
25. Kuczera, G.; Parent, E. Monte Carlo assessment of parameter uncertainty in conceptual catchment models: The Metropolis algorithm. *J. Hydrol.* **1998**, *211*, 69–85. [CrossRef]
26. Montero, R.A.; Schwanenberg, D.; Krahe, P.; Lisniak, D.; Sensoy, A.; Sorman, A.A.; Akkol, B. Moving horizon estimation for assimilating H-SAF remote sensing data into the HBV hydrological model. *Adv. Water Resour.* **2016**, *92*, 248–257. [CrossRef]
27. Yao, H.; Georgakakos, A. Assessment of Folsom Lake response to historical and potential future climate scenarios: 2. Reservoir management. *J. Hydrol.* **2001**, *249*, 176–196. [CrossRef]
28. Zhao, T.; Cai, X.; Yang, D. Effect of streamflow forecast uncertainty on real-time reservoir operation. *Adv. Water Resour.* **2011**, *34*, 495–504. [CrossRef]
29. Liu, Z.; Guo, Y.; Wang, L.; Wang, Q. Streamflow forecast errors and their impacts on forecast-based reservoir flood control. *Water Resour. Manag.* **2015**, *29*, 4557–4572. [CrossRef]
30. Todini, E. Role and treatment of uncertainty in real-time flood forecasting. *Hydrol. Process.* **2004**, *18*, 2743–2746. [CrossRef]
31. Koutsoyiannis, D. The Hurst phenomenon and fractional Gaussian noise made easy. *Hydrol. Sci. J.* **2002**, *47*, 573–595. [CrossRef]
32. Ahmed, J.A.; Sarma, A.K. Artificial neural network model for synthetic streamflow generation. *Water Resour. Manag.* **2007**, *21*, 1015. [CrossRef]

33. Ochoa-Rivera, J.C.; García-Bartual, R.; Andreu, J. Multivariate synthetic streamflow generation using a hybrid model based on artificial neural networks. *Hydrol. Earth Syst. Sci.* **2002**, *6*, 641–654. [CrossRef]
34. Raso, L.; Schwanenberg, D.; van de Giesen, N.C.; van Overloop, P.J. Short-term optimal operation of water systems using ensemble forecasts. *Adv. Water Resour.* **2014**, *71*, 200–208. [CrossRef]
35. Stive, P.M. Performance Assessment of Tree-Based Model Predictive Control. Master's Thesis, Delft University of Technology, Delft, The Netherlands, 2011.
36. Ficchi, A.; Raso, L.; Dorchies, D.; Pianosi, F.; Malaterre, P.O.; van Overloop, P.J.; Jay-Allemand, M. Optimal operation of the multireservoir system in the Seine River Basin using deterministic and ensemble forecasts. *J. Water Resour. Plan. Manag.* **2015**, *142*. [CrossRef]
37. Schwanenberg, D.; Fan, F.M.; Naumann, S.; Kuwajima, J.I.; Montero, R.A.; dos Reis, A.A. Short-term reservoir optimization for flood mitigation under meteorological and hydrological forecast uncertainty. *Water Resour. Manag.* **2015**, *29*, 1635–1651. [CrossRef]
38. Uysal, G.; Şensoy, A.; Şorman, A.A.; Akgün, T.; Gezgin, T. Basin/reservoir system integration for real time reservoir operation. *Water Resour. Manag.* **2016**, *30*, 1653–1668. [CrossRef]
39. Uysal, G.; Schwanenberg, D.; Alvarado-Montero, R.; Şensoy, A. Short term optimal operation of water supply reservoir under flood control stress using model predictive control. *Water Resour. Manag.* **2018**, *32*, 583–597. [CrossRef]
40. Hui, P. *Yellow River Group Project, A Subproject of the China-DC WRE Project*; Final Research Report of Cluster 2; Delft University of Technology: Delft, The Netherlands, 2002.
41. Yan, B.; Guo, S.; Chen, L. Estimation of reservoir flood control operation risks with considering inflow forecasting errors. *Stoch. Environ. Res. Risk Assess.* **2014**, *28*, 359–368. [CrossRef]
42. Datta, B.; Burges, S.J. Short-term, single, multiple-purpose reservoir operation: Importance of loss functions and forecast errors. *Water Resour. Res.* **1984**, *20*, 1167–1176. [CrossRef]
43. Pianosi, F.; Raso, L. Dynamic modeling of predictive uncertainty by regression on absolute errors. *Water Resour. Res.* **2012**, *48*. [CrossRef]
44. Biemans, H.; Hutjes, R.W.A.; Kabat, P.; Strengers, B.J.; Gerten, D.; Rost, S. Effects of precipitation uncertainty on discharge calculations for main river basins. *J. Hydrometeorol.* **2009**, *10*, 1011–1025. [CrossRef]
45. Schwanenberg, D.; Becker, B.P.J.; Xu, M. The open real-time control (RTC)-Tools software framework for modeling RTC in water resources sytems. *J. Hydroinform.* **2015**, *17*, 130–148. [CrossRef]
46. Xu, M.; Schwanenberg, D. Comparison of sequential and simultaneous model predictive control of reservoir systems. In Proceedings of the 10th International Conference on Hydroinformatics, Hamburg, Germany, 14–18 July 2012.
47. Wächter, A.; Biegler, L.T. On the implementation of an interior-point filter line-search algorithm for large-scale nonlinear programming. *Math. Program.* **2006**, *106*, 25–57. [CrossRef]
48. Heitsch, H.; Römisch, W. Scenario reduction algorithms in stochastic programming. *Comput. Optim. Appl.* **2003**, *24*, 187–206. [CrossRef]
49. Nadarajah, S.; Shiau, J.T. Analysis of extreme flood events for the Pachang River, Taiwan. *Water Resour. Manag.* **2005**, *19*, 363–374. [CrossRef]
50. Stedinger, J.R.; Vogel, R.M.; Foufoula-Georgiou, E. Frequency analysis of extreme events. In *Handbook of Hydrology*; Maidment, D.R., Ed.; McGraw-Hill: New York, NY, USA, 1993; Chapter 18.
51. Xiao, Y.; Guo, S.; Liu, P.; Yan, B.; Chen, L. Design flood hydrograph based on multicharacteristic synthesis index method. *J. Hydrol. Eng.* **2009**, *14*, 1359–1364. [CrossRef]
52. General Directorate of State Hydraulic Works (DSI). *Kirazdere Dam Engineering Hydrology and Planning Report*; Prepared by 1st Regional Directorate of State Hydraulics Works; Planning and Science Board: Bursa, Turkey, 1983. (In Turkish)
53. Mediero, L.; Garrote, L.; Martin-Carrasco, F. A probabilistic model to support reservoir operation decisions during flash floods. *Hydrol. Sci. J. ASCE* **2007**, *52*, 523–537. [CrossRef]
54. Jørgensen, J.B. Moving Horizon Estimation and Control. Ph.D. Thesis, Technical University of Denmark, Lyngby, Denmark, 2005.

9

Multiple Climate Change Scenarios and Runoff Response in Biliu River

Xueping Zhu [1,*], Chi Zhang [2], Wei Qi [3,4,*], Wenjun Cai [1], Xuehua Zhao [1] and Xueni Wang [1]

1 College of Water Resources Science and Engineering, Taiyuan University of Technology, Taiyuan 030024, China; caiwenjun62620@163.com (W.C.); zhaoxuehua@tyut.edu.cn (X.Z.); xnwang@mail.dlut.edu.cn (X.W.)
2 School of Hydraulic Engineering, Dalian University of Technology, Dalian 116024, China; czhang@dlut.edu.cn
3 School of Environmental Science and Engineering, South University of Science and Technology of China, Shenzhen 518055, China
4 State Key Laboratory of Water Resource & Hydropower Engineering Science, Wuhan University, Wuhan 430072, China
* Correspondence: xpzhu01@163.com (X.Z.); QiWei_waterresources@hotmail.com (W.Q.)

Abstract: The impacts of temperature and precipitation changes on regional evaporation and runoff characteristics have been investigated for the Biliu River basin, which is located in Liaoning Province, northeast China. Multiple climate change scenarios from phase 3 and phase 5 of the Coupled Model Intercomparison Project (CMIP3 and CMIP5) (21 scenarios in total) were utilized. A calibrated hydrologic model—SWAT model—was used to simulate future discharges based on downscaled climate data through a validated morphing method. Results show that both annual temperature and precipitation increase under most of the CMIP3 and CMIP5 scenarios, and increase more in the far future (2041–2065) than in the near future (2016–2040). These changes in precipitation and temperature lead to an increase in evaporation under 19 scenarios and a decrease in runoff under two-thirds of the selected scenarios. Compared to CMIP3, CMIP5 scenarios show higher temperature and wider ranges of changes in precipitation and runoff. The results provide important information on the impacts of global climate change on water resources availability in the Biliu River basin, which is beneficial for the planning and management of water resources in this region.

Keywords: climate change; CMIP3; CMIP5; downscaling; runoff response; SWAT model

1. Introduction

It has been recognized that climate change could have profound impacts on the global water cycle [1–3]. Therefore, it is important to consider potential impacts of climate change in the planning and management of regional water resources.

To quantitate climate change impacts on water resources, the output of general circulation models (GCMs) is commonly used [4–6], coupled with hydrological models or forcing offline hydrological models [7–10]. The results help to understand the impacts of climate change and develop strategies to adapt to or possibly mitigate these impacts [3]. The Coupled Model Intercomparison Project phase 3 (CMIP3) and phase 5 (CMIP5) have provided abundant climate data (e.g., the IPCC's AR5) [11], and both CMIP3 and CMIP5 climate datasets have been widely utilized globally [12–16]. Several studies have revealed that, compared to CMIP3, CMIP5 ensemble simulations have substantially improved the statistical representation of daily mean precipitation and temperature [17,18]. However, few studies have compared the impacts of CMIP3 and CMIP5 on the design of the planning and management infrastructure of water resources [19–21], especially in China [22].

Northeast China is an important food production area in China and it has suffered from droughts in recent years. It has been reported that the inflow of Biliu River Reservoir (an important river in northeast China) declined significantly over the period of 1990–2005 [23]. Because the reservoir is the most important water source for nearby big cities and also because it plays an important role in cropland irrigation, a water transfer project is currently being constructed to transfer water into the reservoir. Zhang et al. optimized water diversion and supply rules of Biliu River Reservoir and attempt to identify resilient water diversion strategies to mitigate the potential impacts of climate change [10]. However, the design of the water transfer project was based on CMIP3 data, which may inaccurately estimate the severity of a water resource shortage [19,20]. Therefore, it is necessary to compare the availability of simulated water resources using CMIP3 and CMIP5, which can provide important information on the differences in the availability of projected water resources and therefore facilitate the design and evaluation of the water transfer project [19–21].

The overall objective of this study is to compare the projected availability of runoff using datasets from CMIP3 and CMIP5 in the Biliu River basin. Three GCM outputs of CMIP3 for three emission scenarios and six GCM outputs of CMIP5 for two representative concentration pathways are utilized. A 'morphing' downscaling method is utilized to derive local climatic parameters on a daily time scale. The downscaled datasets are utilized as the input of the SWAT hydrological model to simulate future runoff.

2. Methodology

2.1. CMIP3 and CMIP5 Datasets

2.1.1. Climate Models and Emission Scenarios

Several studies have evaluated the precipitation and temperature simulation performance of the CMIP3 and CMIP5 historical experiments and found that CMIP5 improved the simulation of basic atmospheric variables compared to CMIP3 [17,18]. Three CMIP3 models and six CMIP5 models are used in this study to consider the uncertainties of global climate prediction. They cover a representative range of projections of the two experiments and perform well in the statistical representation of daily precipitation and temperature [17,18].

Emission scenarios A1B, A2, and B1 from CMIP3, which can capture the emission uncertainties, have been analyzed by other studies [24,25] and are considered in this study. Three GCMs for each emission scenario are utilized (as shown in Table 1). The baseline scenario (20C3M) for the historical period is also considered. The regional monthly temperature and precipitation of the study area are derived by linear interpolation of the nearest GCM data. Although it is simple, its precision is acceptable [26–28].

The formulation of the long-term (century time-scale) simulations is a new part of CMIP5 when compared with CMIP3. The long-term experiment was initiated from the end of freely evolving simulations of the historical period; the experiment was conducted with atmosphere-ocean global climate models (AOGCMs) which in some cases may be coupled with carbon cycle models [11,29]. The long-term experiment includes the pre-industrial control run, the historical run (1850 to at least 2005), and the future scenario run (~2006–2100, or extended to 2300), etc.

For future climate projections, four emission and concentration scenarios called 'representative concentration pathways' (RCPs) were designed that lead to radiative forcing levels of 8.5, 6, 4.5, and 2.6 W/m^2 around the end of the century. Each of the RCPs covers the period 2006–2100, and extensions have been formulated for periods up to 2300 [30]. RCP4.5 and RCP8.5 are core experiments [29]. RCP4.5 is a medium forcing integration and RCP8.5 is a high radiative forcing case. They correspond to a medium mitigation scenario and a high emissions scenario, respectively. RCP6 and RCP2.6 are two carbon cycle feedback experiments, part of tier 1 experiments. The tier 1 and tier 2 experiments explore various aspects of the core experiments in further detail. The core experiments (i.e., RCP4.5 and

RCP8.5) are utilized in this study. Six climate models (as shown in Table 1) from CMIP5 are considered for comparison.

Table 1. Details of CMIP3 and CMIP5 climate models and scenarios used in this study.

Model	Country	Resolution	Scenarios
BCCR_BCM2.0	Norway	2.81° × 2.81°	A1B, A2, B1 for each model respectively
CSIRO_MK3.0	Australia	1.88° × 1.88°	
MIROC3.2m	Japan	2.81° × 2.81°	
ACCESS1.0	Australia	1.88° × 2.48°	RCP8.5 and RCP4.5 for each model respectively
BCC-CSM1.1(m)	China	1.13° × 1.13°	
CESM1(BGC)	USA	1.3° × 0.9°	
CESM1(CAM5)	USA	1.3° × 0.9°	
CMCC-CM	Italy	0.75° × 0.75°	
MPI-ESM-MR	Germany	1.88° × 1.88°	

2.1.2. Downscaling Method

The coarse resolution climate model predictions need to be downscaled to fine spatial and temporal resolutions to facilitate hydrological simulations. There are many downscaling methods, and they can be divided into two categories: dynamical downscaling and statistical downscaling methods. The statistical downscaling methods are relatively simpler than dynamical downscaling and have been widely used [7–9]. One of the statistical downscaling methods called "morphing" [7] is adopted here. The "morphing" approach has two characteristics. First, the 'baseline climate' is reliable, because it is based on observed climate data. Second, the resulting weather sequence is likely to be meteorologically consistent.

Morphing involves three generic operations: a shift, a linear stretch (scaling factor), and a combination of shift and a stretch. A shift by Δx_m is applied to the present-day climate variable x_0 by $x = x_0 + \Delta x_m$. Here, Δx_m is the absolute changes in monthly mean climate for month m. The monthly variance of climate variables is unchanged. A stretch of α_m is applied by $x = \alpha_m x_0$, where α_m is the fractional change in the monthly-mean value for month m. A combination of shift and stretch is obtained by $x = x_0 + \Delta x_m + \alpha_m \times (x_0 - \overline{x}_m^0)$, in which $\overline{x}_m^0$ is the baseline climatological value for month m, and is calculated as $\overline{x}_m^0 = \frac{1}{24 \times d_m \times Y} \sum_{Y\,year} \sum_{month\,m} x_0$, where Y is the number of years and d_m is the number of days in month m.

Precipitation and temperature are two main climate variables considered in previous research in runoff simulation. Therefore, we analyze runoff responses to changes in precipitation and temperature. The precipitation data is downscaled by a combination of shift and stretch morphing, and the temperature data is downscaled by shift morphing by

$$P_1 = (1 + \alpha_m) \times P_0 = (1 + \frac{\overline{P}_m - \overline{P}_m^b}{\overline{P}_m^b}) \times P_0 \quad (1)$$

$$T_1 = T_0 + \Delta T = T_0 + (\overline{T}_m - \overline{T}_m^b) \quad (2)$$

$$T_1^{\max} = T_0^{\max} + \Delta T_{\max} = T_0^{\max} + (\overline{T}_m^{\max} - \overline{T}_m^{b,\max}) \quad (3)$$

$$T_1^{\min} = T_0^{\min} + \Delta T_{\min} = T_0^{\min} + (\overline{T}_m^{\min} - \overline{T}_m^{b,\min}) \quad (4)$$

where P_1, T_1, $T_1^{\max}$ and $T_1^{\min}$ are downscaled precipitation, daily mean, and maximum and minimum temperature; P_0, T_0, $T_0^{\max}$ and $T_0^{\min}$ are historical observation precipitation, daily mean, and maximum and minimum temperature; $\overline{P}_m$, $\overline{T}_m$, $\overline{T}_m^{\max}$ and $\overline{T}_m^{\min}$ are average values of future precipitation, daily mean, and maximum and minimum temperature over month m; $\overline{P}_m^b$, $\overline{T}_m^b$, $\overline{T}_m^{b,\max}$ and $\overline{T}_m^{b,\min}$ are average values of precipitation, daily mean, and maximum and minimum temperature over month *m* for the historical period from GCMs.

2.2. Hydrological Model

The continuous and physically distributed hydrological model—the Soil and Water Assessment Tool (SWAT) [31]—is utilized in this study. The main components of SWAT include hydrology, climate, nutrient cycling, soil temperature, sediment movement, crop growth, agricultural management, and pesticide dynamics [32]. The model has been widely applied to simulate hydrological processes globally [23,33–35] and it can conveniently consider weather adjustment.

Model calibration and validation are necessary to identify the SWAT model parameters before it can be used for prediction. Sensitive parameters were first identified by the LH-OAT (latin hypercube sampling based on one-factor-at-a-time) method [36] incorporated in SWAT. Manual calibration was then carried out for several of the most sensitive parameters. The performance of SWAT was estimated by three performance metrics: the Nash–Sutcliffe model efficiency (NSE), the average relative error (Re) and the coefficient of determination (R^2). These metrics are defined as

$$\mathrm{NSE} = 1 - \sum_{t=1}^{T} (Q_{mt} - Q_{st})^2 / \sum_{t=1}^{T} (Q_{mt} - Q_{mavg})^2 \tag{5}$$

$$\mathrm{Re} = (Q_{mavg} - Q_{savg}) / Q_{savg} \times 100 \tag{6}$$

$$\mathrm{R}^2 = \sum_{t=1}^{T} (Q_{st} - Q_{savg})(Q_{mt} - Q_{mavg}) / \left\{ [\sum_{t=1}^{T} (Q_{st} - Q_{savg})^2][\sum_{t=1}^{T} (Q_{mt} - Q_{mavg})^2] \right\}^{\frac{1}{2}} \tag{7}$$

where Q_{mt} and Q_{st} are measured and simulated flow at time *t*; Q_{mavg} and Q_{savg} are average values of observed flow and simulated flow; *T* is the total number of time steps. The model performance is considered to be acceptable when NSE > 0.50, R^2 > 0.60, and Re < ±20% according to the study by Hao et al. [37].

3. Study Region and Datasets

3.1. Biliu River Basin

The Biliu River basin is located in the Liaoning Province, northeast China (Figure 1). The study area is about 2085 km^2. The Biliu River Reservoir was built in 1975 and it has a storage capacity of 934×10^6 m^3. The reservoir is the most important water source of nearby big cities and also plays an important role in cropland irrigation. The study area is located in the north temperate zone and is characterized by a moist climate. Forest land and farmland are the main land use types. The main soil types are brown soil and meadow soil. Another reservoir, named Yushi Reservoir, with a drainage area of 313 km^2 and a storage capacity of 89×10^6 m^3, was built upstream in 2001. The reservoir supplies water to the outside of the basin. Therefore, the impact of Yushi Reservoir needs to be considered in the hydrological model. The Biliu River Reservoir has experienced severe water shortage problems recently and therefore future runoff conditions under climate change need to be analyzed for adaptation measures.

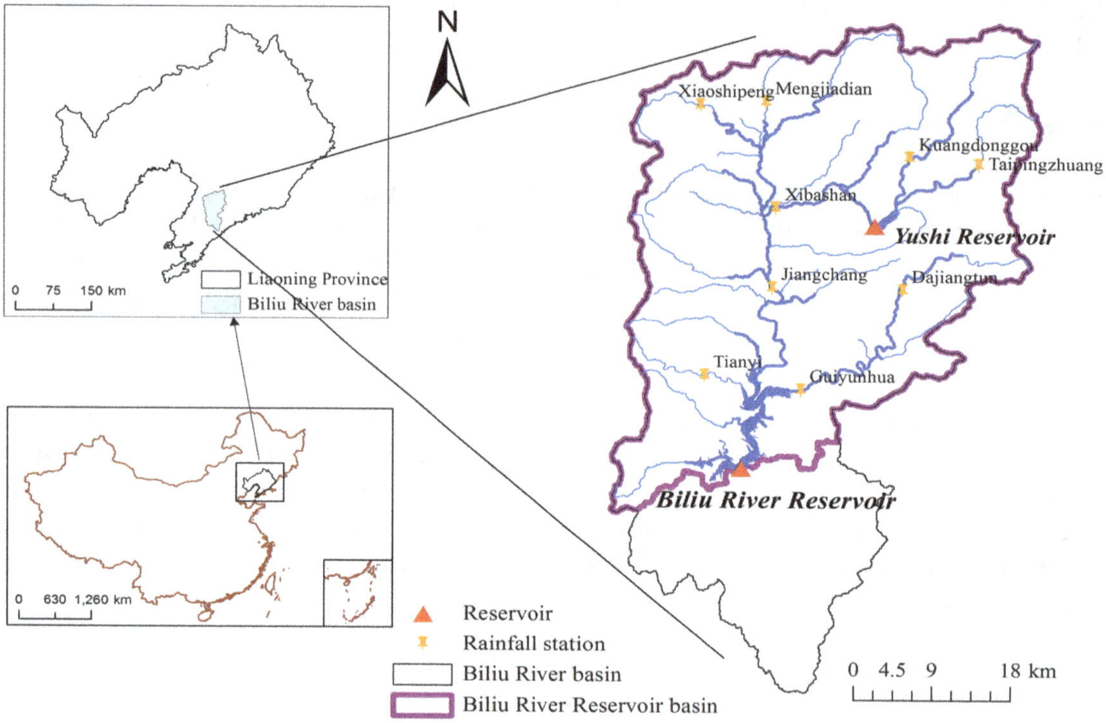

Figure 1. The Biliu River basin.

3.2. Dataset

The climate data in 1901–2099 for A1B, A2, and B1 were downloaded from the National Climate Center (http://ncc.cma.gov.cn). The long-term experiment data of 1850–2100 for the chosen six climate models in CMIP5 were downloaded from the Program for Climate Model Diagnosis and Intercomparison (PCMDI, http://pcmdi3.llnl.gov/esgcet/). The climate data were extracted for 1980–2004 period and two future periods (2016–2040 and 2041–2065). Future precipitation and temperature data output from the climate models are used as the input of the SWAT model to simulate future runoff.

Yearly and monthly precipitation and runoff data in 1958–2011 and daily runoff data in 1978–2004, were obtained from the Biliu River Reservoir administration. Daily precipitation data in 1978–2004 at nine precipitation stations were obtained from the Hydrology Bureau of Liaoning Province. Daily meteorological data—including mean, maximum, and minimum temperature, humidity, wind speed and direction, and solar radiation—were obtained from the China Meteorological Data Sharing Service System (http://cdc.cma.gov.cn/index.jsp). The Digital Elevation Model (DEM) data (90 × 90 m) were obtained from the CGIAR Consortium for Spatial Information (CGIAR-CSI) (http://srtm.csi.cgiar.org). Soil type and land use maps were obtained from the Data Center for Resources and Environmental Sciences, Chinese Academy of Sciences (http://www.resdc.cn/first.asp).

4. Results

This section includes three main parts. First, the SWAT model used is calibrated and validated based on observed data; second, the variations of precipitation and temperature in the study region are studied based on GCMs data from CMIP3 and CMIP5; third, the precipitation and temperature data are used as the input of SWAT model to predict the runoff in this region. In the resulting figures, runoff data are for the location of the Biliu River Reservoir, while precipitation, temperature, and evaporation data are the mean values of the Biliu River Reservoir basin in Figure 1.

4.1. SWAT Model Calibration and Validation Results

The SWAT model is calibrated and validated beforehand. The periods 1980–1994 and 1995–2004 were used as the calibration period and validation period, respectively. Yushi Reservoir runs after 2001 in the model. The simulated and measured monthly runoff are shown in Figure 2. The values of NSE and R^2 exceed 0.90 and 0.95, respectively. Re values are 4.04% and 12.64% in the calibration and validation. Therefore, the SWAT model is applicable for runoff prediction in this area.

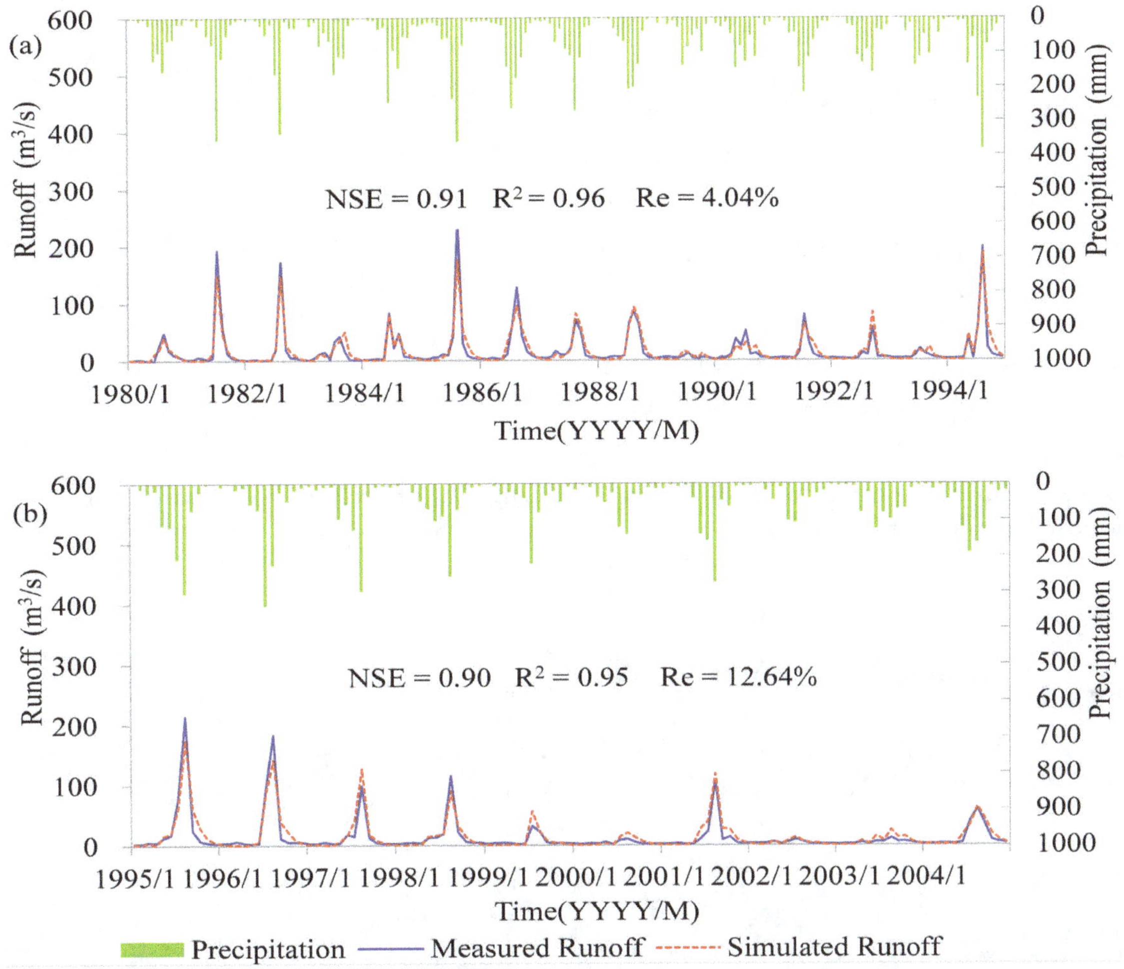

Figure 2. Simulated monthly runoff at Biliu River Reservoir station: (**a**) calibration period of 1980–1989; (**b**) validation period of 1990–1999.

4.2. Precipitation and Temperature Variations

After the 1980s, the Biliu River basin is affected by both climate change and human activities. The influence will exist in the future, and it is difficult to return to the natural state before 1980s. Therefore, the period after 1980s is used for precipitation and temperature downscaling. In addition, daily data are required for hydrological simulation. Thus, 1980–2004 is considered for downscaling since daily scale is not available after 2004. The climate model outputs in 1980–2004 and two future periods, 2016–2040 (near future) and 2041–2065 (far future), are utilized. Precipitation and temperature of the two future periods are downscaled respectively on the basis of 1980–2004 data. The output of the GCMs is bias corrected according to the observation in the historically period, and the same bias correction approach is applied to the output of GCMs in the future.

4.2.1. Temperature Variations

Two time periods, 1985–1994 and 1995–2004, were utilized to validate the morphing method for temperature downscaling. The downscaled temperature of 1995–2004 can be calculated using Equations (2)–(4). The downscaled temperature was compared with the historical observation in Figures 3 and 4. The nine selected models are verified detail by detail because there is no scenario difference in the historical period.

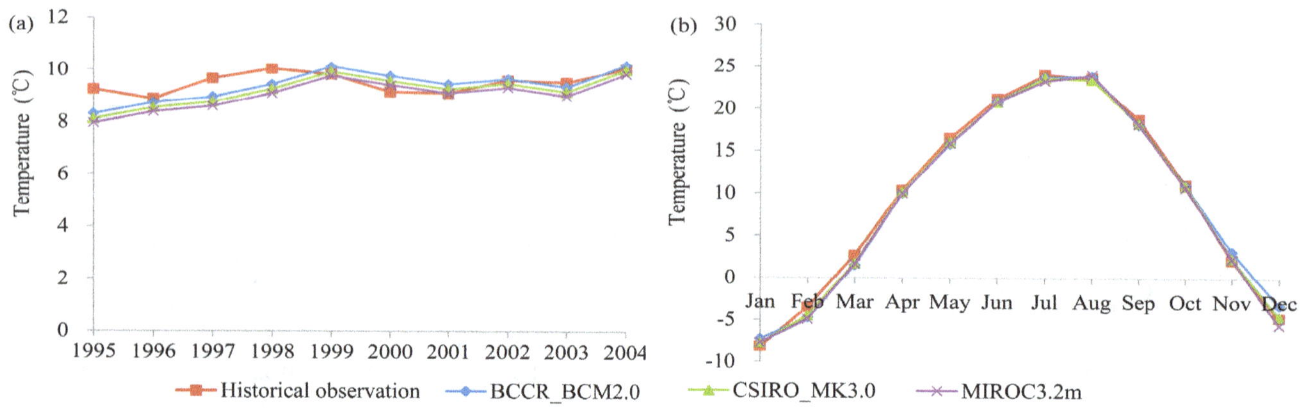

Figure 3. Comparison of the average downscaled temperature from CMIP3 using morphing and historical observation temperature series: (**a**) annual series; (**b**) monthly distribution.

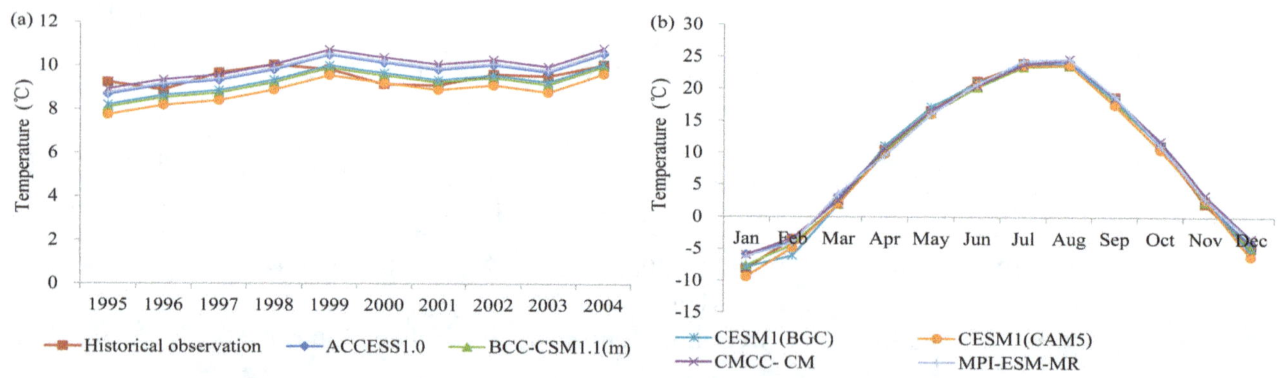

Figure 4. Comparison of the average downscaled temperature from CMIP5 using morphing and historical observation temperature series: (**a**) annual series; (**b**) monthly distribution.

Figure 3a compares the average annual mean temperatures of CMIP3 models. For BCCR_BCM2.0, CSIRO_MK3.0, and MIROC3.2m, the relative errors (Re) are −1.2%, −3.1%, and −4.8% respectively, and correlation coefficients (R^2) all are larger than 0.60. Figure 3b compares monthly temperature variations. It can be seen that they agree well. Figure 4a shows the average annual mean temperature of the six models from CMIP5. Three of them—ACCESS1.0, CMCC-CM and MPI-ESM-MR—slightly overestimate the temperature with Re being 2.7%, 5.2%, and 3.4% respectively. The other three models slightly underestimate the temperature with Re of BCC-CSM1.1(m), CESM1(BGC), and CESM1(CAM5) being −3.4%, −2.3%, and −7.0% respectively. The R^2 of the six models all are larger than 0.60. Figure 4b shows the comparison of monthly temperature variations, and it can be seen that the downscaled temperatures agree well with observation. Overall, the morphing approach shows acceptable performance in the temperature downscaling.

The annual mean temperature changes relative to 1980–2004 period are shown in Figures 5 and 6. The horizontal dash lines refer to the maximum and minimum multi-year mean temperature changes, which show the ranges of change uncertainty, and can be calculated by Equations (8) and (9).

$$z_{MAX} = \max\left\{\frac{1}{Y}\sum_{j=1}^{Y}\Delta z_i^j,\ i = 1,2,\cdots,k\right\} \tag{8}$$

$$z_{MIN} = \min\left\{\frac{1}{Y}\sum_{j=1}^{Y}\Delta z_i^j,\ i = 1,2,\cdots,k\right\} \tag{9}$$

where z_{MAX} and z_{MIN} represent the maximum and minimum multi-year mean changes, respectively; Δz_i^j is the absolute or fractional changes in annual mean climate for year j based on the ith scenario; Y is the number of years and k is the number of scenarios.

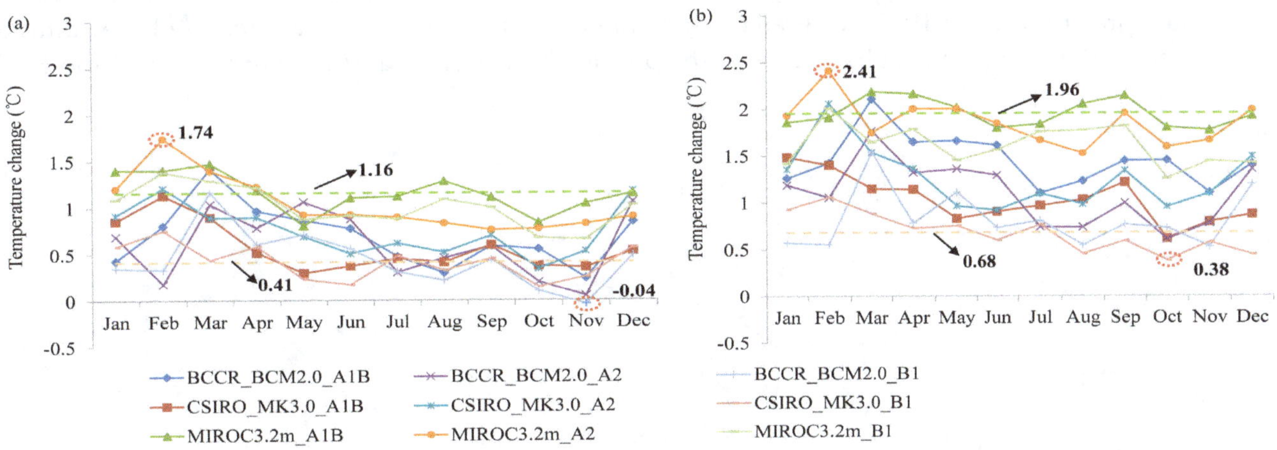

Figure 5. The forecasted temperature differences relative to 1980–2004 period under CMIP3: (**a**) the period 2016–2040; (**b**) the period 2041–2065.

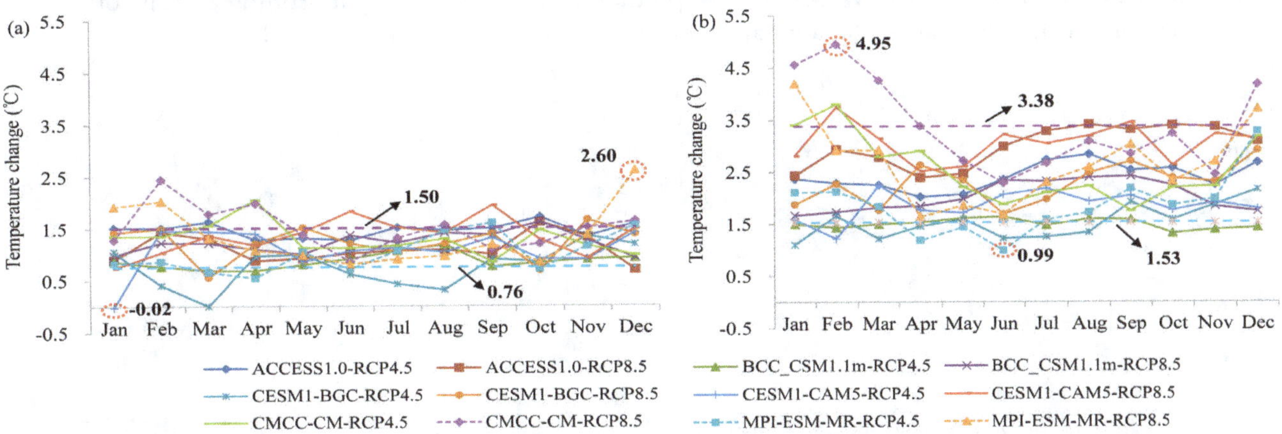

Figure 6. The forecasted temperature differences relative to 1980–2004 period under CMIP5: (**a**) the period 2016–2040; (**b**) the period 2041–2065.

As shown in Figures 5 and 6, the annual mean temperature increases under all of the CMIP3 and CMIP5 scenarios. The annual mean temperature changes range from 0.41 to 1.16 °C under CMIP3 and range from 0.76 to 1.50 °C under CMIP5 in 2016–2040. The temperatures under all of the CMIP3 and CMIP5 scenarios increase even more in 2041–2065 than in 2016–2040, and the increases range from 0.68 to 1.96 °C under CMIP3 and range from 1.53 to 3.38 °C under CMIP5. Overall, future temperature increases range from 0.41 to 1.50 °C in 2016–2040 and 0.68 to 3.38 °C in 2041–2065 relative to 1980–2004 period on the basis of the selected scenarios. The results imply that temperatures are likely to increase in the future and increase more in the far future than in the near future. Moreover, temperature increases more under CMIP5 than under CMIP3.

The temperature differences between emission scenarios are also compared. Among A1B, A2, and B1, the increase in temperature from largest to smallest is A1B > A2 > B1 on the basis of BCCR_BCM2.0 and MIROC3.2m in both future periods, while it is A2 > A1B > B1 based on CSIRO_MK3.0. In addition, it is found that the temperature increases more in RCP8.5 than in RCP4.5 under all the models in both future periods except ACCESS1.0 in 2016–2040. This implies that climate emission scenarios are closely associated with temperature.

4.2.2. Precipitation Variations

Similar to the temperature downscaling validation, two time periods, 1985–1994 and 1995–2004, were used to validate the applicability of the morphing method for precipitation downscaling. Downscaled precipitation of 1995–2004 can be calculated by Equation (1). The downscaled precipitation is compared with the historical observation precipitation to verify its accuracy in Figures 7 and 8.

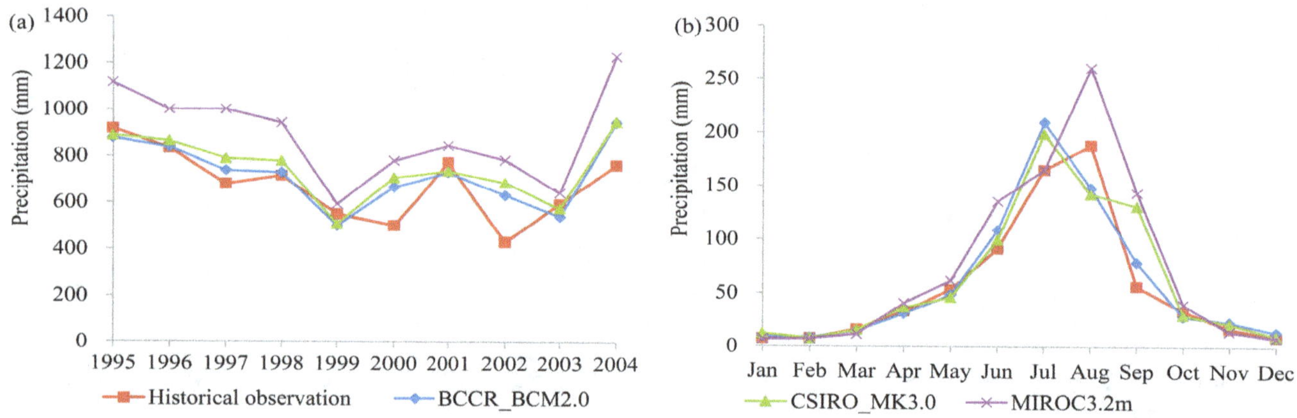

Figure 7. Comparison of the downscaled precipitation from CMIP3 using morphing and historical observation precipitation series: (**a**) annual series; (**b**) monthly distribution.

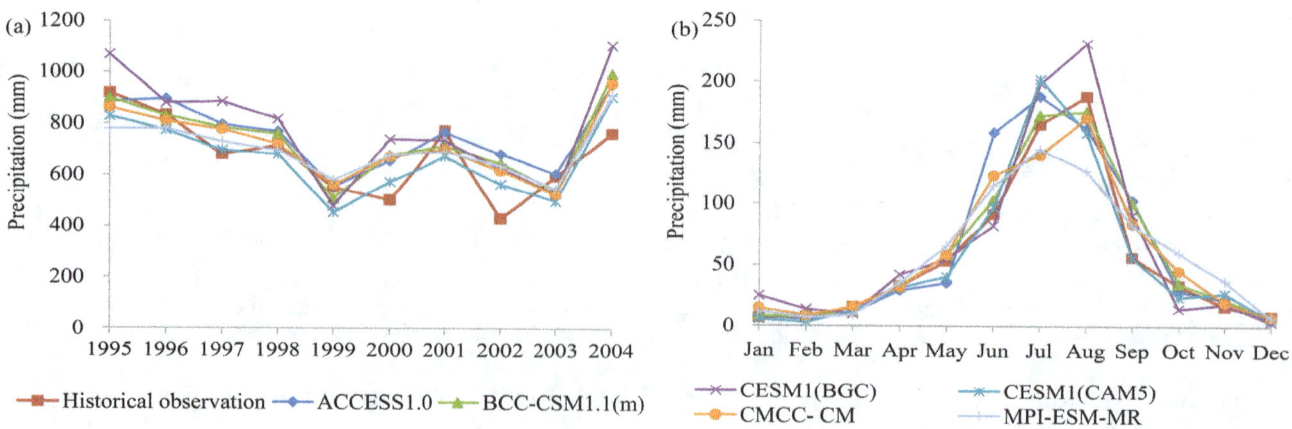

Figure 8. Comparison of the downscaled precipitation from CMIP5 using morphing and historical observation precipitation series: (**a**) annual series; (**b**) monthly distribution.

Figure 7a compares the annual precipitation of CMIP3 models. For BCCR_BCM2.0, CSIRO_MK3.0 and MIROC3.2m, Re are 6.4%, 10.7% and 32.2% respectively, and R^2 are larger than 0.72. Figure 7b compares monthly precipitation variations. It can be seen that they agree well. Figure 8a shows the average annual precipitation of the six models from CMIP5. Only one of them—CESM1(CAM5)—slightly underestimated the precipitation (Re is −1.8%). The other five models all overestimate the precipitation. The Re of ACCESS1.0, BCC-CSM1.1(m), CMCC-CM, CESM1(BGC), and MPI-ESM-MR are 11.9, 8.8, 6.5, 16.3, and 4.0% respectively. The R^2 of the six models all are larger than 0.70. Figure 8b shows a comparison of monthly precipitation variations. The fitting result is accurate: R^2 values are larger than 0.93. Overall, the morphing results basically meet the downscaling requirements, which demonstrates that the morphing method is valid for precipitation downscaling.

The annual mean change percentages relative to 1980–2004 period are shown in Figures 9 and 10. The horizontal dash lines refer to the maximum and minimum multi-year mean precipitation change percentages, which show the ranges of change uncertainty, as calculated by Equations (8) and (9).

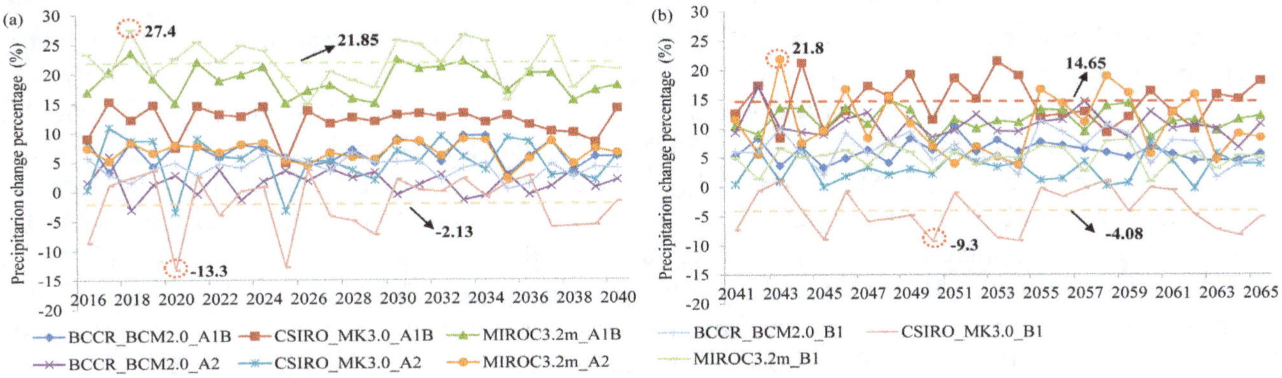

Figure 9. The forecasted precipitation change percentage relative to 1980–2004 period under CMIP3 (**a**) 2016–2040; (**b**) 2041–2065.

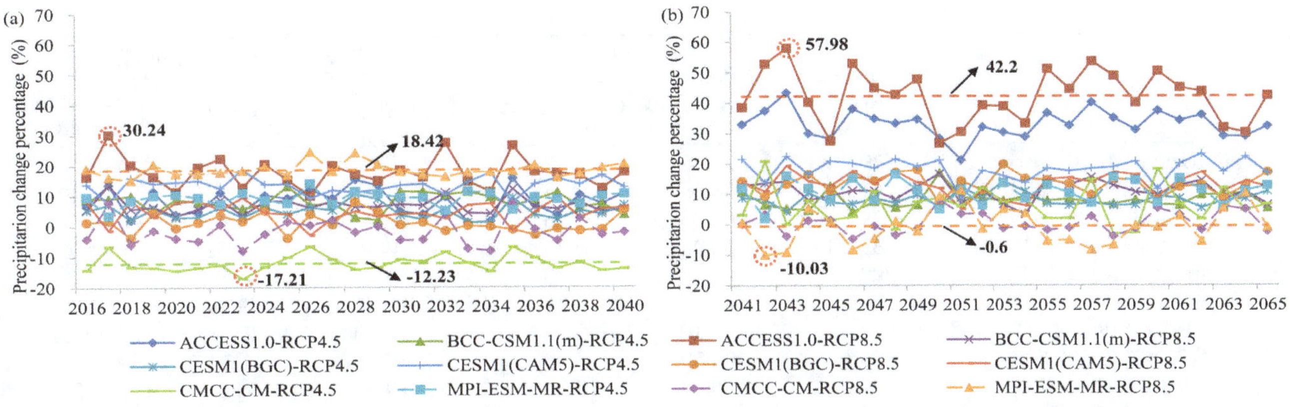

Figure 10. The forecasted precipitation change percentage relative to 1980–2004 period under CMIP5 (**a**) 2016–2040; (**b**) 2041–2065.

As shown in Figure 9, annual precipitation increases under most of the CMIP3 scenarios except CSIRO_MK3.0 (B1). The mean annual change percentage ranges from −2.13% to 21.85% in 2016–2040. The range is slightly smaller in 2041–2065, which ranges from −4.08% to 14.65%. The difference of MIROC3.2m (B1) in the two future periods is the greatest, followed by BCCR_BCM2.0 (A2) and MIROC3.2m (A1B). The other six scenarios differ by less than ±5%. This implies that future precipitation is likely to increase, with similar changes in the near future and far future.

As shown in Figure 10, annual precipitation increases under most of the CMIP5 scenarios except CMCC-CM (RCP4.5 and RCP8.5) in 2016–2040 and MPI-ESM-MR (RCP8.5) in 2041–2065. The annual mean change percentage ranges from −12.23% to 18.42% in 2016–2040. The range is significantly wider in 2041–2065, which ranges from −0.6% to 42.2%. Compared to the results in 2016–2040, the precipitation change percentage in 2041–2065 decreases only under MPI-ESM-MR (RCP8.5), while the other 11 scenarios all increase. This implies that future precipitation is likely to increase. Moreover, precipitation in the far future is likely to increase more significantly than in the near future. On the other hand, future precipitation decreases under a few scenarios, which reveals the uncertainties of future precipitation. Overall, future precipitation changes range from −12.23% to 21.85% in 2016–2040 and −4.08% to 42.2% in 2041–2065 relative to 1980–2004 period on the basis of the selected scenarios.

The precipitation differences between emission scenarios are also compared. Among A1B, A2, and B1, the increase in precipitation from largest to smallest is A1B > A2 > B1 on the basis of CSIRO_MK3.0 in both future periods. However, the pattern is not obvious under the other two models. In addition, it is found that half of the models have larger change percentages under RCP8.5 while the other half have larger change percentages under RCP4.5. This implies that climate emission scenarios are not so closely associated with precipitation.

The historical annual series of 1958–2011 data are considered as the baseline to further analyze the precipitation variations. Compared to the historical mean annual precipitation of the baseline period (739.4 mm), the change percentages of the forecasted long series (2016–2065) are no more than ±10% under CMIP3 scenarios. Specifically, the forecasted mean annual precipitation decreases by 0.56, 1.77, and 8.3% under BCCR_BCM2.0 (B1), CSIRO_MK3.0 (A2), and CSIRO_MK3.0 (B1) respectively, while it increases under the other six CMIP3 scenarios. Among the 12 CMIP5 scenarios, the decline in precipitation only appears in the CMCC-CM model, with declines of 8.67% and 6.56% for RCP4.5 and RCP8.5, respectively. The maximum increase appears in the ACCESS1.0 model, while the other scenarios show smaller increases. The increased proportions of precipitation are mostly below 10%, with two exceptions: the ACCESS1.0 models for RCP4.5 and RCP8.5. A comparison of the results is shown in Table 2.

Table 2. Mean annual precipitation and runoff in the historical period (1958–2011) and future period (2016–2065).

Climate Scenario	Precipitation (mm)	Runoff (mm)	Runoff Coefficient	Change Percentage (%)		
				Precipitation	Runoff	Runoff Coefficient
Historical observation	739.43	275.43	0.37	-	-	-
BCCR_BCM2.0(A1B)	741.76	258.55	0.35	0.31	−6.13	−6.43
BCCR_BCM2.0(A2)	739.93	263.35	0.36	0.07	−4.39	−4.45
BCCR_BCM2.0(B1)	735.28	260.24	0.35	−0.56	−5.52	−4.98
CSIRO_MK3.0(A1B)	790.78	300.37	0.38	6.94	9.06	1.97
CSIRO_MK3.0(A2)	726.37	249.02	0.34	−1.77	−9.59	−7.97
CSIRO_MK3.0(B1)	678.03	217.10	0.32	−8.30	−21.18	−14.04
MIROC3.2m(A1B)	805.40	306.29	0.38	8.92	11.20	2.10
MIROC3.2m(A2)	760.50	276.14	0.36	2.85	0.26	−2.52
MIROC3.2m(B1)	793.18	300.71	0.38	7.27	9.18	1.78
ACCESS1.0 (RCP4.5)	841.15	331.45	0.39	13.76	20.34	5.79
ACCESS1.0 (RCP8.5)	909.90	398.39	0.44	23.05	44.64	17.54
BCC-CSM1.1(m)(RCP4.5)	752.82	261.36	0.35	1.81	−5.11	−6.80
BCC-CSM1.1(m)(RCP8.5)	759.89	269.18	0.35	2.77	−2.27	−4.90
CESM1(BGC) (RCP4.5)	741.93	254.00	0.34	0.34	−7.78	−8.09
CESM1(BGC) (RCP8.5)	748.55	264.06	0.35	1.23	−4.13	−5.30
CESM1(CAM5) (RCP4.5)	807.84	307.95	0.38	9.25	11.81	2.34
CESM1(CAM5) (RCP8.5)	761.03	268.87	0.35	2.92	−2.38	−5.15
CMCC-CM (RCP4.5)	675.31	204.83	0.30	−8.67	−25.63	−18.57
CMCC-CM (RCP8.5)	690.96	204.70	0.30	−6.56	−25.68	−20.47
MPI-ESM-MR (RCP4.5)	766.74	266.63	0.35	3.69	−3.20	−6.64
MPI-ESM-MR (RCP8.5)	756.81	248.18	0.33	2.35	−9.90	−11.97

4.3. *Future Evaporation and Runoff Conditions under Climate Changes*

The annual evaporation and runoff of the 21 scenarios in the two future periods are predicted by the SWAT model using the downscaled daily precipitation and temperature data. The predicted evaporation and runoff percentage differences relative to the 1980–2004 period are shown in Figures 11–14, respectively. The horizontal dashed lines in the figures refer to the maximum and minimum multi-year mean evaporation and runoff change percentages, which can be calculated by Equations (8) and (9), show the ranges of change uncertainty.

Figure 11 shows the annual evaporation changes under CMIP3 scenarios. The annual mean change percentage ranges from −1.09% to 5.50% in 2016–2040. The range is slightly wider in 2041–2065, which ranges from −0.85% to 7.21%. Figure 12 shows the annual evaporation changes under CMIP5 scenarios. The annual mean change percentage ranges from −2.05% to 8.90% in 2016–2040. The range is slightly smaller in 2041–2065, which ranges from 5.10% to 12.55%. Compared to the results in 2016–2040, the annual evaporation change percentages are similar under CMIP3 but much larger under CMIP5. Future annual evaporation all increase except CSIRO_MK3.0 (B1) for two future periods and CMCC-CM (RCP4.5) for 2016–2040. Overall, future evaporation changes range from −2.05% to 8.90% in 2016–2040 and −2.70% to 12.55% in 2041–2065 relative to 1980–2004 period on the basis of the selected scenarios. The increase in future evaporation is mainly caused by the rising temperature.

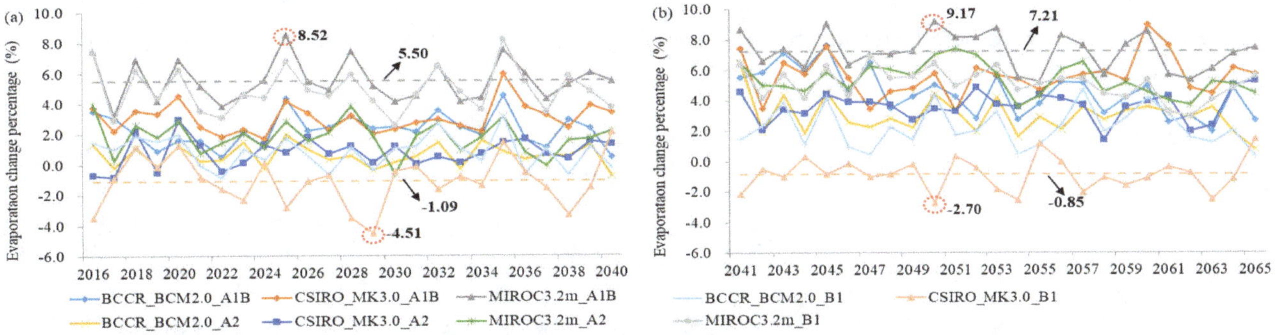

Figure 11. The forecasted evaporation change percentage relative to 1980–2004 period under CMIP3: (**a**) 2016–2040; (**b**) 2041–2065.

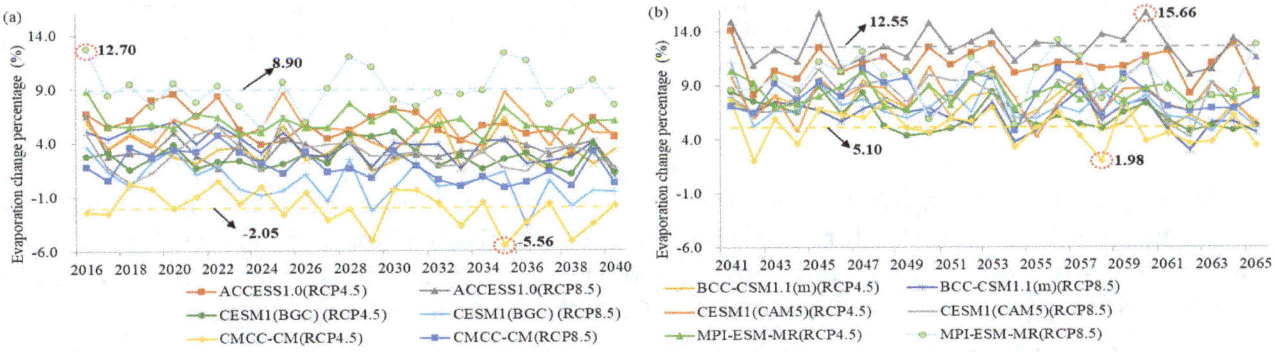

Figure 12. The forecasted evaporation change percentage relative to 1980–2004 period under CMIP5: (**a**) 2016–2040; (**b**) 2041–2065.

Figure 13 shows the annual runoff changes under CMIP3 scenarios. The annual mean change percentage ranges from −13.35% to 46.59% in 2016–2040. The range is slightly smaller in 2041–2065, which ranges from −19.39% to 25.47%. The difference of MIROC3.2m (B1) in the two future periods is the greatest, followed by MIROC3.2m (A1B) and BCCR_BCM2.0 (A2). The other six scenarios differ by less than ±10%.

Figure 14 shows the annual runoff changes under CMIP5 scenarios. The annual mean change percentage ranges from −37.97% to 37.60% in 2016–2040. The range is significantly wider in 2041–2065, which ranges from −24.05% to 90.16%. Compared to the results in 2016–2040, the runoff change percentage in 2041–2065 decreases most under MPI-ESM-MR (RCP8.5), which is also the only scenario under which precipitation decreases. Meanwhile, runoff slightly decreases under BCC-CSM1.1(m) (RCP4.5), CMCC-CM (RCP8.5), and MPI-ESM-MR (RCP4.5), while under the other eight scenarios runoff increases. Overall, future runoff changes range from −37.97% to 46.59% in 2016–2040 and −24.05% to 90.16% in 2041–2065 relative to 1980–2004 period on the basis of the selected scenarios.

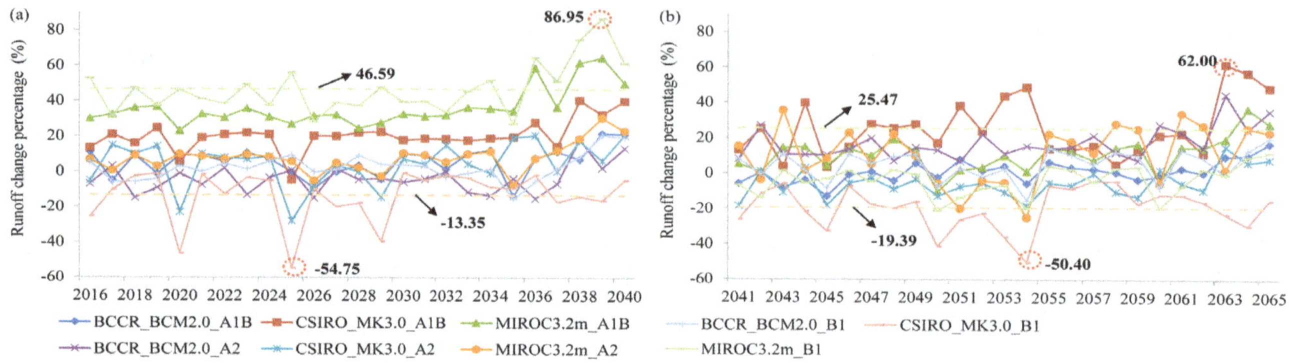

Figure 13. The forecasted runoff change percentage relative to 1980–2004 period under CMIP3: (**a**) 2016–2040; (**b**) 2041–2065.

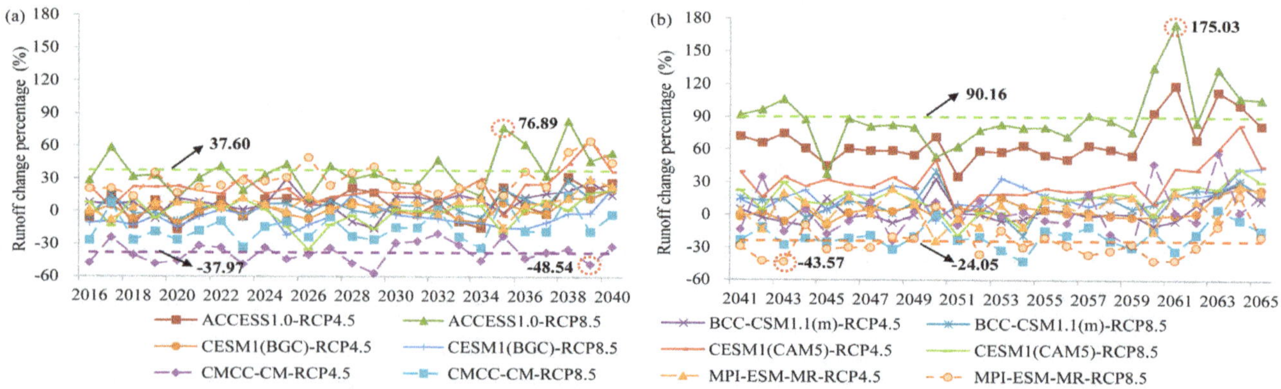

Figure 14. The forecasted runoff change percentage relative to 1980–2004 period under CMIP5: (**a**) 2016–2040; (**b**) 2041–2065.

This implies that changes in future runoff are greater than changes in precipitation under both CMIP3 and CMIP5. However, future runoff changes are relatively similar in the near future and far future under CMIP3, but are much more severe in the far future than in the near future under CMIP5. In addition, the ranges of change in runoff under CMIP5 are much wider than under CMIP3.

Among A1B, A2, and B1, the change of runoff from largest to smallest is A1B > A2 > B1 under CSIRO_MK3.0 in both future periods. However, the pattern is not obvious under the other two models. In addition, it is found that half of the models have larger change percentages under RCP8.5 while the other half have larger change percentages under RCP4.5. The situation is just like that of the precipitation changes, but with different change percentages. This implies that climate emission scenarios are also not so closely associated with runoff.

The forecasted long series (2016–2065) of runoff is compared with the baseline series (1958–2011) in order to thoroughly analyze the future runoff changes which are more practical for decision makers. Compared to the historical mean annual runoff (275.43 mm), the mean annual runoff decreases under five CMIP3 scenarios while it increases under the other four scenarios. Specifically, the scenarios under which mean annual runoff increases include MIROC3.2m for A1B, A2, and B1, and CSIRO_MK3.0 (A1B). The forecasted mean annual runoff increases by up to 11.20% under MIROC3.2m (A1B), while it decreases by up to 21.18% under CSIRO_MK3.0 (B1); this is because the precipitation under these scenarios increases the most and decreases the most respectively. Among the 12 CMIP5 scenarios, runoffs increase in the ACCESS1.0 model for RCP4.5 and RCP8.5, as well as the CESM1 (CAM5) model for RCP4.5, while the other scenarios all have a decline in runoff. Runoff increases by up to 20.34% and 44.64% under the ACCESS1.0 model for RCP4.5 and RCP8.5 respectively, mainly because their precipitation increases the most. On the other hand, runoffs of the CMCC-CM model decrease by up to 25.63% and 25.68% for RCP4.5 and RCP8.5 respectively, which is because the precipitation decreases the most. A comparison of the results is shown in Table 2. Overall, compared to the baseline series, the runoff decreases under two-thirds of the selected scenarios including five CMIP3 scenarios and nine CMIP5 scenarios.

5. Discussion

The morphing results are affected by the characteristics of the 'baseline climate'. However, the morphing results meet the downscaling requirements in this study; in addition, the method is simple and requires only a small computation volume. That is the reason why morphing is utilized in this study.

Compared to 1980–2004 period, both temperature and precipitation increase under most of the scenarios, and they both increase more in the far future than in the near future under both CMIP3 and CMIP5. Moreover, the temperature increases more under CMIP5 than under CMIP3 scenarios. Several studies have shown that CMIP5 presents warmer and wetter predictions [19,20]. Meanwhile, the ranges of change in both precipitation and runoff under CMIP5 are much wider than under CMIP3, which indicates that the selected CMIP5 scenarios are more effective in capturing future runoff uncertainties than the selected CMIP3 scenarios. In addition, the results imply that climate emission scenarios are closely associated with temperature but not so closely associated with precipitation and runoff.

As shown in Table 2, the runoff changes differ from precipitation changes. In most scenarios (15 of the 21 chosen scenarios), the decline in precipitation results in a more significant decline in runoff, while the increase in precipitation results in less increase in runoff. This implies that whether the precipitation increases or decreases, part of the water is failed to form runoff. There are six exceptions when the annual precipitation is more than 790 mm and the increase in runoff is larger than that of precipitation. The six exceptions include three CMIP3 scenarios (CSIRO_MK3.0 (A1B), MIROC3.2m (A1B), and MIROC3.2m (B1)) and three CMIP5 scenarios (the ACCESS1.0 model for RCP4.5, RCP8.5, and CESM1(CAM5) (RCP4.5)), which are the same as the six runoff increase scenarios. Precipitation supplements the soil moisture content when precipitation reaches or exceeds a certain amount. After the soil moisture content is saturated, most of the remaining precipitation forms runoff [31]. The runoff coefficient of precipitation thus increases (as show in Table 2), which may result in the larger increase in runoff than precipitation.

The relatively low flow conditions of the 1980–2004 period may partly result in the decline in future runoff when compared to the baseline series. Three impact factors are further analyzed. First, Yushi Reservoir supplies water to the outside of the basin, about 46 million m^3 each year (about 22.1 mm). This is one cause of the decline in runoff. Second, compared to 1980–2004 period, the simulated future evaporation increased under all the selected scenarios in both future periods except CSIRO_MK3.0 (B1) and CMCC-CM (RCP4.5) with slightly decreases. Evaporation increases greater in the far future than in the near future, which is the same to the temperature increase pattern. The increase in evaporation caused by the rising temperature is another cause of the decline in future runoff. Third, on the basis of water balance in the basin, the upstream water consumption, such as farmers getting water from the stream for irrigation, may be another factor which affects future runoff. Unlike the former two factors, the amount of upstream water consumptions is relatively small, and the impact on runoff is not obvious in these scenarios.

Acknowledging the uncertainty of the simulated runoff using the GCMs in the future [13,20,21,38], it is hard to give an accurate estimate about how much water the reservoir will have. However, the results in this study do show that the runoff in the future could decrease. This decrease could provide important information for the long-term planning of the development of this region. At present, the study region has a few factories that consume much water every year, and the water demand may increase in the future. In addition, the water used for irrigation could also increase in the future. The decreasing runoff (as shown in this study) suggests that this region may suffer from water shortages if industrial and irrigation demands keep increasing in the future. This information is beneficial for regional sustainable development, and indicates that possible adaption measures (e.g., building water transfer projects) should be taken in advance to adapt to climate change.

6. Conclusions

This study investigated the variations of future climate and water resources availability in an important river basin (the Biliu River basin) in northeast China. We found that both temperature and precipitation increase under most of the CMIP3 and CMIP5 scenarios, and CMIP5 shows higher temperature and wider ranges of changes in precipitation and runoff than CMIP3. We also found that the evaporation could increase and the runoff could decrease in the future. The decline in runoff may aggravate water shortages in the Biliu River basin, which may influence the water security in nearby big cities. The results provide very important information for water resources planning and management in this region and could have important implications to the regional sustainable development.

The uncertainties caused by the downscaling method and hydrological model, which are not considered in this article, require further research. In addition, the impact of human activities on future runoff, as well as the comprehensive impact of change conditions, also requires further research. Furthermore, it is necessary to combine these results with the analysis of reservoir regulation rules in order to use this information to make an informed decision about the future water availability for this reservoir.

Acknowledgments: This study was supported by the National Natural Science Foundation of China (grant No. 51509176), Natural Science Foundation of Shanxi Province, China (grant No. 201601D021086 and grant No. 2016011054), and Project of Hydrology Bureau, Shanxi Province (grant No. ZNGZ2015-036). The authors would like to thank the editors and reviewers for their valuable comments and suggestions.

Author Contributions: Xueping Zhu and Chi Zhang conceived and defined the research themes. Xueping Zhu and Wei Qi designed the methods and modeling and conducted the modeling investigations. Xueping Zhu and Wenjun Cai analyzed the data; Wei Qi, Xuehua Zhao, and Xueni Wang contributed to the discussion and analyses. All authors have contributed to the revision and approved the manuscript.

References

1. Menzel, L.; Bürger, G. Climate change scenarios and runoff response in the Mulde catchment (Southern Elbe, Germany). *J. Hydrol.* **2002**, *267*, 53–64. [CrossRef]
2. Hawkins, E.; Sutton, R. The potential to narrow uncertainty in regional climate predictions. *Bull. Am. Meteorol. Soc.* **2009**, *90*, 1095–1107. [CrossRef]
3. Woldemeskel, F.M.; Sharma, A.; Sivakumar, B.; Mehrotra, R. A framework to quantify GCM uncertainties for use in impact assessment studies. *J. Hydrol.* **2014**, *519*, 1453–1465. [CrossRef]
4. Intergovernmental Panel on Climate Change (IPCC). *Climate Change 2007: The Physical Science Basis. Contribution of Working Group I to the Fourth Assessment Report of the Intergovernmental Panel on Climate Change*; Cambridge Universty Press: New York, NY, USA, 2007.
5. Sivakumar, B. Global climate change and its impacts on water resources planning and management: Assessment and challenges. *Stoch. Environ. Res. Risk Assess.* **2011**, *25*, 583–600. [CrossRef]
6. Towler, E.; Rajagopalan, B.; Gilleland, E.; Summers, R.S.; Yates, D.; Katz, R.W. Modeling hydrologic and water quality extremes in a changing climate: A statistical approach based on extreme value theory. *Water Resour. Res.* **2010**, *46*, W11504. [CrossRef]
7. Belcher, S.E.; Hacker, J.N.; Powell, D.S. Constructing design weather data for future climates. *Build. Serv. Eng. Res. Technol.* **2005**, *26*, 49–61. [CrossRef]

8. Milzow, C.; Burg, V.; Kinzelbach, W. Estimating future ecoregion distributions within the Okavango Delta Wetlands based on hydrological simulations and future climate and development scenarios. *J. Hydrol.* **2010**, *381*, 89–100. [CrossRef]
9. Chen, S.T.; Yu, P.S.; Tang, Y.H. Statistical downscaling of daily precipitation using support vector machines and multivariate analysis. *J. Hydrol.* **2010**, *385*, 13–22. [CrossRef]
10. Zhang, C.; Zhu, X.P.; Fu, G.T.; Zhou, H.C.; Wang, H. The impacts of climate change on water diversion strategies into a water deficit reservoir. *J. Hydroinform.* **2014**, *16*, 872–889. [CrossRef]
11. Taylor, K.E.; Stouffer, R.J.; Meehl, G.A. A summary of the CMIP5 experiment design. *Bull. Am. Meteorol. Soc.* **2012**, *93*, 485–498. [CrossRef]
12. Brekke, L.D.; Dettinger, M.D.; Maurer, E.P.; Anderson, M. Significance of model credibility in estimating climate projection distributions for regional hydroclimatological risk assessments. *Clim. Chang.* **2008**, *89*, 371–394. [CrossRef]
13. Hoang, L.P.; Lauri, H.; Kummu, M.; Koponen, J.; van Vliet, M.T.H.; Supit, I.; Leemans, R.; Kabat, P.; Ludwig, F. Mekong River flow and hydrological extremes under climate change. *Hydrol. Earth Syst. Sci.* **2016**, *20*, 3027–3041. [CrossRef]
14. Donat, M.G.; Lowry, A.L.; Alexander, L.V.; O'Gorman, P.A.; Maher, N. More extreme precipitation in the world's dry and wet regions. *Nat. Clim. Chang.* **2016**, *6*, 508–513. [CrossRef]
15. Wang, G.L.; Wang, D.G.; Trenberth, K.E.; Erfanian, A.; Yu, M.; Bosilovich, M.G.; Parr, D.T. The peak structure and future changes of the relationships between extreme precipitation and temperature. *Nat. Clim. Chang.* **2017**, *7*, 268–274. [CrossRef]
16. Wang, X.Y.; Yang, T.; Li, X.L.; Shi, P.F.; Zhou, X.D. Spatio-temporal changes of precipitation and temperature over the Pearl River basin based on CMIP5 multi-model ensemble. *Stoch. Environ. Res. Risk Assess.* **2017**, *31*, 1077–1089. [CrossRef]
17. Koutroulis, A.G.; Grillakis, M.G.; Tsanis, I.K.; Papadimitriou, L. Evaluation of precipitation and temperature simulation performance of the CMIP3 and CMIP5 historical experiments. *Clim. Dyn.* **2016**, *47*, 1881–1898. [CrossRef]
18. Sun, Q.; Miao, C.; Duan, Q. Comparative analysis of CMIP3 and CMIP5 global climate models for simulating the daily mean, maximum, and minimum temperatures and daily precipitation over China. *J. Geophys. Res. Atmos.* **2015**, *120*, 4806–4824. [CrossRef]
19. Ayers, J.; Ficklin, D.L.; Stewart, I.T.; Strunk, M. Comparison of CMIP3 and CMIP5 projected hydrologic conditions over the Upper Colorado River Basin. *Int. J. Climatol.* **2016**, *36*, 3807–3818. [CrossRef]
20. Ficklin, D.L.; Letsinger, S.L.; Stewart, I.T.; Maurer, E.P. Assessing differences in snowmelt-dependent hydrologic projections using CMIP3 and CMIP5 climate forcing data for the western United States. *Hydrol. Res.* **2016**, *47*, 483–500. [CrossRef]
21. Shamir, E.; Megdal, S.B.; Carrillo, C.; Castro, C.L.; Chang, H.I.; Chief, K.; Corkhill, F.E.; Eden, S.; Georgakakos, K.P.; Nelson, K.M.; et al. Climate change and water resources management in the Upper Santa Cruz River, Arizona. *J. Hydrol.* **2015**, *521*, 18–33. [CrossRef]
22. Su, F.; Zhang, L.; Ou, T.; Chen, D.; Yao, T.; Tong, K.; Qi, Y. Hydrological response to future climate changes for the major upstream river basins in the Tibetan Plateau. *Glob. Planet. Chang.* **2016**, *136*, 82–95. [CrossRef]
23. Zhang, C.; Shoemaker, C.A.; Woodbury, J.D.; Cao, M.L.; Zhu, X.P. Impact of human activities on stream flow in the Biliu River basin, China. *Hydrol. Process.* **2013**, *27*, 2509–2523. [CrossRef]
24. Raje, D.; Mujumdar, P.P. Reservoir performance under uncertainty in hydrologic impacts of climate change. *Adv. Water Resour.* **2010**, *33*, 312–326. [CrossRef]
25. Vicuna, S.; Dracup, J.A.; Lund, J.R.; Dale, L.L.; Maurer, E.P. Basin-scale water system operations with uncertain future climate conditions: Methodology and case studies. *Water Resour. Res.* **2010**, *46*, W04505. [CrossRef]
26. Paiva, R.C.D.; Durand, M.T.; Hossain, F. Spatiotemporal interpolation of discharge across a river network by using synthetic SWOT satellite data. *Water Resour. Res.* **2015**, *51*, 430–449. [CrossRef]
27. Matin, M.A.; Bourque, C.P.A. Intra- and inter-annual variations in snow-water storage in data sparse desert-mountain regions assessed from remote sensing. *Remote Sens. Environ.* **2013**, *139*, 18–34. [CrossRef]
28. Glenn, J.; Tonina, D.; Morehead, M.D.; Fiedler, F.; Benjankar, R. Effect of transect location, transect spacing and interpolation methods on river bathymetry accuracy. *Earth Surf. Process. Landf.* **2016**, *41*, 1185–1198. [CrossRef]

29. CLIVAR Exchanges, World Climate Research Programme (WCRP). *WCRP Coupled Model Intercomparison Project—Phase 5 (CMIP5)*; CLIVAR Exchanges No. 56, Vol. 16, No. 2; Indigo Press: Southampton, UK, 2011.
30. Van Vuuren, D.P.; Edmonds, J.; Kainuma, M.; Riahi, K.; Thomson, A.; Hibbard, K.A.; Hurtt, G.C.; Kram, T.; Krey, V.; Lamarque, J.F.; et al. The representative concentration pathways: An overview. *Clim. Chang.* **2011**, *109*, 5–31. [CrossRef]
31. Neitsch, S.L.; Arnold, J.G.; Kiniry, J.R.; Williams, J.R. *Soil and Water Assessment Tool Theoretical Documentation*; Version 2005; US Department of Agriculture (USDA) Agricultural Research Service (ARS): Temple, TX, USA, 2005.
32. Arnold, J.G.; Srinivasan, R.; Muttiah, R.S.; Williams, J.R. Large area hydrologic modeling and assessment. Part 1-Model development. *J. Am. Water Resour. Assoc.* **1998**, *34*, 1–17. [CrossRef]
33. U.S. Environmental Protection Agency (EPA). *Protocols for Developing Nutrient TMDLs*; Office of Water EPA 841-B-99-007; U.S. Environmental Protection Agency (EPA): Washington, DC, USA, 1999.
34. Marshall, E.; Randhir, T.O. Spatial modeling of land cover change and watershed response using Markovian cellular automata and simulation. *Water Resour. Res.* **2008**, *44*, W04423. [CrossRef]
35. Li, Z.; Liu, W.Z.; Zhang, X.C.; Zheng, F.L. Impacts of land use change and climate variability on hydrology in an agricultural catchment on the Loess Plateau of China. *J. Hydrol.* **2009**, *377*, 35–42. [CrossRef]
36. Van Griensven, A.; Meixner, T.; Grunwald, S.; Bishop, T.; Diluzio, M.; Srinivasan, R. A global sensitivity analysis tool for the parameters of multi-variable catchment models. *J. Hydrol.* **2006**, *324*, 10–23. [CrossRef]
37. Hao, F.H.; Cheng, T.G.; Yang, S.T. *Non-Point Source Pollution Model*; China Environmental Science Press: Beijing, China, 2006.
38. Mcsweeney, C.F.; Jones, R.G.; Booth, B.B.B. Selecting ensemble members to provide regional climate change information. *J. Clim.* **2012**, *25*, 7100–7121. [CrossRef]

10

Quantifying the Effects of Near-Bed Concentration on the Sediment Flux after the Operation of the Three Gorges Dam, Yangtze River

Li He

Key Laboratory of Water Cycle and Related Land Surface Processes, Institute of Geographic Sciences and Natural Resources Research, Chinese Academy of Sciences, Beijing 100101, China; heli@igsnrr.ac.cn

Abstract: The regime of sediment transport in the Jingjiang Reach has significantly changed from quasi-equilibrium to sub-saturation since the impoundment of the Three Gorges Dam (TGD), and vertical profiles of suspended sediment concentration (SSC) have changed accordingly. Vertical profiles of SSC data measured at three hydrological stations in the Jingjiang Reach (Zhicheng, Shaishi, and Jianli), before and after the impoundment of TGD, were collected and analyzed. Analytic results indicate a remarkably large concentration in the near-bed zone (within 10% of water depth from the river-bed) in a sub-saturated channel. The maximum measured concentration was up to 15 times the vertical average concentration, while the ratio in quasi-equilibrium channel was less than four times that. Concentrations normalized with reference concentration at the same height, and may decrease with increasing values of suspension index (settling velocity over shear velocity). In addition, concentration near the water surface may be larger than concentration in the near-bed region when the suspension index is smaller than 0.01. Sediment flux transported in the near-bed zone may be up to 35% of the total sediment flux in unsaturated flows. The relationship between deviations of estimating sediment flux when ignoring the near-bed concentration and discharge in flood season and non-flood season are different in unsaturated and quasi-equilibrium channels. Analysis indicates that, in the quasi-equilibrium channel, more attention should be paid to near-bed concentration during non-flood season, the same as measurements during flood season with larger discharge.

Keywords: sediment regime; suspended sediment concentration; vertical profiles of concentration; Jingjiang River Reach; Yangtze River

---

1. Introduction

The majority (>90%) of river-borne flux is closely associated with sediment [1]. A significant proportion of sediments are transported in suspension, as the bed load at river mouths is often less than 1% of the total solid transport [2].

Various vertical profiles of suspended sediment concentration (SSC) have been observed by field observation. For example, these profiles include linear type including linear or quasi-linear (with different gradients), parabolic curve, and mixed linear type [3,4]. For different linear types, the difference between concentration on the water surface and concentration near the bed varies. For linear type, the difference of concentration in the vertical direction increases with increasing slope, and the rate of variation in a vertical direction is constant. Furthermore, the linear types can be observed in estuaries with smaller concentrations [3]. For the parabolic type, the rate of variation in a vertical direction varies, and exists in a relatively larger concentration in the near-bed region. For the parabolic type, the concentration in the near-bed region is a kind of tailing phenomenon, and this kind of vertical profile can be observed in tide water [3]. Various theories, i.e., gravity, diffusion, mixing, energy dissipation, and stochastic models have been applied to simulate the vertical distribution of SSC [5,6]. And various efforts have been focused on the temporal and spatial variation of SSC by field survey, especially the near-surface and near-bed concentration [7,8]. Zuo et al. [3] pointed out that larger concentrations in the near-bed region may be caused by flocculate, saline water and tidal wave. Based on observation in the Jingjiang River Reach after Three Gorges Dam's impoundment, vertical profiles of SSC in channels with a changed sediment regime have revealed a remarkably large concentration in the near-bed region [9]. This indicates that, a larger concentration in the near-bed region can also be observed in a reservoir down-channel [9]. Erosion downstream is a living topic coping with sediment trapping in reservoirs. Varied sediment transport regimes and sediment-related problems in the middle reach (Jingjiang Reach, China) have a profound morphological impact on the lower reach, that is, navigation, pollutant and deposition in channels and ports, water and sediment management in stem channels, social and economic problems and so on [10]. Various studies focus on the changed sediment regime and its influences on the channel downstream (i.e., [11–13]). However, the characteristics of the distinct larger near-bed concentration and its effects on the river reach have not been widely analyzed.

Thus, the study aims to analyze the changed vertical profiles of SSC with changed sediment regimes. Based on vertical profiles of concentration, detailed characteristics of the tailing phenomena are analyzed. Then, the effects of high concentration in the near-bed region on sediment flux are estimated. Finally, the vertical distribution of suspended sediment concentration and its effects after dam operation are compared with data before operation.

2. Materials and Methods

2.1. Study Area and Data

The Yangtze River (YR) is the largest and longest river in China, and the third largest river in the world. The length of the YR is approximately 6.3×10^3 km, and the drainage area is approximately 1.8×10^6 km^2.

The Three Gorges Dam (TGD) is located at the exit of the upper YR, Yichang in Hubei province [14]. The dam is 185 m high and the storage capacity of the reservoir is 3.9×10^8 m^3. The main purposes of the project are flood control, power generation and navigation. It started to impound water in 2003. After 2003, bedload and suspended load from the upper drainage area of the YR are trapped in the reservoir of the TGD. For suspended sediment load (SSL), more than 70% may be trapped during the first 10 years of operation, and approximately half of the SSL may be trapped during the operation over 40–100 years; the ratio may decrease to 15% and 10% after 80 years and 100 years, respectively [15]. Therefore, the SSL entering the middle and lower channel decreases, which leads to erosion of river reach down the dam [16]. During the operation of the TGD, the downstream channel erosion was extensive, and riverbed incision was accelerated [17].

The Jingjiang River is the river reach between Zhicheng and Chenglingji stations, with another two controlling stations of Shashi and Jianli (Figure 1). The length of the Jingjiang River Reach is approximately 348 km, and it is 64 km downstream from the TGD. The length of the upper and the lower part of the Jingjiang River Reach are approximately 172 km and 176 km, respectively [18,19]. During the operation of the TGD, the flow regime and sediment regime of the Jingjiang River Reach have changed [14].

Data measured at these three gauges in the Jingjiang Reach before and after the operation of TGD are collected (Table 1). Observed data are the vertical profiles of concentrations, discharges, wetted area, water depth, temperature, water stages, velocities and corresponding gradations. Data are measured by the Jingjiang Hydrology and Water Resources Surveying Bureau (JHWRSB), and published by the Yangtze River Water Resources Commission (YRWRC).

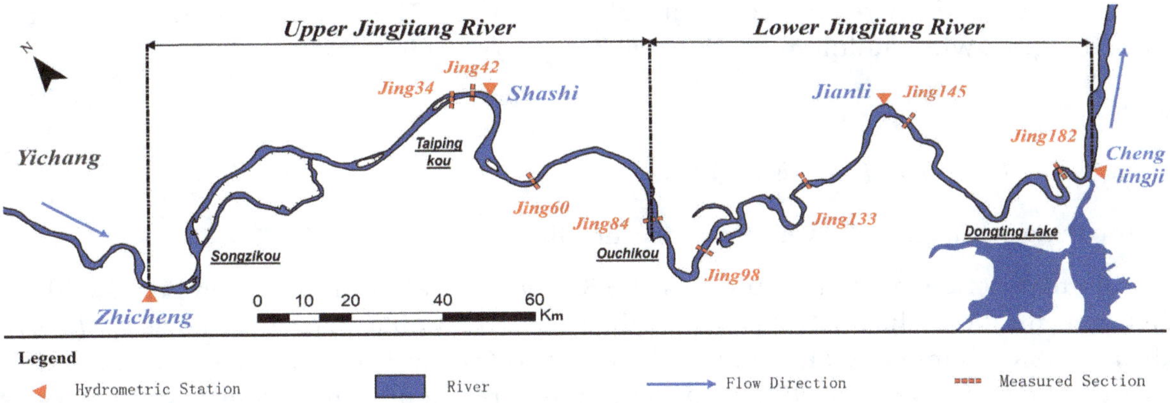

Figure 1. Sketch of the Yangtze River and Jingjiang River Reach.

Table 1. Statics of data measured at the Jingjiang River Reach.

Stations	Measured Year	Groups
Zhicheng	1996, 1998, 2002	55
Shashi	1996, 1998, 2010, 2011, 2012, 2013	130
Jianli	1986, 1998, 2002, 2010, 2011, 2012, 2013	180

Vertical profiles measured before and after 2010 are different. For data measured before dam operation, there are only five measuring points in each vertical line. Typical vertical profiles of data measured before dam operation may be described with relative heights (y/H) of 0.94, 0.8, 0.6, 0.2, and 0.04 (data measured 24 September 2002, at Zhicheng station with distance left of 700 m), where y is the distance of each measured point from the riverbed (m), and H is the averaged water depth (m) of this vertical line. This means that the near-bed region (less than 10% depth) has one measuring point.

For data measured after dam operation, there are seven points in each vertical profile, with typical relative heights (y/H) of 0.98, 0.8, 0.4, 0.2, 0.1, 0.0307, 0.0061 (data measured 11 July 2011, at Jianli station with distance left of 1170 m). The two near-bed points are measured with distance of 0.5 m and 0.1 m from the bed, respectively.

2.2. Equations to Estimate the Effects of Near-Bed Concentration on Sediment Flux

The sediment load by the two near-bed points are compared with the total sediment load of the local vertical profile,

$$R_0 = \frac{Q_{sv(7)} - Q_{sv(5)}}{Q_{sv(7)}} \times 100\% \tag{1}$$

where $Q_{sv(7)}$ and $Q_{sv(5)}$ are vertical sediment flux by seven points and upper five points, respectively.

The result of ignoring these two points can also be estimated. The deviation ratio of sediment transport rate by omitting the two near-bed points can be calculated by forms as:

$$R = \frac{Q_{s(7)} - Q_{s(5)}}{Q_{s(7)}} \times 100\% \tag{2}$$

where $Q_{s(7)}$ and $Q_{s(5)}$ are sediment flux by seven points and five points, respectively. $Q_{s(5)}$ is estimated with two measured near-bed points being omitted artificially. For the cross-sectional estimation, there are only several vertical profiles. The delta-shaped area between the left bank and the first vertical line from the left bank is also considered. The delta-shaped area between the right bank and the last vertical line from the left bank is not considered due to the difficulty of identifying the right bank.

The specific values of data measured during pre- and post-operation may vary in different vertical lines, but the differences are limited. These vertical profiles measured before dam operation missed the near-bed region due to what has been done after operation. They are interpolated and extended to the near-bed region, with minimum relative height (y/H) of 0.006.

3. Results

3.1. Distinct Non-Uniform Vertical Distribution

3.1.1. Comparing with Vertical-Averaged Concentration

In order to describe the vertical profiles of SSC, the vertical profiles are compared with vertical-averaged concentration. The relative height of the vertical coordinate is expressed as y/H, in which y is the distance of each measured point from the riverbed (m), and H is the averaged water depth (m) of this vertical line. The relative concentration in the horizontal coordinate is expressed as S_i/S_{avg}, where S_i is the measured concentration of the vertical profile (kg/m^3), and S_{avg} represents the vertical averaged concentration (kg/m^3). The minimum distance of measuring points from the riverbed can be named as the reference height, and the corresponding concentration is the reference concentration.

Figure 2 compares the vertical distribution of sediment concentrations in the Jingjiang Reach before and after the TGD impounded water in 2003. Figure 2 shows that, after the dam operation, measured vertical profiles of suspended sediment at these two gauges show a distinctly high concentration in the near-bed zone.

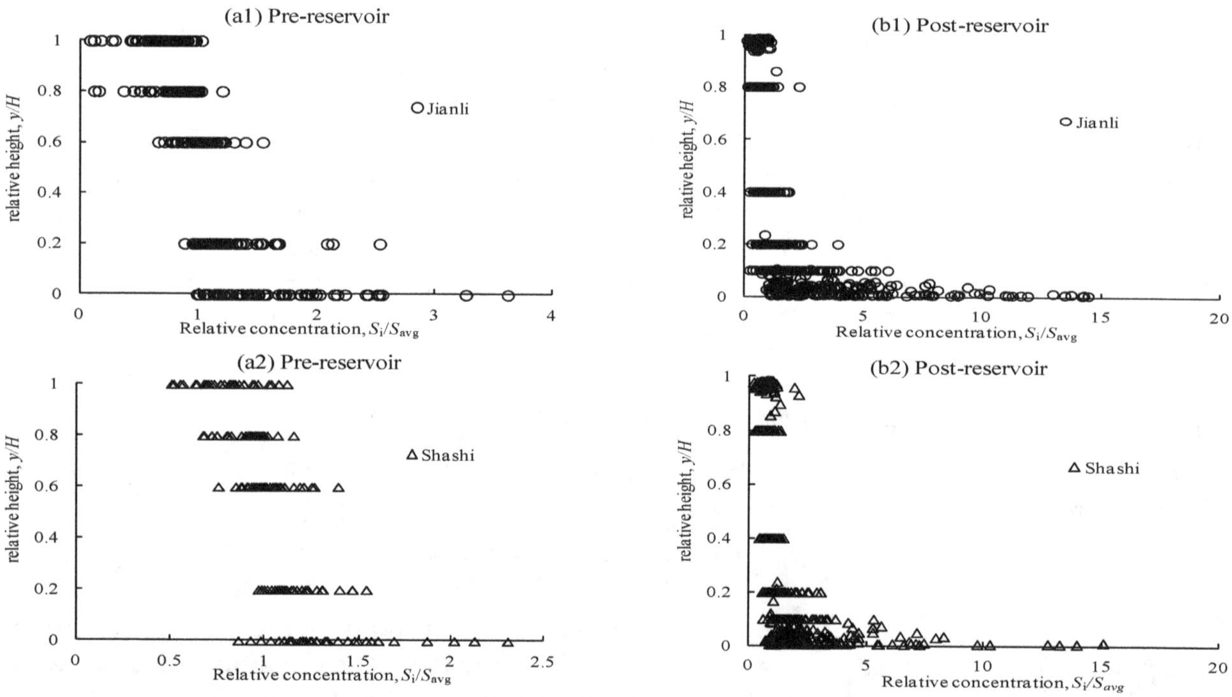

Figure 2. Vertical distributions of sediment concentrations at Shashi and Jianli hydrological stations; data measured before (**a**) and after (**b**) the Three Gorges Dam (TGD) operation in 2003. S_i is the measured concentration of the vertical profile (kg/m^3), and S_{avg} represents the vertical averaged concentration (kg/m^3).

For data measured after operation, the minimum relative height of measured points from the riverbed (y/H) is approximately 0.0051. The maximum relative concentrations (S_i/S_{avg}) are approximately 14.45 (Jianli) and 15.05 (Shashi). This kind of distinctly high concentration in the near-bed zone can be named as "tailing phenomena". It indicates that more sediment is transported in areas near the riverbed; the measured two near-bed points make it much worse. According to the data, average concentrations of these two near-bed points may account for up to 1.86 times of vertical averaged concentration at Shashi hydrological station, and 2.73 times at Jianli hydrological station.

For data measured before operation, the minimum relative height of measured points from the riverbed (y/H) is approximately 0.014 (0.3 m from the riverbed). The maximum relative concentrations (S_i/S_{avg}) are approximately 3.62 (Jianli) and 2.30 (Shashi). The maximum relative concentrations (S_i/S_{avg}) for Zhicheng station is approximately 2.76. For Zhicheng station, there are no data measured after the dam operation. Thus, data measured at Zhicheng are not included in Figure 2. Data measured before dam operation can be viewed as short-tail tailing phenomena.

Relatively high concentrations near the bed region have also been pronounced. Relatively larger near-bed concentrations can also be observed in the estuary of Yangtze River, Qiantangjiang River and Taizhou sea area [20,21]. Figure 3 illustrates relative concentrations (S_i/S_{avg}) measured at several regions, which also demonstrated a kind of tailing phenomena. Relative concentration (S_i/S_{avg}) is estimated with measured concentration over vertical averaged concentration, as in Figure 2. Data measured at Yangtze Estuary are drawn from Zhang [20], data measured at Wuhan and Jiujiang River Reach are drawn from Tang et al. [22], and data at Oujiangkou and Taizhou sea areas are drawn from Dong et al. [21]. The maximum relative concentration (S_i/S_{avg}) is approximately 4.3, and the minimum relative height of measured points from the riverbed (y/H) is approximately 0.03. Therefore, this kind of vertical profile can be viewed as short-tail tailing phenomena. The tailing phenomena which is much apparent in the Middle Jingjiang River Reach can be viewed as long-tail tailing phenomena. The near-bed concentration of the S-type vertical distribution is also relatively larger [23]. The tailing phenomena in the estuary area can be observed during tide periods and non-flood seasons [20].

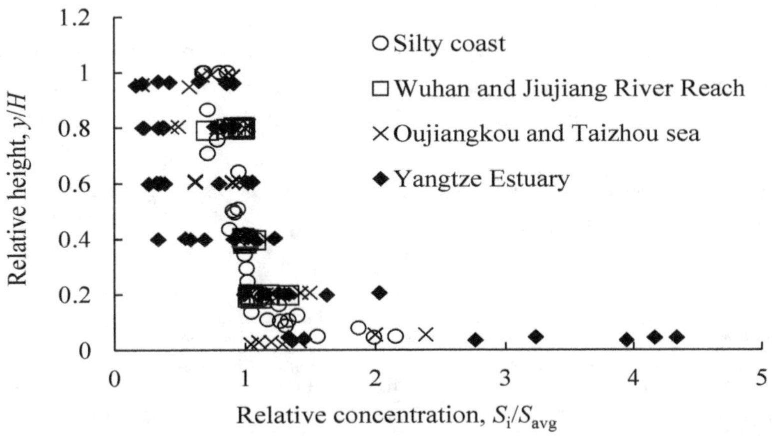

Figure 3. Vertical distribution of sediment concentration measured at different regions. Data at Yangtze Estuary are drawn from Zhang [20], data at Wuhan and Jiujiang River Reach are drawn from Tang et al. [22], data at Oujiangkou and Taizhou sea area are drawn from Dong et al. [21], and data at silty coast are drawn from Xia et al. [24]. S_i is the measured concentration of the vertical profile (kg/m^3), and S_{avg} represents the vertical averaged concentration (kg/m^3).

3.1.2. Comparing with Near-Bed Point

In order to highlight the concentration near the bed, the vertical profiles of suspended sediment load are compared with concentration of the near-bed point. The horizontal coordinate is S_i/S_{ia}, where S_{ia} represents the measured concentration of the near-bed point (kg/m^3). Only data measured after the operation are analyzed.

According to S_i/S_{ia}, the concentration near the bed is not absolutely larger than that of upper points, as shown in Figure 4. Vertical profiles can be grouped into two kinds: the first one is profiles in which concentration of the near-bed point is the largest (Type I), and the second one is profiles in which maximum concentration occurred in the near-bed region (Type II). All the vertical profiles observed in Zhang [20], Tang et al. [22], and Dong et al. [21] can be grouped as Type I with short-tail.

Figure 4 illustrates the vertical profiles measured after dam operation at Shashi and Jianli stations. For data measured after operation, approximately 27% of vertical profiles can be grouped as Type II (Jianli), and the ratio for Shashi station is approximately 48% (Table 2). The suspension index (ω/u_*) in Table 2 is estimated with settling velocity (ω) over shear velocity (u_*). According to Table 2, vertical profiles of Type I have a relatively larger chance of occurring in flood season, while vertical profiles of Type II have a relatively larger chance of occurring in non-flood season.

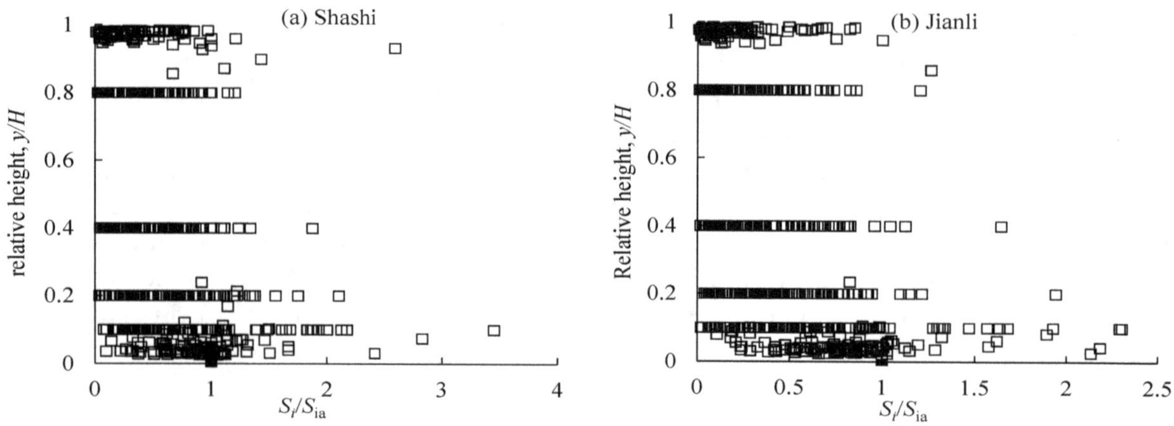

Figure 4. Vertical concentrations comparing with near-bed point concentration at Shashi (**a**) and Jianli (**b**) hydrological stations (data measured after dam operation). S_i = measured concentration of the vertical profile (kg/m^3), and S_{ia} = measured concentration of the near-bed point (kg/m^3).

Table 2. Characteristics of vertical profiles of Type I and Type II measured at Shashi and Jianli hydrological stations (after dam operation).

Stations	Type I						Type II					
	Flood Season			Non-Flood Season			Flood Season			Non-Flood Season		
	Amount of Group	Range of (ω/u_*)	D_{50} (mm)	Amount of Group	Range of (ω/u_*)	D_{50} (mm)	Amount of Group	Range of (ω/u_*)	D_{50} (mm)	Amount of Group	Range of (ω/u_*)	D_{50} (mm)
Shashi	33	0.16–95.38	0.01	17	2.09–108.89	0.06	20	0.16–1.37	0.03	25	1.84–95.67	0.07
Jianli	47	0.14–76.15	0.06	29	2.38–61.64	0.10	10	0.13–32.58	0.06	19	1.38–63.92	0.10

For Type I, five typical profiles are shown in Figure 5a, and five typical profiles of Type II are shown in Figure 5b. This shows that the concentrations of the near-bed region of Type I vary with different coefficients (ω/u_*). For type II, the vertical distribution of concentration is much more complex, while concentration with relative height less than 0.04 is relatively large.

Figure 6 illustrates the vertical distribution of S_i/S_{ia} with different (ω/u_*). In order to make it more readable, the coefficient in Figure 6 is (ω/u_*). It shows that, the concentration of S_i/S_{ia} at the same relative height decreases with increasing coefficients (Figure 6a). When the coefficient decreases to less than 0.01, vertical profiles of Type I change to Type II (Figure 6b). One thing that should be pointed out is that lines in Figure 6 are estimated with data measured in flood seasons with Type I at Shashi station. The lines may vary with different data adopted, but the rules are the same: the concentration of S_i/S_{ia} at the same relative height (y/H) decreases with increasing coefficient (ω/u_*), and Type I may change to Type II with a coefficient less than 0.01.

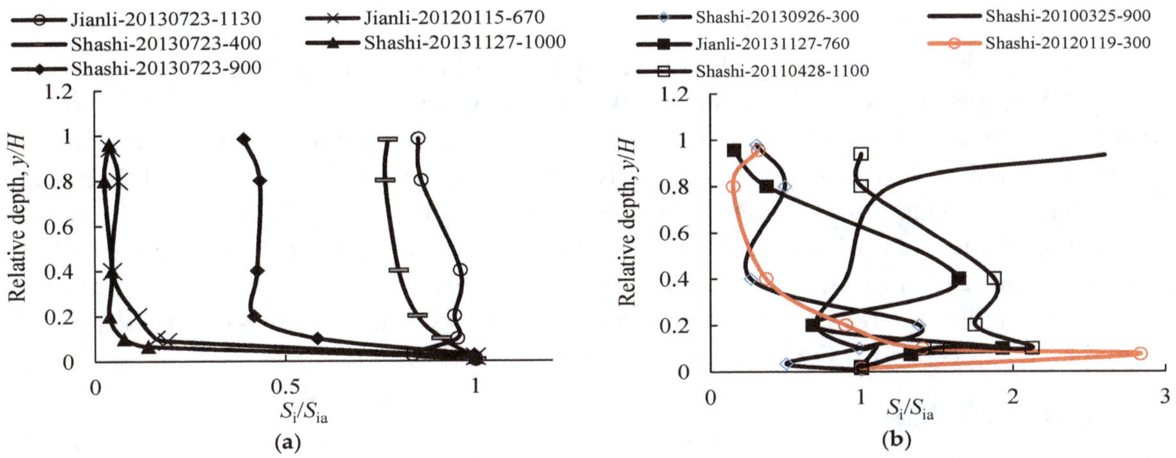

Figure 5. Typical vertical profiles of type I (**a**) and type II (**b**) with tailing phenomena. The legends are composed of station name, date of measurement, and lateral distance of vertical line from left bank. S_i = measured concentration of the vertical profile (kg/m^3), and S_{ia} = measured concentration of the near-bed point (kg/m^3).

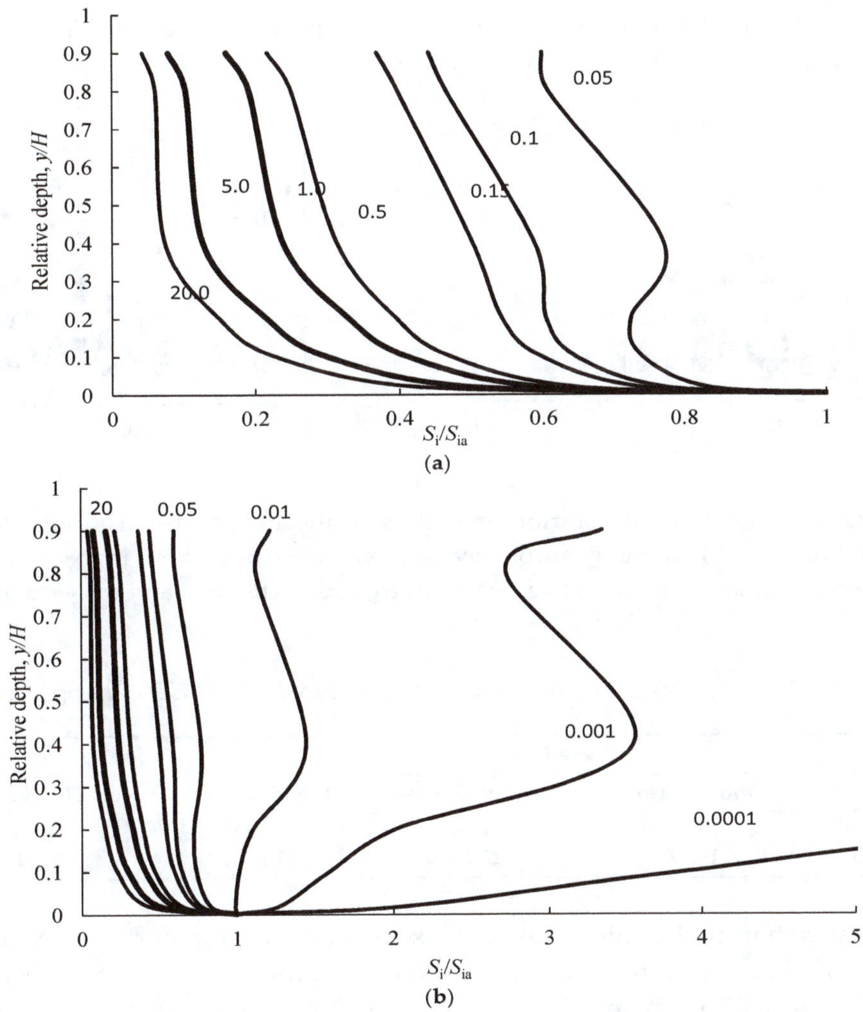

Figure 6. Vertical distribution of S_i/S_{ia} with different ω/u_*, (**a**) for $(\omega/u_*) > 0.05$ and (**b**) for $(\omega/u_*) < 0.0001$. ω = settling velocity, u_* = over shear velocity, S_i = measured concentration of the vertical profile (kg/m^3), and S_{ia} = measured concentration of the near-bed point (kg/m^3).

3.2. Effects on Sediment Flux

Suspended sediment transport is the most significant factor influencing estuaries in emorphodynamics, yet it is often one of the largest unknowns. As to sediment trapped by the TGD and changed sediment regime, the vertical distribution of suspended sediment varies in the Jingjiang River. The estimation of sediment flux of the Jingjiang River may also be influenced, which is important to navigation, channel management of the whole river system, and so on.

The tailing phenomena means that, sediment concentration within 10% of water depth from the river-bed cannot be ignored, otherwise it may lead to different results when estimating erosion or deposition by sediment-transport balance methods and volume methods. Thus, the effects of remarkably larger concentration in the near-bed region are analyzed.

3.2.1. Vertical Sediment Flux

In Figure 7, the B_i is the lateral distance of measured vertical lines from the left bank, and B_{avg} is the maximum width of the cross section. The maximum width of the cross section is assumed to be the maximum lateral distance from the left bank, namely the distance of the last vertical profiles from the left bank. It shows that these two kinds of vertical profiles may occur in the whole lateral cross section. The maximum R_0 value is larger than 47%. The average ratios for these two, type profiles at Shashi station are 18% (Type I) and 13% (Type II). The average ratios for these two, type profiles at Jianli station are 22% (Type I) and 18% (Type II). Table 3 indicates that the R_0-values for non-flood season are relatively larger than those of flood season.

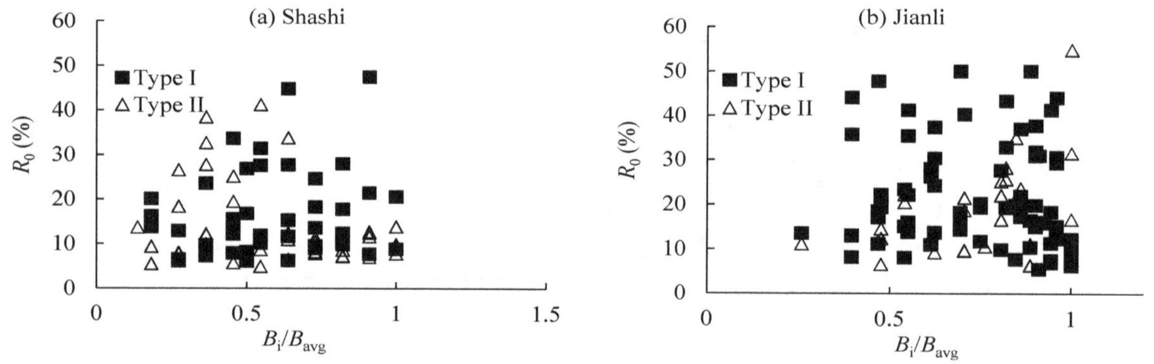

Figure 7. Lateral distribution of contribution ratios of the two near-bed points, (**a**) for Shashi hydrological station, and (**b**) for Jianli hydrological station. R_0 = contribution ratio of the two near-bed points, B_i = lateral distance of measured vertical lines from the left bank, and B_{avg} = maximum width of the cross section.

Table 3. Vertical averaged R_0-values of data measured after TGD dam operation.

Stations	Type I		Type II	
	Flood Season	**Non-Flood Season**	**Flood Season**	**Non-Flood Season**
Shashi	15.67	21.28	12.51	14.10
Jianli	19.57	27.70	16.94	19.61

Figure 8 indicates that the R_0 value may increase with increasing particle size and ω/u_*.

Zhang [20] also pointed out that the near-bed concentration may affect the estimated sediment load when $A = \omega/(\beta k u_*) > 0.15$, where ω is the settling velocity of uniformed particles, k is the Karmen coefficient, u_* is the shear velocity of approaching flow, and β is a coefficient for non-uniform sediment. When suspension index $A = \omega/(\beta k u_*)$ equals two, suspended sediment may concentrate in an area with a distance of 0.2 times the water depth from the riverbed, and the ratio of near-bed transportation over the total transportation (R) is approximately 30% [20]. Tang et al. [22] pointed out that, the near-bed concentration may be more apparent when $\omega/(ku_*) > 5$. The values of ω/u_* are 0.06, 0.8 and 2 (with $k = 0.4$ and $\beta = 1$) by Zhang [20] and Tang et al. [22], respectively. For data illustrated in Figure 8, the values of ω/u_* range from 0.001 to 0.201. Zhang [20] also pointed out that the tailing phenomena is more apparent with larger values of relative particle size (D_i/D_m, particle-size over averaged particle size) of non-uniform sediment. The ratio (R) is approximately 25% when $D_i/D_m > 3$ [20].

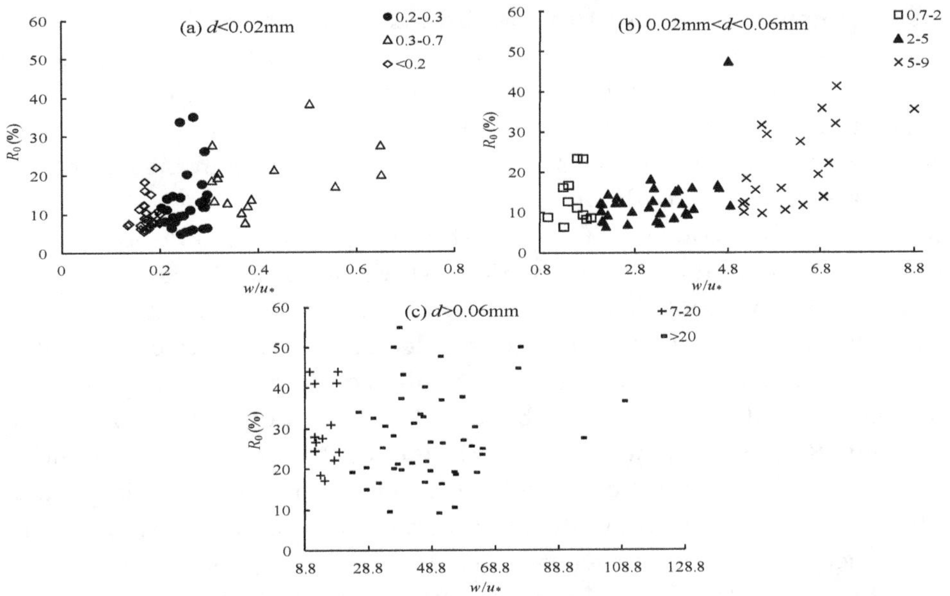

Figure 8. Relationship between R_0, ω/u_* and particle diameters, (**a**) $d < 0.02$ mm, (**b**) 0.02 mm $< d <$ 0.06 mm, and (**c**) $d > 0.06$ mm. The legends are the values of ω/u_*. R_0 = contribution ratio of the two near-bed points, ω = settling velocity, and u_* = shear velocity.

In total, initiation of motion, suspension threshold, and the vertical distribution of sand concentration in the water column all need to be determined for an accurate estimate of sediment transport.

3.2.2. Sectional Sediment Flux

Due to the larger contribution ratio of the two near-bed points (R_0) in Figure 7, the contributions on sectional sediment flux can also be estimated. Figure 9 shows that, if the two near-bed points are not included, the sediment flux may be underestimated by approximately 40%. During flood season, the R-value may increase with increasing discharge, while the largest R-value may occur by measures in non-flood season. In total, the larger average R-value for flood season at Shashi station may be caused by larger discharge (Table 4).

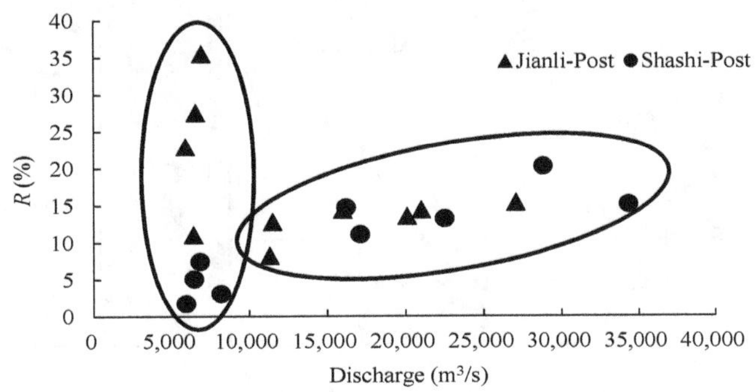

Figure 9. Deviation ratio of sediment transport rate by data measured after TGD dam operation. The two ellipses indicate non-flood season and flood season, respectively. R = contribution ratio of the two near-bed points on vertical sediment flux.

Table 4. Sectional averaged *R*-values of data measured after TGD dam operation.

Stations	Total	Flood Season	Non-Flood Season
Shashi	10.21	14.88	4.36
Jianli	17.56	13.10	24.24

This described situation may result in the different erosion amounts by sediment-transport balance method and volume method [25]. Yuan et al. [26] showed that, the contribution ratio by the near-bed concentration varies by different hydrological years and river reach. For a certain hydrological station, the contribution ratios may also vary with the inflow condition. Therefore, the impact of high concentration near the bed surface must be given more attention in the future, especially during non-flood season and flood season with larger discharge.

The influence of near-bed concentration on estimating the annual deposition by the sediment-balance method has also been analyzed by various studies [26,27]. Based on hydrology data and geometry data measured between 2002 and 2008, contributions of factors that may cause differences were distinguished. For the Yichang–Zhicheng River Reach, there are three kinds of factors that may contribute to deposition estimation with sediment-balance method, namely, sand dredging in the channel, the near-bed concentration, and non-balanced water runoff. For the Zhicheng–Shashi River Reach and Shashi–Jianli River Reach, there are four kinds of contributing factors, namely, sand dredging in the channel, the near-bed concentration, non-balanced water runoff, and simplified measurement.

Figure 10 illustrates the contribution ratios of different factors. The two ratios in Figure 10 are the contribution ratio by near-bed concentration over the total influencing factors, and contribution ratio by total influencing factors on the total deposition estimation. It shows that, for estimating the deposition with sediment-balance method, the contribution by near-bed concentration cannot be ignored, with ratios of 12%, 9% and 18%, respectively. As to its large magnitude of total deposition, the net deposition caused by near-bed concentration is 1.267×10^7 m^3, 9.80×10^9 m^3 and 2.303×10^7 m^3, respectively.

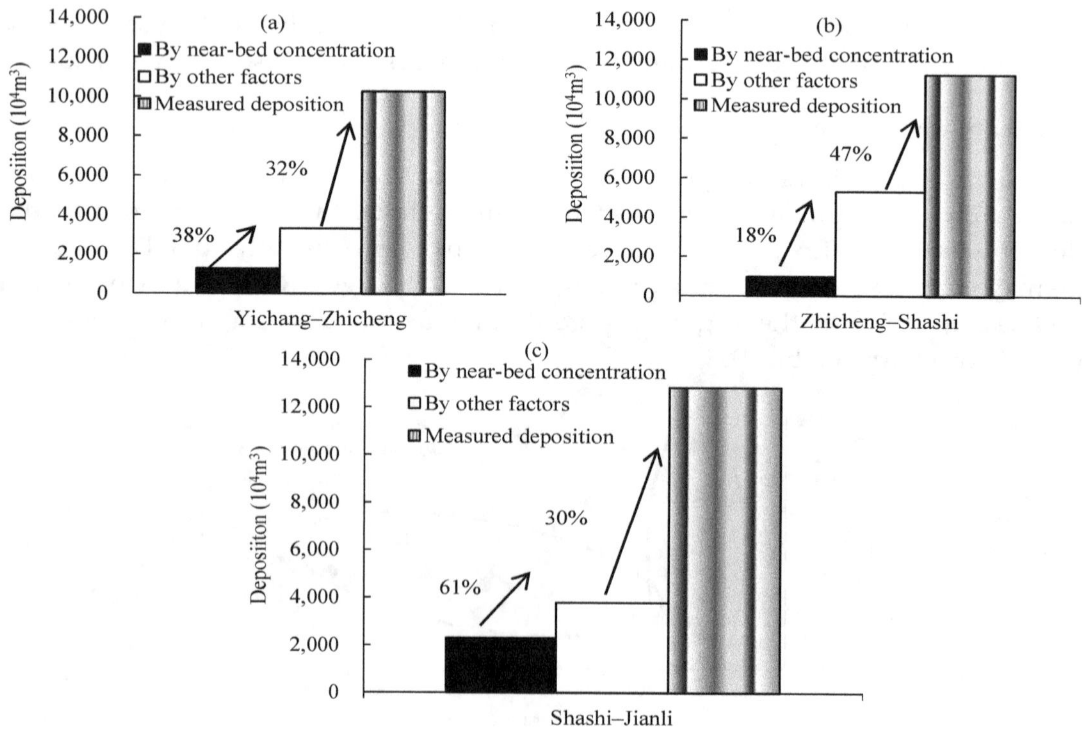

Figure 10. Contributing ratios of near-bed sediment concentration on total deposition estimation, at three sites, (**a**) the Yichang–Zhicheng, (**b**) the Zhicheng–Shashi, and (**c**) the Sahshi–Jianli reaches, based on data measured during 2002–2008 (from Yuan et al. [26]).

Dam construction has primarily contributed to sediment zonation (with different patterns of size distribution) weakening in the Jingjiang reaches (from Yichang to Chenglingji) over the last half-century [28]. The percentage of finer grouped particles (diameter less than 0.01 mm) decreased from 60% (2002) to approximately 40% (2008), and the medium sediment diameter increased from 0.052 mm (2002) to 0.081 mm (2008) [29]. The operation of the TGD and the ensuing decline of SSC was the main reason for the coarsening of the bed sediment in 1977–2003, especially at Shashi station (located 173 km downstream from Yichang station) [28]. It indicates that the coarser particulates had settled in the reservoir, and the river channel between Yichang and Shashi stations was badly eroded by clean water from the reservoir [28]. The coarser eroded sediment gradually settled along the channel between Shashi and Hankou stations due to the decrease in slope and current [28].

4. Discussion

Based on data measured after dam operation, the vertical profiles of SSC in the starved channel show a distinct tailing phenomena. Furthermore, it also influences the estimation of sediment flux. Data measured before dam operation are also collected to make a comparison, between both the vertical distribution and effects on sediment flux estimation.

4.1. Comparing with Data Measured Before Dam Operation

4.1.1. Vertical Profiles of SSC

Tailing phenomena can also be pronounced, however, the tailing phenomena is much more limited (relative concentration of 3.6 for Jianli and 2.3 for Shashi, as shown in Figure 5), the same as verticals by Zhang [20], Tang et al. [22], and Dong et al. [21]. Thus, vertical profiles before operation can be termed as short-tailing.

For data measured before operation, more data are measured in flood season. Approximately 20% (Jianli, Shashi, and Zhicheng stations) of vertical profiles can be grouped as Type II. In total, the occurring frequency of Type II increased after the operation of the TGD, especially at Shashi station (Table 5). In addition, the value of (ω/u_*) increases dramatically.

Table 5. Characteristics of vertical profiles measured at Shashi and Jianli hydrological stations before TGD dam operation.

Stations	Flood Season		Non-Flood Season	
	Amount of Group	Range of (ω/u_*)	Amount of Group	Range of (ω/u_*)
Shashi	27	0.08–0.46	9	0.23–1.34
Jianli	53	0.02–0.78	22	0.05–12.12

Based on theoretical analysis, Zhang and Tan [30] stated that, if the concentration of coarser particles (eight times the finer particles) is larger than 33.5 kg/m^3, the vertical distribution of concentration for finer particles may be different from that of the distribution of uniform particles, with $dS_f/dy > 0$ (Figure 11). It also stated that this kind of distribution of particles finer than 0.01 mm can be observed in Tongguan and Huayuankou stations, Yellow River.

Figures 12 and 13 illustrate the coarseness of suspended sediment and bed load before and after dam operation. For Shashi station, the percentage of particles coarser than 0.008 mm increases after the dam operation (Figure 12). For Jianli station, the percentage of particles coarser than 0.062 mm increases after the dam operation (Figure 12). The coarsening of bed materials after operation is apparent, especially at Shashi station (Figure 13). The coarsened suspended sediment and bed materials lead to the occurrence of remarkably large concentration in the near-bed region.

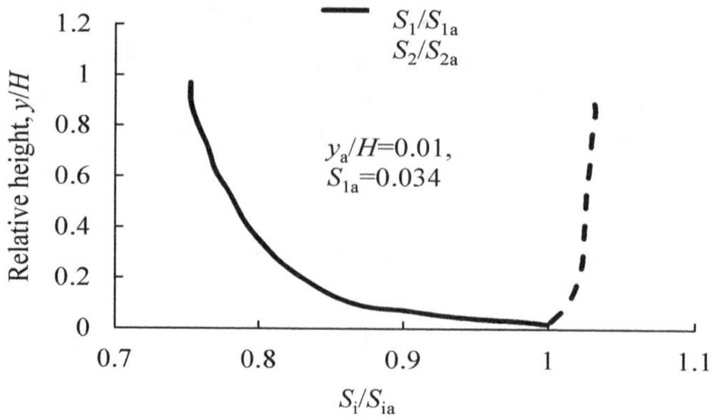

Figure 11. Calculated vertical sediment concentrations comparing with near-bed point concentration (After Zhang and Tan [30]). S_i = measured concentration of the vertical profile (kg/m^3), and S_{ia} = measured concentration of the near-bed point (kg/m^3).

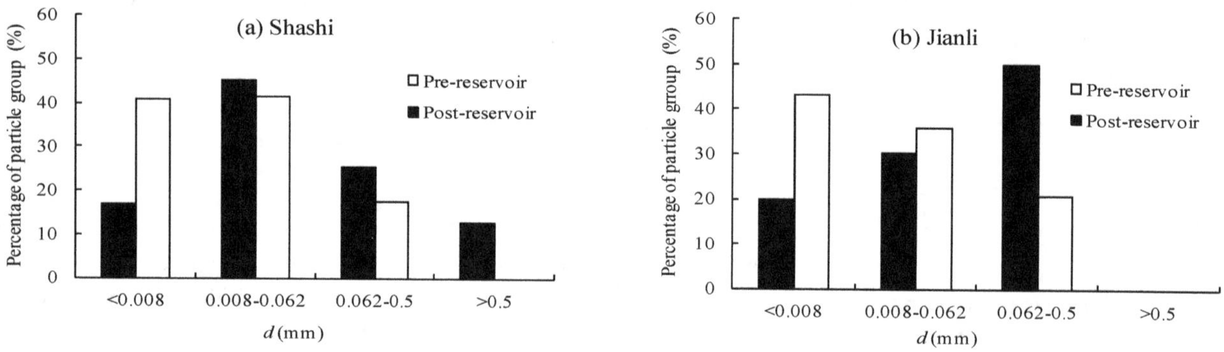

Figure 12. Percentages of particle groups before and after TGD dam operation at Shashi (**a**) and Jianli (**b**) hydrological stations. Data for pre-reservoir are average values between 1991 and 2002, and data for post-reservoir are average values between 2003 and 2016.

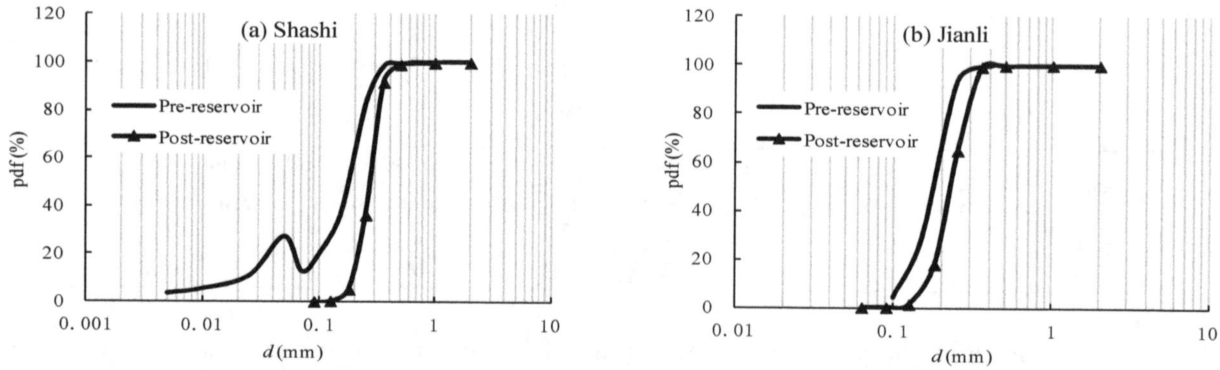

Figure 13. Bed load of Shashi (**a**) and Jianli (**b**) hydrological stations. Data in (**a**) were measured 9 July 1986 at Xinchang hydrological station, and 6 June 2013 at Shashi hydrological station; data in (**b**) were measured 26 May 1986 and 13 March 2013 at Jianli hydrological station.

4.1.2. Sediment Flux

The interpolation made for data measured before dam operation has extended five point profiles to seven point profiles (with almost the same relative height of data after operation). Then, the deviations of flux estimation can also be calculated during the cross-sectional estimation. The maximum ratio may be nearly 35% (Figure 14). The maximum R-values may have occurred during non-flood season, and more attention should be paid on measuring during non-flood season (Table 6). The variances of R-value before and after dam operation are almost the same. However, the R-values during flood season may remain constant with increasing discharge for data measured before operation.

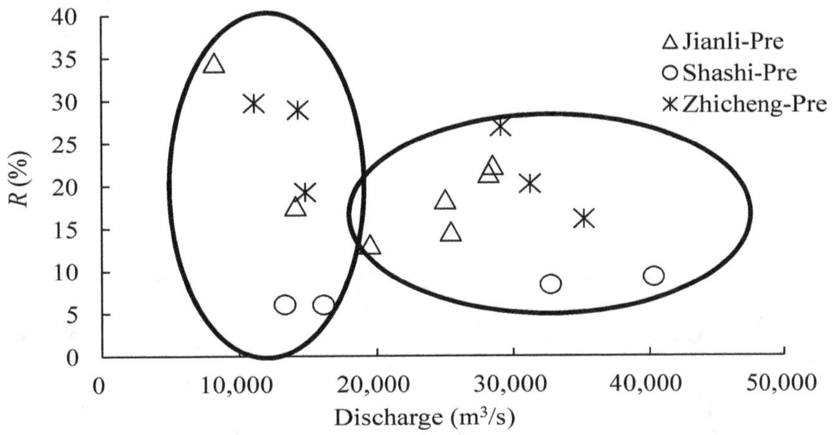

Figure 14. Deviation ratio of sediment transport rate by data measured before TGD dam operation. The two ellipses indicate non-flood season and flood season, respectively. R = contribution ratio of the two near-bed points on vertical sediment flux.

Table 6. Sectional averaged R-values of data measured before TGD dam operation.

Stations	Total	Flood Season	Non-Flood Season
Zhicheng	23.50	21.04	25.86
Shashi	7.43	8.81	6.04
Jianli	32.95	17.96	45.93

During the estimation of sectional averaged sediment flux, there is an assumption that may lead to uncertainties of estimation. For data measured before operation, the extension and interpolation may also lead to uncertainties. This kind of discrepancy can be verified by comparison between estimated and measured sectional discharge and concentration, as shown in Figure 15.

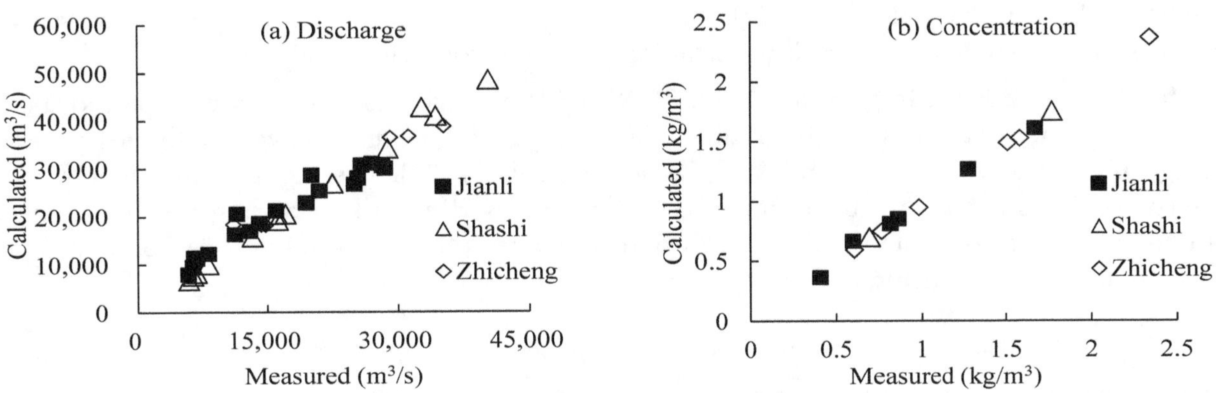

Figure 15. Comparison between measured and calculated sectional discharge (**a**) and concentration (**b**).

4.2. Contribution of Large Floods

The characteristics of suspended sediment concentration and yield at the event scale have been widely analyzed in various environments worldwide [31–33]. Floods are relevant for most of the suspended sediment load [34]. While sediment transport by floods correlated with flood regime, different flood regimes lead to different sediment transportation. For instance, analysis of a typical agro-catchment of the Loess Plateau shows that, the contribution of accumulative total sediment yield by the different flood regimes to the summed sediment yield of all of the examined 158 events are 4%, 13%, 6%, 21%, and 56% for Regimes A to E, respectively [35]. Thus, the flood regimes are analyzed and shown in Figure 16.

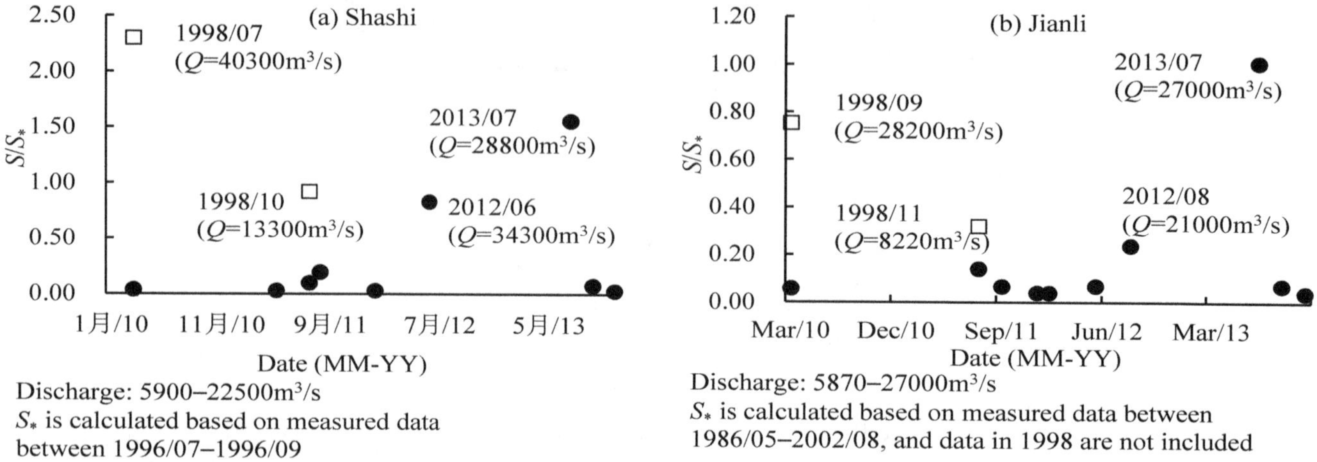

Figure 16. Sediment regimes of floods occurring before and after the operation of the TGD, the Jingjiang River Reach. (**a**) for Shashi Hydrological Station, and (**b**) for Jianli Hydrological Station, S = average concentration, and S_* = average carrying capacity.

Measured data during flood period, such as the floods in 1998, 2013 and 2012, show an over-saturated state, and after the peak flood, the regime may change to pre-equilibrium or starved (Figure 16, Table 7). Figure 16 shows that, the relationships between discharge and concentration measured in floods with high discharges are different from those of floods with low discharges.

Shifts in the discharge-rating curves after three extreme flood events have also been observed by Whitaker et al. [36]. This indicates that the Jingjiang Reach of the YR has changed its sediment regime from pre-equilibrium to starved of sediment because of the operation of the TGD, and the flood regime may be different from the normal sediment regime. Shifts in the discharge-rating curves caused a period of channel instability [36]. Higher suspended sediment concentration may reduce the channel capacity and lead to channel erosion [37]. Finally, the fluvial erosion intensity during flood seasons is highly correlated with a corresponding incoming sediment coefficient [19].

The 1998 flood is the largest flood in the YR's history since China's economy entered the high-speed development period at the beginning of the 1980s, and it caused severe damage to the basin, causing losses exceeding 2×10^{10} RMB (Chinese currency) [38]. The higher water levels also led to overbank flows in the middle Yangtze River [39]. The sediment flux during this flood has been estimated [40]. The water stage in the 1998 flood was higher than that in the 1954 flood, despite the smaller total discharge during the 1998 flood.

Table 7. Characteristics of major floods in China in 1998 (Data adapted from Zhou and Chen [41]).

Stations	Flood Volumes 10^9 m^3		Maximum Peak Flow Discharge		Maximum Flood Stage	
	In 30 Days	In 60 Days	Discharge (m^3/s)	Date	Stage (m)	Date
Yichang	137.9	254.5	63,600	16/8	54.49	16/8
Shashi	–	–	53,700	17/8	45.22	17/8
Jianli	–	–	45,200	17/8	38.31	17/8
Luoshan	–	–	68,600	27/7	34.95	20/8
Hankou	175.4	336.5	72,300	19/8–20/8	29.43	20/8
Jiujiang	–	–	–	–	23.03	2/8
Datong	202.7	395.1	82,300	–	16.32	2/8
Nanjing	–	–	–	–	10.14	29/7

The variability in the relationship between sediment concentration and water discharge (namely hysteretic patterns) has been used to explain the variability of sediment sources from one flood to another [42]. Whitaker et al. [36] compared the estimated sediment yield for an unsampled flood event, a single-event surrogate SSC–Q rating curve, and a long-term SSC–Q rating curve. Analysis revealed that, the long-term SSC–Q rating curve is estimated to be approximately 37% greater than the single-event SSC–Q estimates, which indicates a high degree of uncertainty. Analysis shows a high variability in the rating curve between similar floods [36]. Thus, large floods may also contribute to the occurrence of remarkably large concentration in the near-bed region (Type I), and the deviation of estimated sediment flux with sediment-balance method.

4.3. Connection with Dongting Lake

Reservoir sedimentation would disrupt previous flow and sediment delivery systems [43], especially channels down the dam. The TGD is the world's biggest dam built on the largest river in China, the YR [14]. With impoundment of the TGD, channel reach downstream of the reservoir has adjusted to a changed flow and sediment situation, especially in the Jingjiang reach [29,44].

The Dongting Lake is connected with the Jingjiang River Reach with three distributaries' channels and one outlet. It receives water from the main stream of the Yangtze River via three distributaries' channels, and it discharges into the main stream of the Yangtze River at Chenglingji (the outlet of Dongting lake, Hunan Province, China). The annual runoff received from the Yangtze River is approximately 92.3 km^3 through the Sankou distributary channels [45]. Li et al. [46] pointed out that, the channel changes serve as the primary factor in facilitating the decrease in the discharge diversion ratio, but not the main factor for the decreased amount of the discharge diversion. The occurring frequency of Type II increases after the operation of the TGD, especially at Shashi station, while two of the three distributary channels (Songzikou, Jingzhou City, Hubei Province, China and Taipingkou, Jingzhou City, Hubei Province, China) are located in the reach of upper Shashi station. Thus, vertical profiles of Type II may contribute to this relationship with much concentration transported into the lake.

4.4. Uncertainties

Turbidity monitoring is developed and applied attempting to better predict the continuous variability of suspended sediment concentration during a flood event, and in turn the total sediment yield [47,48]. The accurate in situ quantification of suspended sediment concentration is essential for model calibration and to identify sediment pathways and sediment flux [49].

The two near-bed points are measured with distances of 0.5 m and 0.1 m from the bed, respectively. As to the near-bed measuring, the small distance from the riverbed may be doubtful, as it is difficult to distinguish the riverbed. Equipment used in the field survey includes real-time kinematic Global Navigation Satellite System (RTK–GNSS), Total Station, vessel-mounted Acoustic Doppler Current Profilers (ADCP), Digital Level, Laser Particle Size Analyzer (LPSA), and so on. All the equipment and criteria have been strictly handled and executed by professional surveyors in JHWRSB. Several standards from the technical manual by JHWRSB [25] are listed in Table 8.

The main advantage of a RTK–GNSS survey is that it is faster, cheaper and easier to use during the surveying phase than classical topographic surveying [50]. Pagliari et al. [50] compared the differences between coordinates of the ground-controlled points estimated using RTK–GNSS, and those computed from the classical topographic measurements, during processing photogrammetric block. It pointed out that a RTK–GNSS survey may be sufficient to determine the requested tolerance [50]. Thus, in recent years, miniaturized components (GNSS receivers) have been adopted in field surveys [51].

Table 8. Instruments, methods and criteria adopted in the field survey.

Measuring	Instruments or Methods	Criteria or Error
Positioning of instrument	Total station Digital Level	Positioning error: ±0.3 m (vertical), ±1.5 m (vertical lines)
Vertical lines for measuring velocity	RTK–GNSS (antenna and receiver)	Position errors of antenna: 0.5° RMS (orientation) and 0.5–3 m RMS (position). Position errors of receiver: ±10 mm + 1 ppm (horizontal), ±20 mm + 1 ppm (vertical), where ppm means additional error per km of baseline.
Velocity	Acoustic Doppler Current Profilers (ADCP)	The deviation between each measured discharge and the averaged discharge should be less than ±5%, otherwise, data should be re-measured.
Water depth	ADCP	Verified with fish lead. Frequency: 600 KHz Sounding range: 0.7–75 m Velocity measurement: ±20 m/s Resolution: 0.1 cm/s
Sampling Suspended sediment	Bottom-touched automatic-closing sampler	The dropping speed of the sampler is reduced when approaching the riverbed. The sampler may close automatically when it is brought into contact with the riverbed.
Suspended sediment gradation	Sieving method	Field (2 mm, 5 mm, 10 mm, 25 mm, 50 mm, 75 mm, 100 mm, 150 mm, 200 mm, 250 mm, and 300 mm) Lab (0.002 mm, 0.004 mm, 0.008 mm, 0.016 mm, 0.032 mm, 0.062 mm, 0.125 mm, 0.25 mm, 0.5 mm, 1.0 mm, and 2.0 mm)
Near-bed concentration	Double-checked after sampling and grain-size analysis	The measured concentration is reliable when the vertical profiles of concentration of particles with d_{50} < 0.062 mm has no changing point in near-bed region. Otherwise, related measurements should be omitted.

ADCP is relatively easy to use, enables gathering various data types at once, can be attached to a RC platform, and has rapid three-dimensional flow structure measuring [52,53]. However, disadvantages of ADCP include snap-shot measurement of flow-change to errors, being time-consuming over large areas, and not being able to measure flow or depth in depths under 0.2 m [53]. As it is rather expensive, ADCP is relatively adopted to measure flow structure, flow discharge, bathymetry, bed load discharge, suspended load and channel change in field surveys [53]. Zhang et al. [54] pointed out that the error of ADCP in measuring cross-sections in the middle and lower YR channel is ±1% and the random uncertainty is 2–5%, and the integrated system (GPS and ASCP) delivered positioning with <15 cm accuracy vertically and <30 cm horizontally.

For sediment grain-size analysis, the results from LPSA and sieving are comparable when particles have a spherical shape [55]. Using a sieving method for coarser particles is more acceptable than for finer ones [55]. A sieving method is used in field surveys by JHWRSB [25]. According to the backscatter intensity of ADCP beams, suspended sediment concentration can also be estimated with inversion [56].

A major focus of research has been the development and application of turbidity monitoring, in an attempt to better predict the continuous variability of suspended sediment concentration during a flood event, and in turn the total sediment yield [47]. The concentration by sampling can be validated by echo intensity of the ADCP for future analysis.

5. Conclusions

The sediment regime of the Jingjiang Reach of the Yangtze River has changed dramatically because of the operation of the TGD. Based on vertical distribution data at the three controlling stations (Zhicheng, Shashi, and Jianli) on the Jingjiang Reach, the characteristics of remarkably large concentration in near-bed region and its implications are analyzed in this study. Our conclusions can be summarized as:

(1) In sub-saturated channels, vertical distribution of suspended sediment concentration (SSC) is characterized by a remarkably large concentration in the near-bed zone (within 10% of water

depth from the river-bed). The maximum measured concentration may be up to 15 times of vertical average concentration, while the ratio in quasi-equilibrium channels is less than four.

(2) Concentrations normalized with reference concentration at the same height may decrease with increasing values of suspension index (ω/u_*). Additionally, concentration near water surface may be larger than concentration in near-bed region when suspension index is smaller than 0.01.

(3) After the dam operation, ignoring the near-bed concentration may cause up to 35% deviation when applied to estimate the sediment flux in the unsaturated flows. Deviations may increase with increasing discharge in flood season, while the maximum value may occur in non-flood season. Deviations in quasi-equilibrium channel may not increase with increasing discharge in flood season, while the maximum value may also occur in non-flood season. Deviations may increase with increasing particle size and suspension index.

Analytic results indicate that, in sub-saturated channels, more attention should be paid to near-bed concentration during non-flood season, the same as measurements during flood season with larger discharge.

Acknowledgments: This work was supported by the National Program on Key Basic Research Project (No. 2016YFC0402303); and the National Natural Science Foundation of China (Nos. 51579230, 51109198). We wish to express our deep gratitude to the Jingjiang Hydrology and Water Resources Surveying Bureau, Yangtze River Water Resources Commission for permission to access the hydrological data.

References

1. Ludwig, W.; AmiotteSuchet, P.; Probst, J.L. River discharge of carbon to the world's oceans: Determining local inputs of alkalinity and of dissolved and particulate organic carbon. *Comptes Rendus de L Academie des Sciences Serie II Fascicule A-Sciences de la Terre et des Planetes* **1996**, *323*, 1007–1014.
2. Asselman, N.E.M. Fitting and interpretation of sediment rating curves. *J. Hydrol.* **2000**, *234*, 228–248. [CrossRef]
3. Zuo, S.H.; Li, J.F.; Wan, X.N.; Shen, H.T.; Fu, G. Characteristics of temporal and spatial variation of suspended sediment concentration in the Changjiang Estuary. *J. Sediment Res.* **2006**, *3*, 68–75. (In Chinese with English Abstract)
4. Pal, D.; Ghoshal, K. Hydrodynamic interaction in suspended sediment distribution of open channel turbulent flow. *Appl. Math. Modell.* **2017**, *49*, 630–646. [CrossRef]
5. Higgins, A.; Restrepo, J.C.; Otero, L.J.; Ortiz, J.C.; Conde, M. Vertical distribution of suspended sediment in the mouth area of the Magdalena River, Colombia. *Lat. Am. J. Aquat. Res.* **2017**, *45*, 724–736. [CrossRef]
6. Nir, S.Q.; Sun, H.G.; Zhang, Y.; Chen, D.; Chen, W.; Chen, L.; Schaefer, S. Vertical distribution of suspended sediment under steady flows: Existing theories and fractional derivative model. *Discret. Dyn. Nat. Soc.* **2017**, *2017*, 1–11.
7. Sahin, C.; Verney, R.; Sheremet, A.; Voulgaris, G. Acoustic backscatter by suspended cohesive sediment: Field observations, Seine Estuary, France. *Cont. Shelf Res.* **2017**, *134*, 39–51. [CrossRef]
8. Armijos, E.; Crave, A.; Espinoza, R.; Fraizy, P.; Dos Santos, A.L.M.R.; Sampaio, F.; De Oliveria, E.; Santini, W.; Martinez, J.M.; Autin, P.; et al. Measuring and modelling vertical gradients in suspended sediments in the Solimões/Amazon River. *Hydrol. Process.* **2017**, *31*, 654–667. [CrossRef]
9. He, L.; Chen, D.; Duan, G.L.; Zhu, Z.H.; Zhang, S.Y. Sediment suspension in the "starving" Jingjiang reach, downstream from the Three Gorges Dam, China. In Proceedings of the World Environmental and Water Resources Congress, West Palm Beach, FL, USA, 22–26 May 2016; pp. 303–313.
10. Fang, H.Y.; Cai, Q.G.; Chen, H.; Li, Q.Y. Temporal changes in suspended sediment transport in a gullied loess basin: The lower Chabagou Creek on the Loess Plateau in China. *Earth Surf. Process. Landf.* **2008**, *33*, 1977–1992. [CrossRef]
11. Zhang, R.; Zhang, S.H.; Xu, W.; Wang, B.D.; Wang, H. Flow regime of the three outlets on the south bank of Jingjiang River, China: An impact assessment of the Three Gorges Reservoir for 2003–2010. *Stoch. Environ. Res. Risk Assess.* **2015**, *29*, 2047–2060. [CrossRef]
12. Fan, P.; Li, J.C.; Liu, Q.Q.; Singh, V.P. Case Study: Influence of morphological changes on flooding in Jingjiang River. *J. Hydraul. Eng.* **2008**, *134*, 1757–1766. [CrossRef]

13. Yang, C.; Cai, X.B.; Wang, X.L.; Yan, R.R.; Zhang, T.; Zhang, Q.; Lu, X.R. Remotely sensed trajectory analysis of channel migration in lower Jingjiang Reach during the period of 1983–2013. *Remote Sens.* **2015**, *7*, 16241–16256. [CrossRef]
14. Gao, B.; Yang, D.W.; Yang, H.B. Impact of the three Gorges Dam on flow regime in the middle and lower Yangtze River. *Quat. Int.* **2013**, *304*, 43–50. [CrossRef]
15. Yang, H.; Tang, R. *Study on the Evolution of Jingjiang Reach of Middle Yangtze River*; China Water Power Press: Beijing, China, 1999. (In Chinese)
16. Dai, Z.J.; Fagherazzi, S.; Mei, X.F.; Gao, J.J. Decline in suspended sediment concentration delivered by the Changjiang (Yangtze) River into the East China Sea between 1956 and 2013. *Geomorphology* **2016**, *268*, 123–132. [CrossRef]
17. Xu, K.H.; Milliman, J.D. Seasonal variations of sediment discharge from the Yangtze River before and after impoundment of the Three Gorges Dam. *Geomorphology* **2009**, *104*, 276–283. [CrossRef]
18. Xia, J.Q.; Zong, Q.L.; Deng, S.S.; Xu, Q.X.; Lu, J.Y. Seasonal variations in composite riverbank stability in the Lower Jingjiang Reach, China. *J. Hydrol.* **2014**, *519*, 3664–3673. [CrossRef]
19. Xia, J.Q.; Zong, Q.L.; Zhang, Y.; Xu, Q.X.; Li, X.J. Prediction of recent bank retreat processes at typical sections in the Jingjiang Reach. *Sci. China* **2014**, *57*, 1490–1499. [CrossRef]
20. Zhang, C.F. Vertical Profile of Non-Uniform Suspended Sediment Concentration in Yangtze Estuary. Master's Thesis, Zhejiang University, Hangzhou, China, 2016. (In Chinese with English Abstract)
21. Dong, X.T.; Li, R.J.; Fu, G.C.; Zhang, H.C. Relationship between vertical distribution of velocity and sediment concentration of sediment-laden flow. *J. Hohai Univ. Natl. Sci.* **2015**, *43*, 371–376. (In Chinese with English Abstract)
22. Tang, M.L.; Cao, H.Q.; Li, Q.Y.; Zhai, W.L. Vertical distribution of suspended sediment in middle reaches of Yangtze River. *Yangtze River* **2017**, *48*, 6–11. (In Chinese with English Abstract)
23. Feng, Q.; Xiao, Q.L. A new S-type vertical distribution of suspended sediment concentration. *J. Sediment Res.* **2015**, *1*, 19–24. (In Chinese with English Abstract)
24. Xia, Y.F.; Xu, H.; Chen, Z.; Wu, D.W.; Zhang, S.Z. Experimental study on suspended sediment concentration and its vertical distribution under spilling breaking wave actions in silty coast. *China Ocean Eng.* **2011**, *25*, 565–575. [CrossRef]
25. JHWRSB (Jingjiang Hydrology & Water Resources Surveying Bureau, Hydrology Bureau). *Technical Report: Observation of Unbalanced Sediment Transport in the Lower Reaches of the Three Gorges Reservoir*; Changjiang Water Resources Commission: Wuhan, Hubei, China, 2016. (In Chinese)
26. Yuan, Y.; Zhang, X.F.; Duan, G.L. Modifying the channel erosion and deposition amount of Yichang-Jianli reach calculated by sediment discharge method. *J. Hydroelectr. Eng.* **2014**, *33*, 163–169. (In Chinese with English Abstract)
27. Duan, G.L.; Peng, Y.B.; Guo, M.J. Comparative analysis on riverbed erosion and deposition amount calculated by different methods. *J. Yangtze River Sci. Res. Inst.* **2014**, *31*, 108–118. (In Chinese with English Abstract)
28. Dai, S.B.; Lu, X.X. Sediment load change in the Yangtze River (Changjiang): A review. *Geomorphology* **2014**, *215*, 60–70. [CrossRef]
29. Zhao, G.S.; Lu, J.Y.; Visser, P.J. Fluvial river regime in disturbed river systems: A case study of evolution of the Middle Yangtze River in post-TGD (Three Gorges Dam), China. *J. Geol. Geophys.* **2015**, *4*, 6. [CrossRef]
30. Zhang, X.F.; Tan, G.M. Characteristics of vertical concentration distribution of non-uniform particles. *J. Hydraul. Eng.* **1992**, *10*, 48–52. (In Chinese with English Abstract)
31. Lenzi, M.A.; Mao, L.; Comiti, F. Interannual variation of suspended sediment load and sediment yield in an alpine catchment. *Hydrol. Sci. J.* **2003**, *48*, 899–915. [CrossRef]
32. Zabaleta, A.; Martinez, M.; Uriarte, J.A.; Antiguedad, I. Factors controlling suspended sediment yield during runoff events in small headwater catchments of the Basque Country. *Catena* **2007**, *71*, 179–190. [CrossRef]
33. Langlois, J.L.; Johnson, D.W.; Mehuys, G.R. Suspended sediment dynamics associated with snowmelt runoff in a small mountain stream of Lake Tahoe (Nevada). *Hydrol. Process.* **2005**, *19*, 3569–3580. [CrossRef]
34. Tena, T.; Batalla, R.J.; Vericat, D. Reach-scale suspended sediment balance downstream from dams in a large Mediterranean River. *Hydrol. Sci. J.* **2012**, *57*, 831–849. [CrossRef]
35. Zhang, W.; Yuan, J.; Han, J.Q.; Huang, C.T.; Li, M. Impact of the Three Gorges Dam on sediment deposition and erosion in the middle Yangtze River: A case study of the Shashi Reach. *Hydrol. Res.* **2016**, 175–186. [CrossRef]

36. Whitaker, A.C.; Sato, H.; Sugiyama, H. Changing suspended sediment dynamics due to extreme flood events in a small pluvial-nival system in northern Japan. In *Sediment Dynamics in Changing Environments*; IAHS-AISH Publication: Christchurch, New Zealand, 2008; Volume 325, pp. 192–199.
37. Li, L.Q.; Lu, X.X.; Chen, Z.Y. River channel change during the last 50 years in the middle Yangtze River, the Jianli reach. *Geomorphology* **2007**, *85*, 185–196. [CrossRef]
38. Zhu, Y.H.; Fan, B.L.; Yao, S.M.; Sun, G.Z.; Li, F.Z. Propagation features of the 1998 big floods in the Jinagjiang Reach of the Yangtze River. In *Advances in Water Resources and Hydraulic Engineering*; Springer: Berlin/Heidelberg, Germany, 2009; pp. 1032–1037.
39. Zong, Y.Q.; Chen, X.Q. The 1998 flood on the Yangtze, China. *Nat. Hazards* **2000**, *22*, 165–184. [CrossRef]
40. Xu, K.Q.; Chen, Z.Y.; Zhao, Y.W.; Wang, Z.H.; Zhang, J.Q.; Hayashi, S.J.; Murakami, S.; Watanabe, M. Simulated sediment flux during 1998 big-flood of the Yangtze (Changjiang) River, China. *J. Hydrol.* **2005**, *313*, 221–223. [CrossRef]
41. Zhou, W.H.; Chen, Y.H. The 1998 flood of the Yangtze River. *Int. J. Sediment Res.* **1999**, *14*, 61–66.
42. Seeger, M.; Errea, M.-P.; Begueria, S.; Arnaez, J.; Marti, C.; Garcia-Ruiz, J.M. Catchment soil moisture and rainfall characteristics as determinant factors for discharge/suspended sediment hysteretic loops in a small headwater catchment in the Spanish Pyrenees. *J. Hydrol.* **2004**, *288*, 299–311. [CrossRef]
43. Ran, L.; Lu, X.; Xin, Z.; Yang, X. Cumulative sediment trapping by reservoirs in large river basins: A case study of the Yellow River basin. *Glob. Planet. Chang.* **2013**, *100*, 308–319. [CrossRef]
44. Xia, J.Q.; Deng, S.S.; Lu, J.Y.; Xu, Q.X.; Zong, Q.L.; Tan, G.M. Dynamic channel adjustments in the Jingjiang reach of the middle Yangtze River. *Sci. Rep.* **2016**, *6*, 22802. [CrossRef] [PubMed]
45. Dou, H.; Jiang, J. *Dongting Lake*; Press of University of Science and Technology of China: Hefei, China, 2000.
46. Li, N.; Wang, L.C.; Zeng, C.F.; Wang, D.; Liu, D.F.; Wu, X.T. Variations of runoff and sediment load in the Middle and Lower Reaches of the Yangtze River, China (1950–2013). *PLoS ONE* **2016**, *11*, e160154. [CrossRef] [PubMed]
47. Orwin, J.F.; Smart, C.C. An inexpensive turbidimeter for monitoring suspended sediment. *Geomorphology* **2005**, *68*, 3–15. [CrossRef]
48. Pfannkuche, J.; Schmidt, A. Determination of suspended particulate matter concentration from turbidity measurements: Particle size effects and calibration procedures. *Hydrol. Process.* **2003**, *17*, 1951–1963. [CrossRef]
49. Stutter, M.; Dawson, J.J.C.; Glendell, M.; Napier, F.; Potts, J.M.; Sample, J.; Vinten, A.; Watson, H. Evaluating the use of in-situ turbidity measurements to quantify fluvial sediment and phosphorus concentration and flux in agricultural streams. *Sci. Total Environ.* **2017**, *607–608*, 391–402. [CrossRef] [PubMed]
50. Pagliari, D.; Rossi, L.; Passoni, D.; Pinto, L.; De Michele, C.; Avanzi, F. Measuring the volume of flushed sediments in a reservoir using multi-temporal images acquired with UAS. *Geomat. Nat. Hazards Risk* **2017**, *8*, 150–166. [CrossRef]
51. Bandini, F.; Jakobsen, J.; Olesen, D.; Reyna-Gutierrez, J.A.; Bauer-Gottwein, P. Measuring water level in rivers and lakes from lightweight unmanned aerial vehicles. *J. Hydrol.* **2017**, *548*, 237–250. [CrossRef]
52. Togneri, M.; Lewis, M.; Neill, S.; Masters, I. Comparison of ADCP observations and 3D model simulations of turbulence at a tidal energy site. *Renew. Energy* **2017**, *114*, 273–282. [CrossRef]
53. Kasvi, E.; Hooke, J.; Kurkela, M.; Vaaja, M.T.; Virtanen, J.P.; Hyyppa, H.; Alho, P. Modern empirical and modelling study approaches in fluvial geomorphology to elucidate sub-bend-scale meander dynamics. *Prog. Phys. Geogr.* **2017**, *41*, 533–569. [CrossRef]
54. Zhang, Q.; Shi, Y.F.; Xiong, M. Geometric properties of river cross sections and associated hydrodynamic implications in Wuhan-Jiujiang river reach, the Yangtze River. *J. Geogr. Sci.* **2009**, *19*, 58–66. [CrossRef]
55. Li, W.K.; Wu, Y.X.; Huang, Z.M.; Fan, R.; Lv, J.F. Measurement results comparison between Laser Particle Analyzer and Sieving method in particle size distribution. *Zhongguo Fenti Jishu* **2007**, *5*, 10–13. (In Chinese with English Abstract)
56. Jin, W.F.; Liang, C.J.; Zhou, B.F.; Li, J.D. Measurement and analysis of high suspended sediment concentration in Jintang channel with vessel-mounted ADCP. *J. Mar. Sci.* **2009**, *27*, 31–39. (In Chinese with English Abstract)

11

Multi-Objective Optimization of Resilient Design of the Multipurpose Reservoir in Conditions of Uncertain Climate Change

Stanislav Paseka [1,*], Zoran Kapelan [2] and Daniel Marton [1]

[1] Institute of Landscape Water Management, Faculty of Civil Engineering, Brno University of Technology, 602 00 Brno, Czech Republic; marton.d@fce.vutbr.cz

[2] Centre for Water Systems, College of Engineering, Mathematics and Physical Sciences, University of Exeter, Exeter EX4 4QF, UK; Z.Kapelan@exeter.ac.uk

* Correspondence: paseka.s@fce.vutbr.cz

Abstract: This paper presents and assesses a new approach to decision-making methods for the design of new reservoirs due in times of decreasing water resources. The methods used in this case are decision theory, Resilience and Robustness method. The methods have been selected primarily to analyze different design parameters of a new dam, mainly dam heights leading to different reservoir volumes. The study presents a novel approach to the optimal design of a multipurpose reservoir that would provide enough water for downstream environmental flow, residential and industrial water supply, agricultural water supply, and hydropower generation in the current conditions of climate uncertainty. Uncertainties are interpreted as possible future changes in the climate system using outputs from regional climatic models. In the case study, a simulation model was developed which is able to quantify long-term water balance and use this data to quantify resilience and robustness of its water supply. The simulation model was correlated to the GANetXL software in order to perform Genetic Algorithms based optimization of the reservoir's operation. The simulation–optimization model was then applied to a real-life case study in the Czech Republic, in the Morava River Basin where a new dam with the multipurpose reservoir is planned to be built in the future. The results obtained in this way were analyzed in detail to identify the overall best solution consist of dam height and the total reservoir monthly outflow and new operational rules for the analyzed multipurpose reservoir.

Keywords: design and operation of the multipurpose reservoir; water deficit; reservoir simulation model; climate change; multi-objective optimization NSGA II; resilience and robustness; costs and benefits; water energy

---

1. Introduction

In Czechia, climate change has caused change of the hydrological cycle due to redistribution of precipitation during the hydrological year in the last decade. This phenomenon results in more frequent occurrences of extremes in the form of floods and droughts. But the most urgent problem is that the values of the long-term mean flows are decreasing in rivers as well as the yield of groundwater sources.

Moreover, the years 2015 and 2017 were considered extremely dry in many regions of Czechia. It is becoming a trend of the last couple of years that meteorologists start to call the previous year as the warmest year in the history of meteorological measurement, this confirms WMO Statement on the State of the Global Climate 2017 [1].

The above-mentioned factors create pressure on the effective management of the surface water resources which will become stronger as well as the Czech Republic is sometimes called "roof of

Central Europe", as there is no major inflow to the territory, but only outflows. The limited availability of water resources in this country requires the government to prepare new climate change adaptations and focus on increasing the retention capacity of water in the landscape, long-term water management planning, and environmental politics on global level.

The government of the Czech Republic has issued several documents on adaptation strategies to climate change [2–5], which have been developed by the research institutes: VÚV TGM, v.v.i; EKOTOXA s.r.o.; ČHMÚ in cooperation with Ministry of the Environment of the Czech Republic. These documents indicate many measures to achieve the objectives of protection against the negative impacts of drought namely adaptive measures leading to saving drinking water, increasing of water retention capacity of the soil, stabilizing of water regimes in river basins, restoration of small water reservoirs and increasing their reliability, optimization of the existing reservoir volumes, and importantly, identification of other locations suitable for new water reservoirs and strategic technical option is allocating new water resources like open water reservoirs. Under conditions of climate change uncertainties currently there are several locations surveyed in the process of feasibility studies. These new multipurpose reservoirs will be primarily ensuring drinking water supply, keep ecological flows in the rivers, ensuring water supply for industry and agriculture, and may help to resist the drought problems.

The overall objective of this paper is to introduce a novel approach to the multipurpose reservoir design. The evolutionary multi-objective optimization method was utilized to find the optimal design parameter values of the dam, mainly dam height leading to different reservoir volumes. For the design of dam height and reservoir management, a new level so-called water excess level (I) and the critical threshold level (II) were introduced. In order to complete the main goal, it is necessary that the design must follow the individual objectives:

(a) the goal is to apply a suitable metric which is defined by the resilience of reservoir storage capacity and find optimal economic solutions (to maximize resilience and benefits from the reservoir) using the multi-objective optimization method with required robustness,
(b) the objective of this method is to analyze a range of the dam heights and volumes of the multipurpose dam in conditions of uncertain climate change on the basis of given future climate scenarios using hydrological outputs from 15 regional climate models and potential demand scenarios.

This new approach responding to the actual demands of water under climate change can be a benefit for the design of the multipurpose reservoir under similar conditions. The methods are applied and tested on the case study in the Hanusovice dam profile.

2. Background

In the Czech Republic water reservoirs have always been designed according to the historical hydrological or generated time series. Czech National Standard [6] classifies open water reservoirs based on their strategic importance. Strategic classification is evaluated by temporal reliability R_T which is defined as the ratio of the months without fault and the number of all months for the given time series [7,8]. The class A is the most strategic (A $\geq$ 99.5% R_T) and D is the least strategic (D $\geq$ 95.0% R_T). The water deficit is defined when the active storage capacity is in an unsatisfactory state. Water supply reservoirs have generally been designed and operated by a set of predetermined rules, not taking into consideration climate uncertainty. These rules were formulated based on hydrology, the ability of storage capacity, and yield criteria. Reliability has been the main objective in the sense of achieving the long-term target demand and the prevention against floods. The practice for the design and operation of the reservoir used reliability and the concepts of resilience, which experimental setting is different, were not used. The problem is that this approach does not take into consideration the consequences of possible future hydrological and climate changes when there is an increased social demand, increased needs of agriculture, and long-term decreasing flows. Climate change is creating a potentially large range of uncertainty of possible futures that could threaten the reliability of water supplies.

When modeling water resources systems, we need to determine reliability, vulnerability, resilience, and risk. First Kritskiy and Menkel [7] described the definition of evaluation criteria of water resources; reliability. Hashimoto et al. [8] were among the first to propose the use of the terms of water resource system performance evaluation. These performance criteria refer to (i) how probable is it that a system can get into a unsatisfactory or failure state (reliability), (ii) how severe the consequences of failure might be (vulnerability), and (iii) how quickly it can bounce back, which is the recovery from a failure (resilience) [8]. Using multi-objective linear programming, the trade-offs between reliability, vulnerability, and resilience were researched in water supply reservoir operation in a previous study [9].

Resilience for water resources has generally been quantified as the duration of time (maximum or average) when temporary restrictions are in place due to low supply availability. But its calculation is varied in the literature. Relationships among resilience, reliability, and vulnerability of water supply using many-objective analysis were explored in a past paper [10]. For example, resilience was used as a performance criterion metric [11], in a past work [12] resilience is calculated as the average duration of time when a system is under a temporary restriction, or in another previous work [13] resilience was calculated as a fraction of the total future time a system is under an unsatisfactory state. Roach [14,15] characterized and tested several potential resilience metrics and looked for the resilience of the water system and robustness of the water supply based on adaptation strategies. Watts et al. [16] and Amarasinghe et al. [17,18] have undertaken studies on the resilience of water supply systems. The reservoir resilience as a function of time (static and dynamic) for a multipurpose reservoir operation was compared a past study [19]. The resilience along with robustness to system design wastewater treatment plant control were used previously [20].

The risk and uncertainty were first described by Knight [21], who was the first to distinguish between risk and uncertainty. The risk assessment and risk analysis for water management purposes were described by Kaplan in [22]. Uncertainties used in hydrology have been presented by Beven and Binley [23]. Uncertainties can be categorized by the generation of multiple future scenarios that represent alternative likely conditions under different assumptions [24]. Uncertainties are the factors which can affect reservoir design or operation, as well as the accuracy of results. Uncertainty analysis focusing on reservoir capacity performance with various sizes and types of input uncertainties were tested for the design of the active storage capacity and reliability in the articles [25–28]. Based on the achieved results it is possible that both values active storage capacities and reliability, determined without considering input data uncertainty, may be undervalued. This can lead to failure of storage capacity.

Using Monte Carlo experiments the Markov model reproduced relationships between resilience and reliability for a wide class of water supply systems and provided a general theoretical foundation for understanding the trade-offs among reservoir system storage, yield, reliability, and resilience [29]. In a past paper [30] the behavior of statistical performance indices (reliability, resilience, and vulnerability for a multipurpose storage reservoir) using simulations Monte Carlo on a real reservoir in India was performed, and it was confirmed that traditional reliability for reservoir design and operation cannot completely describe the strengths and weaknesses of a given issue. A past paper [31] showed the optimized operation of the largest multipurpose reservoir in Vietnam using the complex evolution algorithm and the MIKE 11 simulation model. The optimization puts focus on the trade-off between flood control and hydropower generation for the reservoir operation in the flood season and the reservoir level at the beginning of the dry season. In another past paper [32] a novel multi-objective optimization modeling framework for the operation of multipurpose simple reservoirs was presented. This model uses genetic algorithms as optimization techniques. The main objective function minimizes the cost of the annual water shortage for irrigation and the secondary objective maximizes energy production. A previous work [33] presents an evolutionary multi-objective optimization approach NSGA-II for the study of multiple water usages in multiple interlinked reservoirs. The case studies involve, primarily, the objectives of power generation and navigability on the river.

One of the newest statistical concepts could be used to evaluate results under uncertainty is robustness. Robustness can be defined as the degree to which a water supply system performs at a satisfactory level across a range of future scenarios or conditions [34]. One of the types of satisficing criteria may be based on the proportion of possible future conditions under which system performs correctly. Developing robust long-term water resources plans described by the above definition have generally been assessed previously [35]. The robustness water resources system under deep uncertainty has been used in many other cases and the results from past papers [20,36–42] show that the using of the robustness concept is beneficial for the analysis of results under uncertainty.

For the case study, the optimizing method Non-dominated Sorting Genetic Algorithm II (NSGA II) described previously [43] was used. The NSGA II was successfully used in handling multi-objective optimization problems in water management in several past works [44–50]. The NSGA II was developed from the multi-objective algorithm NSGA [51], which was formulated based on the suggestions made by Goldberg in 1989 [52]. The relation between risk, reliability, water shortages, and hydropower energy in an open reservoir using the NSGA II optimization was tested previously [53]. The NSGA II algorithm and reservoir simulation model were applied in GENetXL [54]. The GENetXL was presented by a simple water supply hypothetical reservoir operation model with two objectives: maximize yield and maximize recreational benefit, and a large combinatorial optimization problem of pump scheduling in water distribution systems.

There is not much research on optimal reservoir design and management under uncertainty. In the world, complex water management planning and optimization of extensive water resources systems in the context of climatic uncertainties are mostly solved [40,55–61]. Given that case studies of water management systems are solved complexly and on a large scale on the other hand, we focus on the individual study in small scale but in more detail. A solution, in the sense of the whole reservoir capacity under climate change, has not been well-studied as an individual target in the design of reservoirs. In addition, the increasing water demand and distribution of water are not sufficiently safeguarded today. Current issues lead to use the uncertainty of climate change, multi-objective optimization, and concepts of resilience and robustness. All of these creating optimal effective design and new operational rules of a multipurpose water reservoir.

3. Case Study

The case study is based on the intended profile in Hanusovice in the Czech Republic, 200 km east of the capital city Prague, see Figure 1. The new dam is planned in the Morava River Basin, maintained by the manager of the Morava River Basin, State Enterprise (PMO), and the Krupa River is the main inflow into the reservoir. The long-term mean river flow Q_a is 2.12 m^3 s^{-1}. The elevation volume and area curves for a given profile were determined by a digital terrain model using GIS software. In this software, Figure 1 has been also generated showing one of the possible new dam locations.

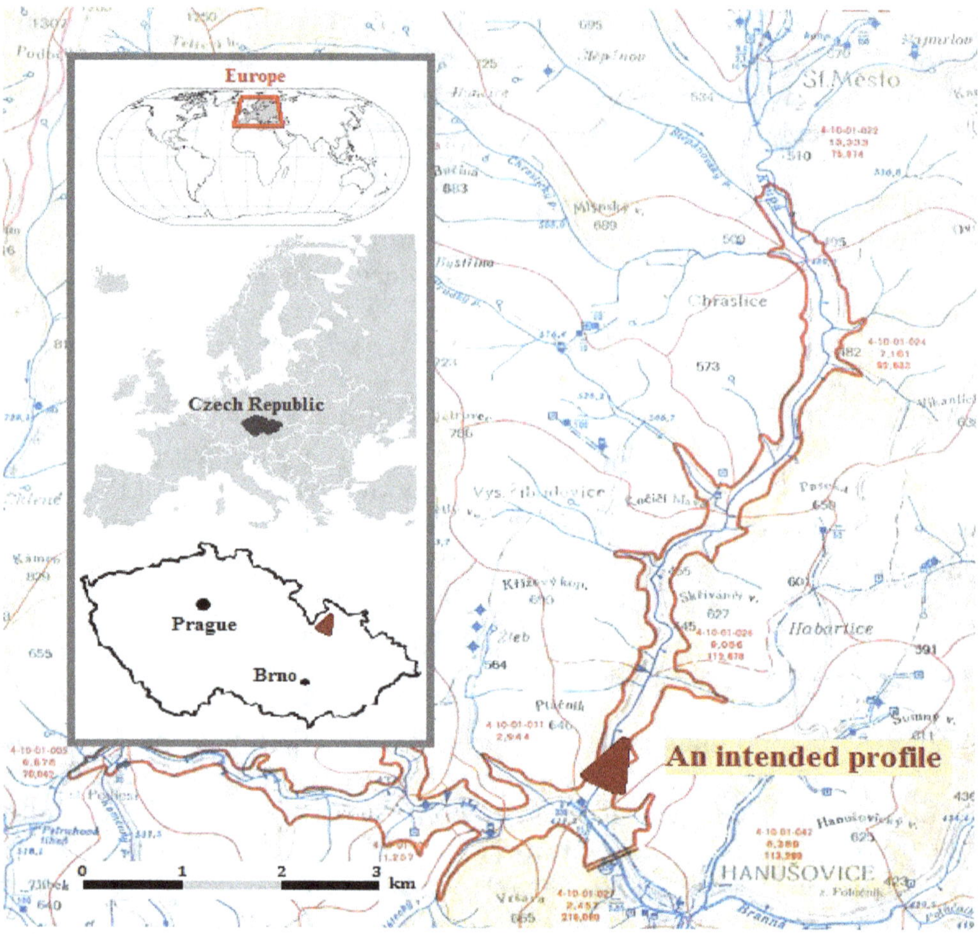

Figure 1. The locality of the plan reservoir to be built in the future.

Hydrologic data in the form of water inflows were derived from 15 regional climate models for the period 2015 to 2100. These are scenarios with a resolution of 25 km × 25 km which are controlled by one of the four global climate models. The results of the climate models are based on the IPCC SRES AR4 [62], only for the conservative emission scenario A1B which represents a very fast future economy growth and development of new technologies with a balance in the use of all fuels as energy sources. In the paper, each regional model represents 1 input of climate scenario. Data were taken from the project RSCN VUV [63]. The data were modified from hydrometric profile Morava/Raskov to hydrometric profile Krupa/Habartice using the analogy method for the river flows were determined.

Total water demand was determined as a sum of the Q_{ECO} (ecological outflow to the river), $Q_{WS:drink}$ (drinking water supply), and $Q_{WS:ind.+agric.}$ (water supply for industry and agriculture). Q_{ECO} = 0.54 m^3 s^{-1} and is constant for all months, $Q_{WS:drink.}$ and $Q_{WS:ind.+agric.}$ are based on 25% of total current demand for the part of the Olomouc Region from the plan for the development of water supply [64]. Mean annual demand was divided to monthly demands, see Table 1. Potential monthly redistribution of water demand was set according to the data from the PMO.

Table 1. Potential monthly water demands.

Demands [m^3 s^{-1}]	January	February	March	April	May	June	July	August	September	October	November	December
Q_{ECO}	0.54	0.54	0.54	0.54	0.54	0.54	0.54	0.54	0.54	0.54	0.54	0.54
$Q_{WS:drink.}$	0.54	0.56	0.56	0.58	0.58	0.60	0.56	0.55	0.55	0.56	0.55	0.53
$Q_{WS:ind.+agric.}$	0.35	0.35	0.36	0.37	0.38	0.40	0.40	0.40	0.38	0.36	0.35	0.35
Total Demand	**1.43**	**1.45**	**1.46**	**1.49**	**1.50**	**1.54**	**1.50**	**1.49**	**1.47**	**1.46**	**1.44**	**1.42**

The permanent storage capacity V_P was simplified and evaluated to height 15.0 m and corresponds to 0.44 mil. m^3. The flood storage capacity V_F was calculated for flood hydrograph $Q_{10.000}$. The Klemes method [65] was used to transform the controlled flood and to evaluate V_F. The total flood storage capacity V_F = 7.72 mil. m^3. The safety reserve 2.0 m was used to prevent overflow of the dam.

The location of the power plant is situated just below the dam and water supply for industry and agriculture is taken downstream the power plant, while drinking water supply is taken in the reservoir itself.

For the case study, an earth dam was used. Parameters of the dam body are; width of the dam crown is 5.0 m, bottom length is 100.0 m, an upstream slope is 1:3, downstream slope is 1:2, and terrain slopes are 1:1 (left side) and 1:1.8 (right side). The height of dam was calculated ranging from 80.0 to 100.0 m.

The target levels of resilience were set for three targets the duration of the longest water deficit period is up to 3 months (main target), up to 5 months, and 0 months. Two target levels of robustness 80% and more and 90% and more were tested.

All prices in Equations (4)–(6) corresponded to the present trade prices in Euro. The $PRICE_{WATER}$ was 0.245 € m^{-3} charged by PMO last year and the $PRICE_{RED,ELE}$ was 0.10 € kWh^{-1}. The discount rate r according to a past paper [66] for 2015–2044 is presumed 0.035, 2045–2089 is presumed 0.03, and 2090–2100 is presumed 0.025. The total coefficient of efficiency η for the calculation of the energy from the hydropower in Equation (7) is 0.80. The cost of the dam ZCU in Equation (10), respectively (9), is around 26.2 € m^{-3} for the earth dam in this location.

The following configuration of parameters NSGA II optimization in GENetXL was used. Population size was 100 genes. The number of generations was 200. Selection the crowded tournament was set up. The crossover was set up the simple one point and the rate was 0.90. Mutation type was set up as simple with mutation rate 0.05. The number of chromosomes (decision variables) was 12 for $i = 1, \ldots, 12$ months of year with boundary conditions $Q_{EXC,i} \in \langle 0.0\ m^3\ s^{-1}; 2.0\ m^3\ s^{-1} \rangle$.

4. Methodology

4.1. Problem Formulation

Where designing the new multipurpose dam, the objective functions have to be defined to satisfy all requests at the same time and lead to a new approach for using a multi-objective optimization.

The resilience of storage capacity must be set to a target level to be considered as acceptable under given future climate scenarios and water demand scenario, as well as robustness. That means if the target levels of resilience and robustness are satisfactory then the designed solutions are acceptable.

In the analysis, 12 decision variables were created. Each of them was described as month excess water Q_{EXC} that was added to the total water demand.

In total, two objective functions were used, net present value and resilience. The results obtained from the 12 decision variables for four chosen climate scenarios. Then the robustness was evaluated based on decision variables and resilience.

4.2. Resilience and Robustness

In this case every kind of potential resilience (RES) [month] metrics, defined previously [14], were tested, and finally, was traditional metric of resilience was selected with the duration of the longest water deficit period:

$$RES = \max(\text{duration of water deficit}), \tag{1}$$

The robustness (ROB) [%] is defined in this case as the percentage ratio of future supply and demand scenarios [36,39]:

$$ROB = \frac{S_A}{S_T} \times 100, \tag{2}$$

where S_A is the number of scenarios in which the reservoir simulation model performs at an acceptable level of resilience and S_T is the total number of scenario combinations of supply and demand.

4.3. Costs and Benefit

One of the objectives is to minimize the cost of dam construction and maximize benefits from the dam. This is expressed by the net present value (*NPV*) [€], which is calculated according to the following Equation (3):

$$NPV = \sum_{t=1}^{T} \left(\frac{BEN_{WS:drink,t}}{(1+r)^{t-1}} + \frac{BEN_{WS:ind.+agric.,t}}{(1+r)^{t-1}} + \frac{BEN_{HPP,t}}{(1+r)^{t-1}} \right) - C_{DAM}, \tag{3}$$

where *BEN* [mil €] are benefits from the utilization of water for hydropower and sales of water, C_{DAM} is the cost of the dam construction, r is discount rate and $t = 1, \ldots, T$ for $T = 86$ (the total number of years until 2100).

We have three possible benefits: drinking water supply Equation (4), water supply for industry and agriculture Equation (5), and benefits from hydropower plant Equation (6):

$$BEN_{WS:drink,t} = \sum_{i=1}^{n} (Q_{WS:drink,i}) \times PRICE_{WATER}, \tag{4}$$

$$BEN_{WS:ind.+agric.,t} = \sum_{i=1}^{n} \left(Q_{WS:ind.+agric.,i} \right) \times PRICE_{WATER}, \tag{5}$$

$$BEN_{HPP,t} = E_{HPP,t} \times PRICE_{RED,ELE}, \tag{6}$$

where $Q_{WS:drink.,i}$ is drinking water for residential water supply and $Q_{WS:ind.+agric.,i}$ is water for industrial and agricultural water supply; $PRICE_{WATER}$ is the price for water consumption. $PRICE_{RED,ELE}$ is the redemption price of electricity from the hydropower according to the Czech Energy Regulatory Office, $i = 1, \ldots, n$ for $n = 12$ (number months of the t-year) and $E_{HPP,t}$ [Wh·month^{-1}] is generated energy from hydropower for a year Equation (7):

$$E_{HPP,t} = \sum_{i=1}^{n} (\rho \times g \times Q_{HPP,i} \times H_i \times \eta) \times \Delta t, \tag{7}$$

where ρ is density of water, g is gravity, H_i is the height of water level in given i-month, η is total coefficient of efficiency, $i = 1, \ldots,$ n for $n = 12$, Δt is one year and $Q_{HPP,i}$ [m^3 s^{-1}] is flowed through a hydropower in given i-month and is calculated according to the Equation (8):

$$Q_{HPP,i} = Q_{ECO,i} + Q_{WS:ind.+agric.,i} + (Q_{EXC,i}), \tag{8}$$

where $Q_{ECO,i}$ is ecological outflow for downstream environmental flow, $Q_{WS:ind.+agric.,i}$ is water supply for industry, and $Q_{EXC,i}$ is the water excess for hydropower and to a river in given i-month.

The last parameter of Equation (3) is C_{DAM} [€], which describes the cost of the dam construction and is determined by following Equation (9):

$$C_{DAM} = V_{DAM} \times ZCU, \tag{9}$$

where V_{DAM} is the volume of the dam structure and ZCU [€m^{-3}] is the cost of the dam calculated according to Czech property valuation decree [67]:

$$ZCU = ZC \times K_5 \times K_i, \tag{10}$$

where ZC is a cost according to the type of material of the dam and K are coefficients according to the location of the dam.

4.4. Reservoir Simulation Model

A reservoir simulation model was created based on emptying and filling of reservoir storage capacity. The newly introduced water excess level (I) and critical threshold (II) were used to calculate the resilience. This principle can be described by the inequality (11) which proceeds from the mass balance equation [68]:

$$Water\ lvl.(0) \leq \sum_{i=0}^{n-1}(Q_{OUT,i} - Q_{IN,i})\Delta t + (Q_{OUT,i+1} - Q_{IN,i+1})\Delta t \leq Water\ lvl.(I),(II),(III), \quad (11)$$

where $Q_{IN,i}$ [$m^3\ s^{-1}$] is inflow water to the reservoir for $i = 0, \ldots, n-1$. $Q_{IN,i}$ is generated by future scenarios of regional climate models from [63]. $Q_{OUT,i}$ [$m^3\ s^{-1}$] is the reservoir outflow for $i = 0, \ldots, n - 1$. The $Q_{OUT,i}$ is formulated by the sum of the $Q_{ECO,i}$, $Q_{WS:drink.,i}$, $Q_{WS:ind.+agric.,i}$, and $Q_{EXC,i}$ (water excess for hydropower). Δt is the time step of calculation (one month). For $i = 0$ it is necessary to enter the initial condition (full active storage capacity). Inequality (11) is limited from the left side by water level (0) which characterizes full active storage capacity and from the right as active storage capacity divided by level (I–III) as shown in Figure 2. Each level of active storage capacity is characterized by given outflow $Q_{OUT,i}$.

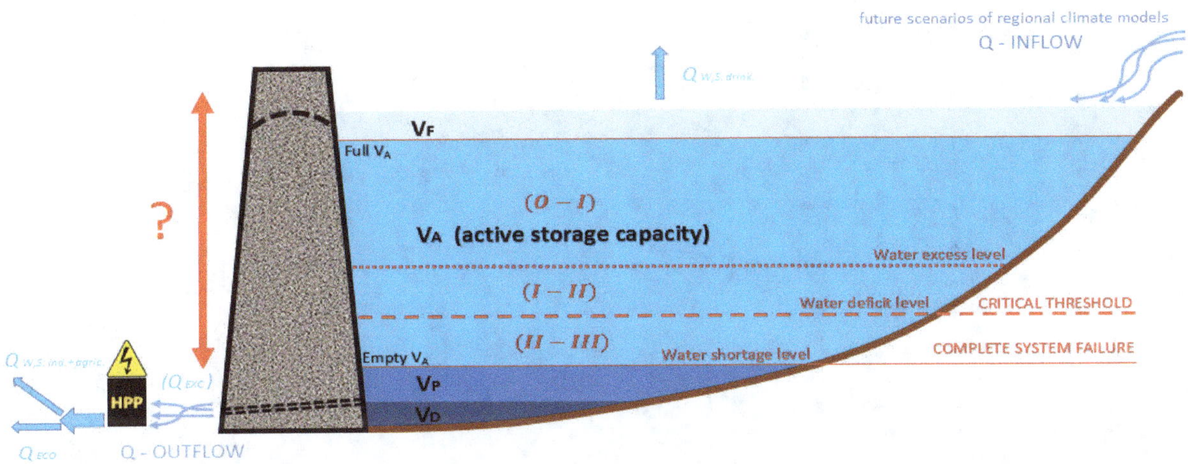

Figure 2. The scheme of the model set up.

The resulting scheme of dam water storage in Figure 2 shows variants of water levels, where V_F is flood storage capacity, V_A is active storage capacity, V_P is permanent storage capacity, and V_D characterizes dead storage capacity.

When the current step i emptying and filling of storage capacity is going up from level (0–III) in (11), then the $Q_{OUT,i}$ [$m^3\ s^{-1}$] is equal to the $Q_{IN,i}$, [$m^3\ s^{-1}$] otherwise new operating rules and restrictions are created according to the following control Equations (12)–(14):

$$\text{0–I:} \quad Q_{OUT,i} = Q_{ECO,i} + Q_{WS:drink,i} + Q_{WS:ind.+agric.,i} + Q_{EXC,i}, \quad (12)$$

$$\text{I–II:} \quad Q_{OUT,i} = Q_{ECO,i} + Q_{WS:drink,i} + Q_{WS:ind.+agric.,i}, \quad (13)$$

$$\text{II–III:} \quad Q_{OUT,i} = Q_{ECO,i} + (Q_{WS:drink,i}) \times 0.7 + \left(Q_{WS:ind.+agric.,i}\right) \times 0.3, \quad (14)$$

As mentioned above, the water level (0) which characterizes full active storage capacity and water level (III) is empty active storage capacity. The volume between levels I–III is set up 11.6 mil. m^3. This volume corresponds to three months of estimated total demand. The volume between levels II–III correspond to the total demand in given month of the year.

When the volume of water is between level I–II, water excess $Q_{EXC,i}$ is null. When the volume of water drops down the critical threshold (II) the deficit begins. At this moment, the restriction of reservoir management by priority roles begins. The main rule is to guarantee Q_{ECO} and next rule is to ensure $Q_{WS:drink.}$ and $Q_{WS:ind.+agric.}$ in the ratio of 70:30, see Equation (14).

4.5. Optimization Method

For the case study, the model in cooperation with the NSGA II algorithm was used in order to create decision variables population of an excess of water $Q_{EXC,i}$.

The key parameters for NSGA II are possible decision variables, dam heights and analyzes of reservoir performance across all climate scenarios in the simulation model. Data output of the optimization method are presented in a form of Pareto sets showing relations between net present value and resilience of the reservoir.

5. Result and Discussion

The obtained relation between resilience and *NPV* in the form of Pareto sets is in Figure 3. For each dam height, 15 climate scenarios from 2015 to 2100 were used from [63] for the planned profile.

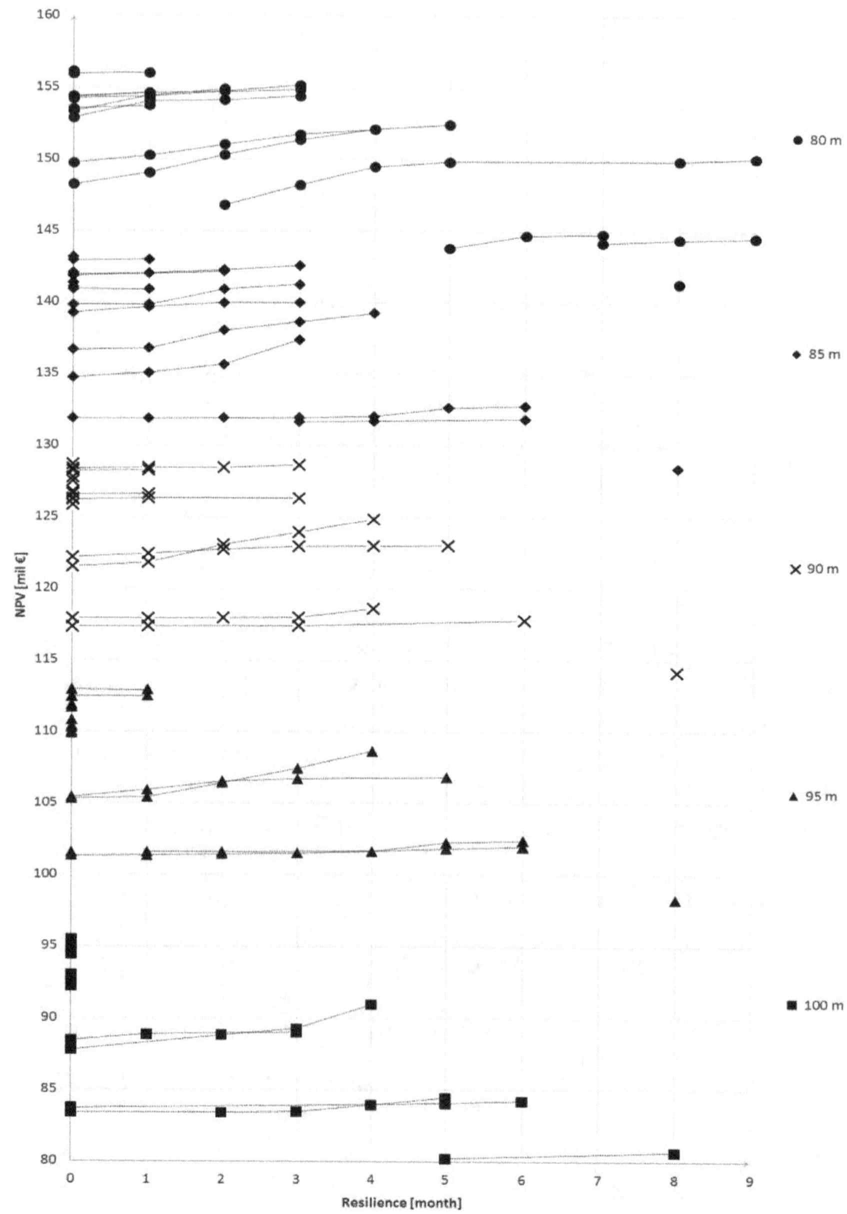

Figure 3. The Pareto optimal sets for resilience vs. *NPV* for varying dam heights and all scenarios.

Then each obtained point of these Pareto optimal sets presents the best solution and same a possible value of excess water Q_{EXC} (decision variables) or the total reservoir outflow Q_{OUT}. All these values can be applied to the design and operation of the dam.

Figure 3 shows increasing resilience (short deficit) in relation to reducing *NPV*. That means less resilient solution (longer deficit duration) representing higher *NPV* (higher benefits of the dam). Pareto sets are relatively flat (only slightly curved), this is mainly due to the setting of present prices for electric power and water supply. It could be expected that future water price increasing would make the Pareto sets more curved because the benefits will increase. Another factor that influences the shape of the Pareto sets could be the dam type.

As it can be seen from Figure 3, the results are better for the lower dam heights, as there are lower costs for construction of the dam. It is obvious that the increasing dam height increases the cost for construction of the dam, but benefits are lower than costs. Lower dam heights bring better results. In addition, building a bigger dam generally induces higher environmental costs, but it was not considered in this case study due to the difficult price quantification

Only Pareto points with 0, 3, and 5 months of resilience, for the lower suitable dam heights of the 80 m, 85 m, and 90 m and only 4 from 15 climate scenarios with different mean annual flows Q_a: as (Sc1) the most water (Q_a = 2.29 m^3 s^{-1}), as (Sc2) more water (Q_a = 2.17 m^3 s^{-1}), as (Sc3) less water (Q_a = 2.09 m^3 s^{-1}), and as (Sc4) the least water (Q_a = 1.84 m^3 s^{-1}) were chosen, see Figure 4.

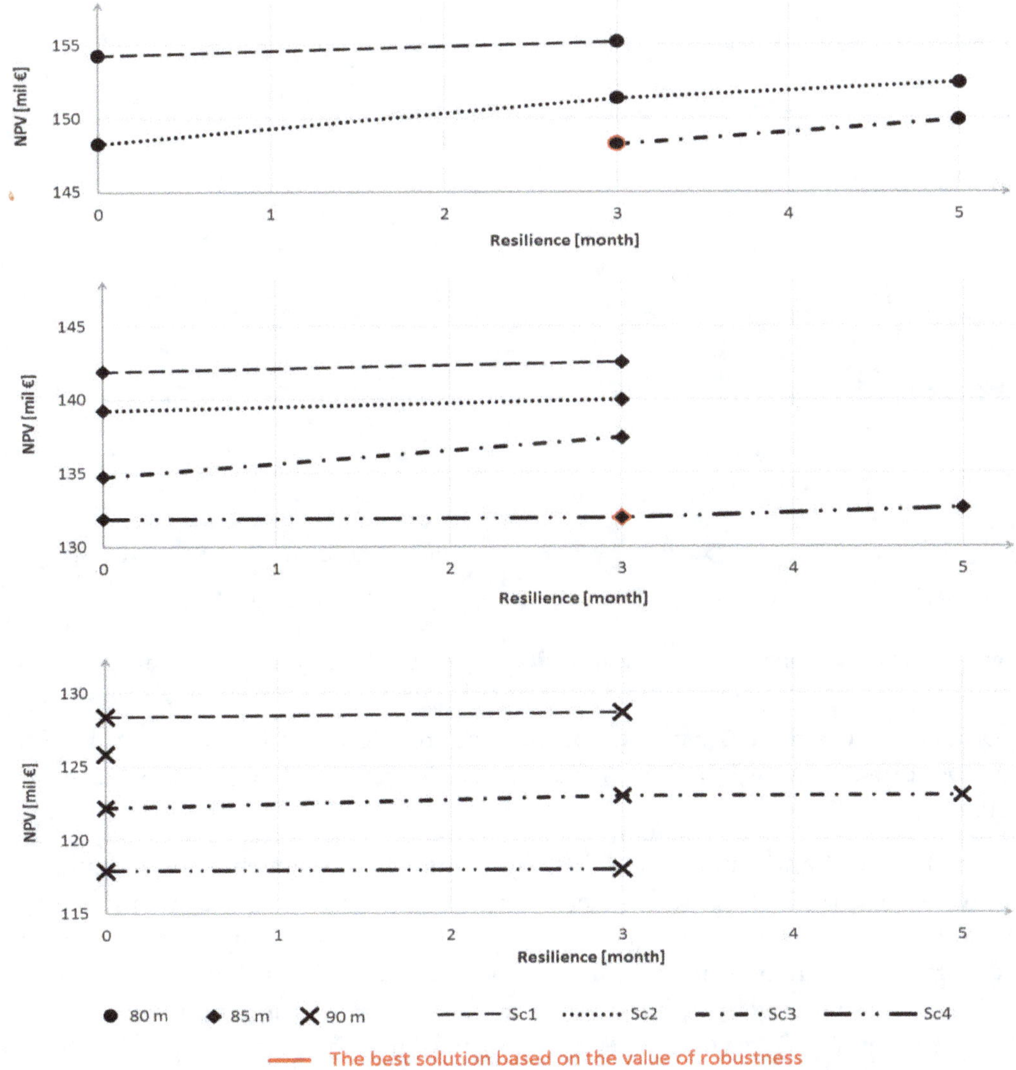

Figure 4. The Pareto optimal sets from four selected climatic scenarios for resilience vs. *NPV*.

Figure 4 shows the Pareto sets for lower dam heights of Figure 3 and only for selected climate scenarios Sc1, Sc2, Sc3, and Sc4. It is clear, that the more water scenarios (Sc1 and Sc2) produce higher the *NPV*, because the benefits are higher for the same cost and *RES*.

Figure 5 specifically shows the relation between the present value of benefit *PV*(*benefit*) and the present value of cost *PV*(*cost*) for the selected four climate scenarios with different mean annual flows Q_a, resilience of the dam from 0 to 5 months and heights of dam 80, 85, and 90 m. As mentioned above, Figure 5 confirms that the more water scenarios (Sc1 and Sc2) cause higher the *NPV* (higher benefit) than less water scenarios (Sc3 and Sc4).

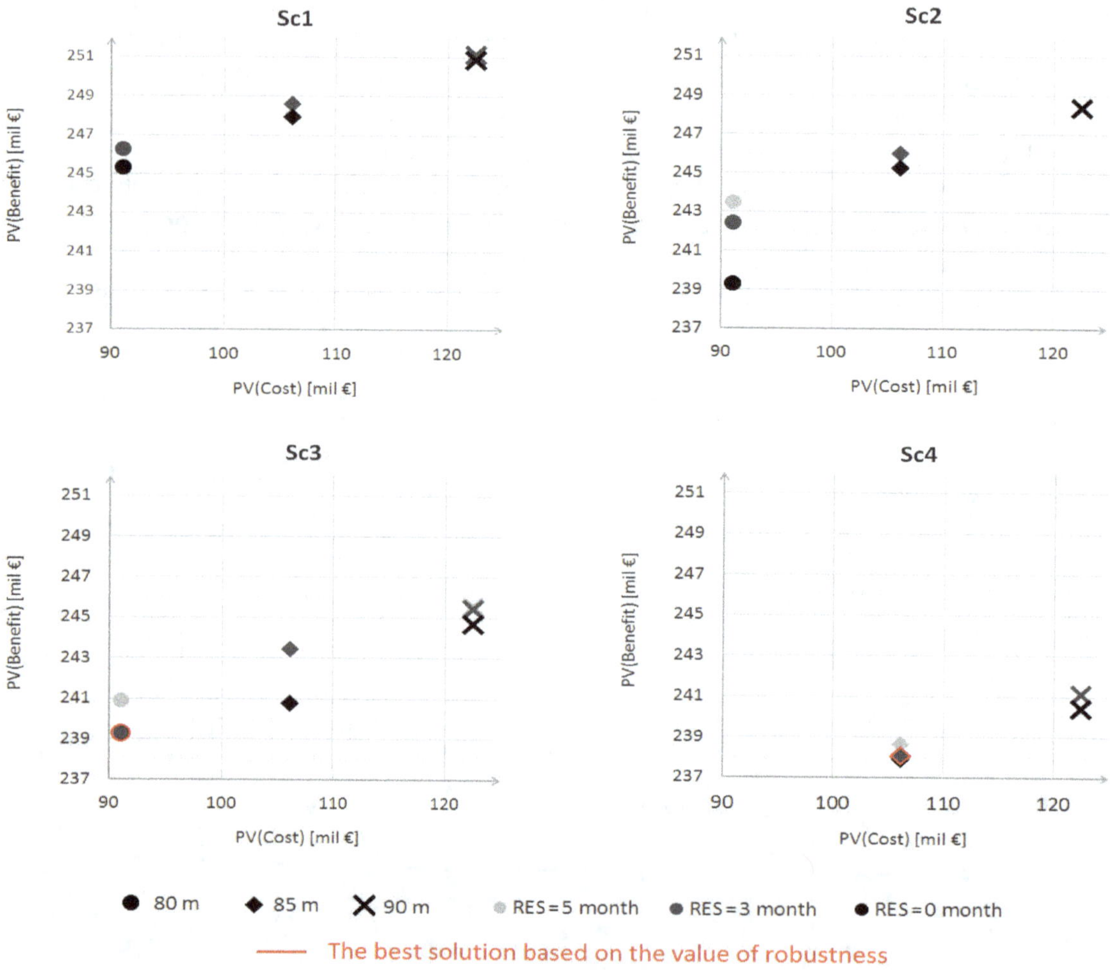

Figure 5. The relations between *PV* (*benefit*) and *PV* (*cost*) for selected climate scenarios.

Lower points in Figure 5 are the points for the most resilient design of the dam (*RES* = 0 months) and the lowest benefits. The higher points refer to the lowest resilience (*RES* = 5 months) and the highest benefits.

The points in Figures 4 and 5 mark the selected results which were chosen for testing of robustness. The results were in the form of decision variables (Q_{EXC}) achieved by optimization for four climatic scenarios. Then the decision variables were tested in the reservoir simulation model for 15 climatic scenarios and different dam heights variants to determine the resilience. After that robustness evaluation was finally made. The robustness was calculated as the fraction number of scenarios with *RES* = 3 months or less (5 months or less, 0 months) and the total number of climate scenarios, according to Equation (2). In Table 2, we can see the resulting robustness, and *NPV* for the points from Figures 4 and 5 (*RES* = 0, 3, and 5 months or less).

Table 2. Robustness for selected points.

		Height 80.0 m				Height 85.0 m				Height 90.0 m			
	C_{DAM} [mil. €] V_A [m^3]	91.05 34,480,000				106.03 44,940,000				122.51 57,270,000			
RES [month]	Scenario:	Sc1	Sc2	Sc3	Sc4	Sc1	Sc2	Sc3	Sc4	Sc1	Sc2	Sc3	Sc4
0	*ROB* [%]	66.7	73.3	-	-	66.7	73.3	**80.0**	**93.3**	66.7	66.7	**80.0**	**93.3**
	⌀ *NPV* [mil. €]	151.4	147.3	-	-	138.0	137.2	134.3	132.2	123.5	123.4	121.6	118.5
3	*ROB* [%]	53.3	73.3	**80.0**	-	60.0	66.7	73.3	**93.3**	66.7	-	73.3	**93.3**
	⌀ *NPV* [mil. €]	151.0	149.7	147.8	-	137.9	138.1	137.1	132.5	123.5	-	122.8	118.6
5	*ROB* [%]	-	60.0	60.0	-	-	-	-	80.0	-	-	66.7	-
	⌀ *NPV* [mil. €]	-	150.9	149.8	-	-	-	-	134.3	-	-	123.1	-

In Table 2 we can see, if the decision variables for the less water scenarios (Sc3 and Sc4) are selected, than robustness for given solutions will be higher, because decision variables (Q_{EXC}) were lower than decision variables from the more water scenarios. According to Table 2, it is obvious that one of the best results is for the decision variables in Sc3, resilience $RES \leq 3$ months, robustness $ROB \geq 80.0\%$, and then the dam height is 80.0 m. The average *NPV* is 147.8 mil. €, the cost of the construction of dam is 91.05 mil. €, and the benefits from the hydropower and water supply are on average 238.8 mil. €. The active storage capacity V_A is 34.48 mil. m^3 and total volume of the dam is 42.63 mil. m^3.

Another optimal solution is for the decision variables in Sc4, $RES \leq 3$ months and $ROB \geq 90.0\%$, and then the dam height is 85.0 m. The average *NPV* is 132.5 mil. €, the C_{DAM} = 106.03 mil. €, and the *BEN* are on average 238.5 mil. €. The V_A is 44.94 mil. m^3 and total volume of the dam is 53.09 mil. m^3. If we compare chosen results cost for construction, the dam in more robust solution will be about 15 mil. € (16.5%) more expensive than a less robust solution. Benefits are 0.3 mil. € (0.1%) lower and the *NPV* is lower about 15.3 mil. € (11.5%).

The results of the decision variables in form monthly water excess for hydropower Q_{EXC} and following the total reservoir outflow Q_{OUT} for each month and for *ROB* = 80.0% and 90.0% are presented in Table 3.

Table 3. Selected optimal results of the monthly water excess Q_{EXC} and the total reservoir monthly outflow Q_{OUT} for two target levels of *ROB*.

		January	February	March	April	May	June	July	August	September	October	November	December
$ROB \geq 80\%$	Q_{EXC} [m^3 s^{-1}]	0.93	1.12	0.06	0.00	0.13	0.00	0.00	0.01	0.00	0.00	0.00	0.02
(h = 80 m)	Q_{OUT} [m^3 s^{-1}]	2.36	2.57	1.52	1.49	1.63	1.54	1.50	1.50	1.47	1.46	1.44	1.44
$ROB \geq 90\%$	Q_{EXC} [m^3 s^{-1}]	0.18	0.01	0.00	0.01	0.00	0.00	0.00	0.00	0.00	0.00	0.09	0.02
(h = 85 m)	Q_{OUT} [m^3 s^{-1}]	1.61	1.46	1.46	1.50	1.50	1.54	1.50	1.49	1.47	1.46	1.53	1.44

In Table 3 we can see that for $ROB \geq 80\%$, the dam height is only 80.0 m and water excess Q_{EXC}, and thus total reservoir outflow Q_{OUT} are higher than for the $ROB \geq 90\%$.

The decision variables (water excess Q_{EXC}) could increase in the future by the rise of water and electricity prices. The question is what the future evolution of water and electricity demand will be. The model is using only the long-term scenarios for hydrology changes and current prices for water and electricity. There is a huge amount of uncertainties in the prediction of water and electricity demands as well as market prices of these commodities in the upcoming decades. Thus, in future, it would be useful to add several future demands or socioeconomic scenarios. It should be noted that the climate scenarios in this study are based on only one conservative emission scenario A1B. Further, it could be very interesting to test the robustness effect on several percentage sets of an increase in water and electricity demand and increase in water and electricity prices for a given result.

6. Conclusions

This article describes a new possible attitude to support decision-making in the future for a new reservoir management within a decrease water resources scenario. A simulation model has been coupled with a cost model and an NSGA II multi-objective optimization algorithm to quantify resilience and robustness under a range of uncertain future climate supply scenarios and one possible demand scenario. The results obtained lead to the following key conclusions:

- The analysis of the different dam heights produced different recommendations for the multipurpose reservoir design. This approach has recommended a specific design of the dam height of 80 m for $ROB = 80\%$ and $RES = 3$ months or 85 m for $ROB = 90\%$ and $RES = 3$ months.
- As a result, recommended dam height, and also recommended total reservoir monthly outflows that, in combination with the control equations, determine new operational rules of a multipurpose water reservoir.
- The new operating rules have been created: The water management in the reservoir should be set according to the current month and recommended reservoir monthly outflows Q_{OUT} in Table 3. In the case of dry periods when the water level falls below level (I) or more as it is shown in Figure 1, the Q_{OUT} have to be restricted according to the control Equation (13) or (14).
- All potential resilience metrics are based on those in a past paper [14] which were tested in this analysis and gave similar shapes of Pareto sets, and therefore, the traditional metric of resilience with duration of the longest water deficit period was selected.
- The model clearly shows that the higher dam heights increase the cost for construction of the dam, but the benefits are lower than costs, therefore the results are better for lower dam height.
- Although, in the summer months, due to the higher water demands and lower flow in the river, the water excess is minimal or null, the minimal demand is guaranteed for targets resilience and robustness.
- A more robust solution generally produced lower benefits respectively lower *NPV* due to a lower water excess. Under the current conditions, it may seem as little profitable, but in the future, with the development of climate change uncertainty, the price of water is expected to increase, and the benefits will be higher.
- The key conclusions, based on the results obtained, serve only as recommendations, but a final decision on the safeness and economy of a new reservoir is on-site of decision-makers.

Based on the case study including climate scenarios, we can recommend using the created universal model considering drought prevention. This new approach to design and operation of the multipurpose reservoir in conditions of uncertain climate change has been tested on the case study and can be applicable to cases in similar conditions with the problem of decreasing water sources. However, several points of the methodological section are specific to the case study and cannot be generalized to all dams.

Author Contributions: S.P. performed the design, collected the data, made the calculations and analysis, and wrote the paper. Z.K. set the objectives of the optimization method, provided useful advice, contributed to the interpretation of results, and made a revision. D.M. helped with setting the reservoir simulation model, contributed to the interpretation of results, and text corrections.

References

1. World Meteorological Organization. *WMO Statement on the State of the Global Climate in 2017*; Chairperson, Publications Board: Geneva, Switzerland, 2018; ISBN 978-92-63-11212-5.
2. Czech Document: Příprava Realizace Opatření pro Zmírnění Negativních Dopadů Sucha a Nedostatku Vody. *Meziresortní Komise VODA-SUCHO*. 2015. Available online: http://eagri.cz/public/web/file/417667/_3_material_VLADA.pdf (accessed on 1 April 2018).

3. Czech Government Document: Komplexní Studie Dopadů, Zranitelnosti a Zdrojů Rizik Souvisejících se Změnou Klimatu v ČR. 2015. EKOTOXA s.r.o.. Available online: https://www.mzp.cz/C1257458002F0DC7/cz/studie_dopadu_zmena_klimatu/$FILE/OEOK-Komplexni_studie_dopady_klima-20151201.pdf (accessed on 10 April 2018).
4. Czech Government Document: Národní Akční Plán Adaptace na Změnu Klimatu. Ministerstvo Životního Prostředí. 2015. Available online: https://www.mzp.cz/C1257458002F0DC7/cz/narodni_akcni_plan_zmena_klimatu/$FILE/OEOK-NAP_cely_20170127.pdf (accessed on 3 April 2018).
5. Czech Government Document: Strategie Přizpůsobení se Změně Klimatu v Podmínkách ČR. Ministerstvo Životního Prostředí. 2015. Available online: http://www.mzp.cz/C1257458002F0DC7/cz/zmena_klimatu_adaptacni_strategie/$FILE/OEOK-Adaptacni_strategie-20151029.pdf (accessed on 1 April 2018).
6. Czech Technical Standard ČSN 75 2405 Reservoir Storage Capacity Analysis, ICS 93.160. Available online: http://seznamcsn.unmz.cz/Detailnormy.aspx?k=69792 (accessed on 3 April 2018).
7. Kritskiy, S.N.; Menkel, M.F. *Water Management Computations*; GIMIZ: Leningrad, Russia, 1952. (In Russian)
8. Hashimoto, T.; Stedinger, J.R.; Loucks, D.P. Reliability, resiliency, and vulnerability criteria for water resource system performance evaluation. *Water Resour. Res.* **1982**, *18*, 14–20. [CrossRef]
9. Moy, W.-S.; Cohon, J.L.; ReVelle, C.S. A Programming Model for Analysis of the Reliability, Resilience, and Vulnerability of a Water Supply. *Water Resour. Res.* **1986**, *22*, 489–498. [CrossRef]
10. Zhang, C.; Xu, B.; Li, Y.; Fu, G. Exploring the relationships among reliability, resilience, and vulnerability of water supply using many-objective analysis. *J. Water Resour. Plan. Manag.* **2017**, *143*. [CrossRef]
11. Matrosov, E.S.; Padula, S.; Harou, J.J. Selecting portfolios of water supply and demand management strategies under uncertainty-contrasting economic optimisation and "Robust decision making" approaches. *Water Resour. Manag.* **2012**, *27*, 1123–1148. [CrossRef]
12. Paton, F.L.; Dandy, G.C.; Maier, H.R. Integrated framework for assessing urban water supply security of systems with non-traditional sources under climate change. *Environ. Model. Softw.* **2014**, *60*, 302–319. [CrossRef]
13. Fowler, H.J.; Kilsby, C.G.; O'Connell, P.E. Modeling the impacts of climatic change and variability on the reliability, resilience, and vulnerability of a water resource system. *Water Resour. Res.* **2003**, *39*, 1222. [CrossRef]
14. Roach, T. Decision Making Methods for Water Resources Management Under Deep Uncertainty. Ph.D. Theses, University of Exeter, Exeter, UK, 2016.
15. Roach, T.; Kapelan, Z.; Ledbetter, R. Comparison of Info-gap and Robust Optimisation Methods for Integrated Water Resource Management under Severe Uncertainty. *Procedia Eng.* **2015**, *119*, 874–883. [CrossRef]
16. Watts, G.; Christierson, B.; Hannaford, J.; Lonsdale, K. Testing the resilience of water supply systems to long droughts. *J. Hydraul.* **2012**, *414–415*, 255–267. [CrossRef]
17. Amarasinghe, P.; Liu, A.; Egodawatta, P.; Bernes, P.; McGree, J.; Goonetilleke, A. Quantitative assessment of resilience of a water supply system under rainfall reduction due to climate change. *J. Hydraul.* **2016**, *540*, 1043–1052. [CrossRef]
18. Li, Y.; Lence, B.J. Estimating resilience for water resources systems. *Water Resour. Res.* **2007**, *43*, W07422. [CrossRef]
19. Simonovic, S.P.; Arunkumar, R. Comparison of static and dynamic resilience for a multipurpose reservoir operation. *Water Resour. Res.* **2016**, *52*, 8630–8649. [CrossRef]
20. Sweetapple, C.; Fu, G.; Butler, D. Reliable, Robust, and Resilient System Design Framework with Application to Wastewater-Treatment Plant Control. *J. Environ. Eng.* **2017**, *143*, 04016086. [CrossRef]
21. Knight, F. *Risk, Uncertainty and Profit*; Hart, Schaffner & Marx: Boston, MA, USA; Houghton Mifflin Co.: Boston, MA, USA, 1921.
22. Kaplan, S. *Risk Assessment and Risk Management—Basic Concepts and Terminology. Risk Management: Expanding Horizons in Nuclear Power and other Industries*; Hemisphere Publ. Corp.: Boston, MA, USA, 1991; pp. 11–28.
23. Beven, K.J.; Binley, A.M. The future of distributed models: Model calibration and uncertainty prediction. *Hydrol. Proceaaes* **1992**, *6*, 279–298. [CrossRef]
24. Mahmoud, M.; Liu, Y.; Hartmann, H.; Stewart, S.; Wagener, T.; Semmens, D.; Stewart, R.; Gupta, H.; Dominguez, D.; Dominguez, F.; et al. A formal framework for scenario development in support of environmental decision-making. *Environ. Model. Softw.* **2009**, *24*, 798–808. [CrossRef]

25. Marton, D.; Paseka, S. Uncertainty Impact on Water Management Analysis of Open Water Reservoir. *Environments* **2017**, *4*, 10. [CrossRef]
26. Marton, D.; Starý, M.; Menšík, P. Analysis of the influence of input data uncertainties on determining the reliability of reservoir storage capacity. *J. Hydrol. Hydromech.* **2015**, *63*, 287–294. [CrossRef]
27. Marton, D.; Starý, M.; Menšík, P.; Paseka, S. Hydrological Reliability Assessment of Water Management Solution of Reservoir Storage Capacity in Conditions of Uncertainty. In *Drought: Research and Science-Policy Interfacing*; CRC Press Taylor & Francis Group: Leiden, The Netherlands, 2015; pp. 377–382. ISBN 978-1-138-02779-4.
28. Paseka, S.; Marton, D.; Menšík, P. Uncertainties of reservoir storage capacity during low water period. In Proceedings of the SGEM International Multidisciplinary Geoconference: Hydrology and Water Resources; STEF92 Technology Ltd.: Sofia, Bulgaria, 2016; pp. 789–796, ISBN 978-619-7105-61-2.
29. Vogel, R.M.; Bolognes, A. Storage-reliability-resilience-yield relations for over-year water supply systems. *Water Resour. Res.* **1995**, *31*, 645–654. [CrossRef]
30. Jain, K.S.; Bhunya, K.P. Reliability, resilience and vulnerability of a multipurpose storage reservoir. *Hydrol. Sci. J.* **2008**, *53*, 434–447. [CrossRef]
31. Ngo, L.L.; Madsen, H.; Rosbjerg, D. Simulation and optimisation modelling approach for operation of the Hoa Binh reservoir, Vietnam. *J. Hydrol.* **2007**, *336*, 269–281. [CrossRef]
32. Villa, I.R.; Rodríguez, J.B.M.; Molina, J.L.; Tarragó, J.C.P. Multiobjective Optimization Modeling Approach for Multipurpose Single Reservoir Operation. *Water* **2018**, *10*, 427. [CrossRef]
33. Scola, L.A.; Takahashi, R.H.C.; Cerqueira, S.A.A.G. Multipurpose Water Reservoir Management: An Evolutionary Multiobjective Optimization Approach. *Math. Probl. Eng.* **2014**, *2014*, 638259. [CrossRef]
34. Groves, D.G.; Yates, D.; Tebaldi, C. Developing and applying uncertain global climate change projections for regional water management planning. *Water Resour. Res.* **2008**, *44*, W12413. [CrossRef]
35. Herman, J.D.; Zeff, H.B.; Reed, P.M.; Characklis, G.W. Beyond optimality: Multistakeholder robustness tradeoffs for regional water portfolio planning under deep uncertainty. *Water Resour. Res.* **2014**, *50*, 7692–7713. [CrossRef]
36. Beh, E.H.Y.; Maier, H.R.; Dandy, G.C. Adaptive, multiobjective optimal sequencing approach for urban water supply augmentation under deep uncertainty. *Water Resour. Res.* **2015**, *51*, 1529–1551. [CrossRef]
37. Haasnoot, M.; Kwakkel, J.H.; Walker, W.E.; Maat, T.J. Dynamic adaptive policy pathways: A method for crafting robust decisions for a deeply uncertain world. *Glob. Environ. Chang.* **2013**, *23*, 485–498. [CrossRef]
38. Jeuland, M.; Whittington, D. Water resources planning under climate change: Assessing the robustness of real options for the Blue Nile. *Water Resour. Res.* **2014**, *50*, 2086–2107. [CrossRef]
39. Paton, F.L.; Maier, H.R.; Dandy, G.C. Including adaptation and mitigation responses to climate change in a multiobjective evolutionary algorithm framework for urban water supply systems incorporating GHG emissions. *Water Resour. Res.* **2014**, *50*, 6285–6304. [CrossRef]
40. Roach, T.; Kapelan, Z.; Ledbetter, R.; Ledbetter, M. Comparison of Robust Optimization and Info-Gap Methods for Water Resource Management under Deep Uncertainty. *J. Water Resour. Plan. Manag.* **2017**, *143*. [CrossRef]
41. Whateley, S.; Steinschneider, S.; Brown, C. A climate change range-based method for estimating robustness for water resources supply. *Water Resour. Res.* **2014**, *50*. [CrossRef]
42. Borgomeo, E.; Mortazavi-Naeini, M.; Hall, J.W.; Guillod, B.P. Risk, Robustness and Water Resources Planning Under Uncertainty. *Earth's Future* **2018**, *6*. [CrossRef]
43. Deb, K.; Agrawal, S.; Pratap, A.; Meyarivan, T.; Amrit, P.; Meyarivan, T. *A Fast Elitist Non-Dominated Sorting Genetic Algorithm for Multi-Objective Optimization: NSGA-II*; Schoenauer, M., Deb, K., Rudolph, G., Yao, X., Lutton, E., Merelo, J.J., Schwefel, H.-P., Eds.; Parallel Problem Solving from Nature PPSN VI; Springer: Berlin/Heidelberg, Germany, 2000; pp. 849–858.
44. Kollat, J.B.; Reed, P.M. Comparing state-of-the-art evolutionary multiobjective algorithms for long-term groundwater monitoring design. *Adv. Water Res.* **2006**, *29*, 792–807. [CrossRef]
45. Nicklow, J.; Reed, P.; Savic, D.; Dessalegne, T.; Harrell, L.; Chan-Hilton, A.; Karamouz, M.; Minsker, B.; Ostfeld, A.; Singh, A.; et al. State of the art for genetic algorithms and beyond in Water resources planning and management. *J. Water Resour. Plan. Manag.* **2010**, *136*, 412–432. [CrossRef]
46. Tukimat, N.N.A.; Harun, S. Optimization of water supply reservoir in the framework of climate variation. Inter. *J. Softw. Eng. Appl.* **2014**, *8*, 361–378. [CrossRef]

47. Fu, G.; Kapelan, Z.; Kasprzyk, J.R.; Reed, P. Optimal Design of Water Distribution Systems Using Many-Objective Visual Analytics. *J. Water Resour. Plan. Manag.* **2013**, *139*. [CrossRef]
48. Perelman, L.; Ostfeld, A.; Salomons, E. Cross Entropy multiobjective optimization for water distribution systems design. *Water Resour. Res.* **2008**, *44*. [CrossRef]
49. Zheng, F.; Simpson, A.R.; Zecchin, A.C. An efficient hybrid approach for multiobjective optimization of water distribution systems. *Water Resour. Res.* **2014**, *50*. [CrossRef]
50. Yang, G.; Guo, S.; Liu, P.; Li, L.; Liu, Z. Multiobjective Cascade Reservoir Operation Rules and Uncertainty Analysis Based on PA-DDS Algorithm. *J. Water Resour. Plan. Manag.* **2017**, *143*. [CrossRef]
51. Srinivas, N.; Deb, K. Muiltiobjective Optimization Using Nondominated Sorting in Genetic Algorithms. *Evol. Comput.* **1994**, *2*, 221–248. [CrossRef]
52. Goldberg, D.E. *Genetic Algorithms in Search, Optimization, and Machine Learning*; Addison-Wesley Publishing Company: Reading, MA, USA, 1989; ISBN 0-201-15767-5.
53. Marton, D.; Kapelan, Z. Risk and reliability analysis of open reservoir water shortages using optimization. *Procedia Eng.* **2014**, *89*, 1478–1485. [CrossRef]
54. Savić, D.A.; Bicik, J.; Morley, M.S. A DSS Generator for Multiobjective Optimisation of Spreadsheet-Based Models. *Environ. Model. Softw.* **2011**, *26*, 551–561. [CrossRef]
55. Borgomeo, E.; Hall, J.W.; Fung, F.; Watts, G.; Colquhoun, K.; Lambert, C. Risk-based water resources planning: Incorporating probabilistic nonstationary climate uncertainties. *Water Resour. Res.* **2014**, *50*, 6850–6873. [CrossRef]
56. Deepashree, R.; Mujumdar, P.P. Reservoir performance under uncertainty in hydrologic impacts of climate change. *Adv. Water Res.* **2010**, *33*, 312–326. [CrossRef]
57. Kling, H.; Stanzel, P.; Preishuber, M. Impact modelling of water resources development and climate scenarios on Zambezi River discharge. *J. Hydrol.* **2014**, *1*, 17–43. [CrossRef]
58. Majone, B.; Bovolo, C.I.; Bellin, A.; Blenkinsop, S.; Fowler, H.J. Modeling the impacts of future climate change on water resources for the Gállego river basin (Spain). *Water Resour. Res.* **2012**, *48*, W01512. [CrossRef]
59. Mantel, S.K.; Hughes, D.A.; Slaughter, A.S. Water resources management in the context of future climate and development changes: A South African case study. *J. Water Clim. Chang.* **2015**, *6*, 772–786. [CrossRef]
60. Maiolo, M.; Mendicino, G.; Senatore, A.; Pantusa, D. Optimization of drinking water distribution systems in relation to the effects of climate change. *Water* **2017**, *9*, 803. [CrossRef]
61. Kapelan, Z.; Savić, D.; Mahmoud, H. A Response Methodology for Reducing Impacts of Failure Events in Water Distribution Networks. *Procedia Eng.* **2017**, *186*, 218–227. [CrossRef]
62. IPCC Secretariat. *Climate Change 2007: Impacts, Adaptation and Vulnerability: Summary for Policymakers*; IPCC Secretariat: Geneva, Switzerland, 2007; ISBN 92-9169-121-6.
63. Czech Project. *Development of Information and Data Support for Design of Adaptation Measures and Long-Term Planning of Water Resources Considering the Climate Change Effects*; T. G. Masaryk Water Research Institute: Praha, Czech Republic; Czech Hydro-Meteorological Institute: Praha, Czech Republic; Water Management Development, RSCN: Jubeiha, Jordan, 2012–2014.
64. Czech Document PRVKUK. *The Plan for the Development of Water Supply and Sewerage Systems in the Olomouc Region*; The Regional Office of the Olomouc Region, Ministry of Agriculture of the Czech Republic: Olomouc, Czech Republic, 2006.
65. Klemeš, V. A simlified solution of the flood-routing problem. *Vodohosp. Časopis* **1960**, *8*, 317–326.
66. Government Finance Function. *The Green Book: Appraisal and Evaluation in Central Government*; Government Finance Function: London, UK, 2016.
67. Czech Property Valuation Decree. *Vyhlaska č. 441/2013 Sb., Vyhlaska k Provedeni Zakona o Ocenovani Majetku*; Ministry of Finance of the Czech Republic: Praha, Czech Republic, 2013.
68. Starý, M. *Reservoir and Reservoir System*; Education Tutorial, Faculty of Civil Engineering, Brno University of Technology: Brno, Czech Republic, 2006.

12

Sensitivity Analysis for the Inverted Siphon in a Long Distance Water Transfer Project: An Integrated System Modeling Perspective

Sifan Jin [1], Haixing Liu [1], Wei Ding [1,*], Hua Shang [2] and Guoli Wang [1]

[1] School of Hydraulic Engineering, Dalian University of Technology, Dalian 116024, China; sfjin@mail.dlut.edu.cn (S.J.); hliu@dlut.edu.cn (H.L.); wanggl@dlut.edu.cn (G.W.)

[2] Faculty of Management and Economics, Dalian University of Technology, Dalian 116024, China; shanghua@dlut.edu.cn

* Correspondence: weiding@dlut.edu.cn

Abstract: Long distance water diversion projects are developed to alleviate the conflicts between supply and demand of water resources across different watersheds. However, the significant scale water diversion projects bring new challenges for the water supply security. This paper presents the flood risk of inverted siphon structure which is used for crossing transversally in the water diversion project through sensitivity analysis. Soboĺ and regionalized sensitivity analysis are used to investigate the sensitive parameters of the integrated model and the sensitive range of the parameters, respectively. The integrated system model consists of the hydrologic model, the sediment transport model and the siphon hydraulic model to determine the flood overtopping duration and volume, which are used to quantify flood risk in this study. The flood overtopping duration and volume indicators are used to quantify flood risk in the sensitivity analysis. The South to North Water Diversion Project in China is used as a case study. The results show the mean rainfall and roughness coefficient of the pipe are the most sensitive parameters in the integrated models, while the sensitive range of these two parameters are distinct. The sensitivity analysis of the inverted siphon provides an insight into the significant contributions to the flood risk. The analysis can provide the guidance for the system operation security.

Keywords: long distance water diversion; inverted siphon; sensitivity analysis; integrated supply system modeling

---

1. Introduction

Due to urbanization and the uneven distribution of water resources in time and space, long-distance water transfer projects are constructed to alleviate the shortage of water resource and meet the increasing demands [1–3]. Water transfer projects often involve huge capital investment and pose complex security problems [4–6]. For example, canals usually cross hundreds of kilometers of complicated terrain, which leads to water quality deterioration, temperature variation, and long operation response period. Moreover, critical hydraulic structures are essential for ensuring the operation security of the projects, e.g., gate/valve, pump station, and intersection structures. Here, we focus on the inverted siphon structure, which is linked to the river to drain water by gravity that is collected from a relatively small hydrographic basin, so that the river can cross transversally a large artificial open channel carrying fresh water for supply.

Inverted siphon structures are prone to be impacted by the hydraulic transient process and structure stability. The potential failures of inverted siphons can be divided into two categories: (1) structure failure; and (2) operation failure. Structure failure indicates the structure identity is destroyed or collapsed due to structure aging and external forces. Operation failure refers to the flows exceeding the design standard of inverted siphon. That is because the designed flood and flood design standard that are used to guide the structure design are subject to the hydrologic parameter uncertainties (e.g., rainfall, soil moisture, and land use) and the hydraulic uncertainties (e.g., sediment deposition). The extra upstream flow of inverted siphon is retained at the entry and a large flood can potentially enter the canal and contaminate the quality of source water in the canal. Therefore, there is a need to investigate which uncertain parameters (e.g., rainfall, sediments, and pipe roughness coefficient) can significantly contribute to the flooding incidents. This paper utilizes an inverted siphon structure that passes underneath a long distance transfer project to illustrate the flood overtopping risk. The sensitivity analysis with respect to inverted siphons is of vital importance to guarantee the safe operation of the long-distance transfer project.

Sensitivity Analysis (SA) has gathered plenty of attention for describing the sensitivity of parameters in terms of the contributions to the model output [7]. Global SA method, in contrast to local SA, is capable of accounting for the whole range of input parameter variation to avoid the subjective judgment and case-specific characteristics on the parameter range. The global SA can deal with the non-linear and non-monotonic models [8–10]. Another type of SA is regression-related, e.g., Multivariate Adaptive Regression Splines (MARS), Gaussian Process (GP) and Radial Basis Function [11]. They use linear or non-linear models to refit the original model and investigate which parameters can give the relative large reduction in goodness-of-fit. The large reduction represents the sensitive parameters [12]. The regression sensitivity methods have the advantage of less computational effort. Moreover, Soboĺ SA [13] is typically a variance based method. It decomposes the response variances for the specific order SA indices. The Soboĺ SA method can calculate the interactions among input parameters, but it should be noted that the higher order Soboĺ analysis could significantly increase the computational burden with the increase in the input number [14–16]. Regional Sensitivity Analysis (RSA) was developed by Spear and Hornberger [17] and improved by Beven and Binley [18], which is also a global SA method. RSA elaborates the sensitivity variation over the full range for a given parameter. The cumulative possibility of behavioral sets is investigated in the RSA to reflect the parameter interaction implicitly. Both Soboĺ and RSA methods are employed in this paper to ascertain the sensitive parameters in the inverted siphon flooding model [19].

This paper aims to address the sensitivities of any value over the domain of the parameters and further identifies the sensitive parameters in the inverted siphon models. The sensitive parameter screening and sensitive range identification of the parameters are implemented by Soboĺ SA and RSA methods, respectively. The upstream runoff of the siphon is simulated by a local hydrologic model, and the sediment transportation and pipe transmission model are used to calculate the inverted siphon flows. The evaluation indexes (flood overtopping duration and volume) are set up to demonstrate the flood risk of the inverted siphon. The sensitivity analysis results are demonstrated based on an inverted siphon structure across the South to North Water Diversion project.

2. Methods

The sensitivity analysis methods including Soboĺ SA and RSA method are introduced to evaluate the sensitivity of parameters in the integrated system modeling. The integrated system model is set up by integrating hydrologic model, sediment transport model and inverted siphon hydraulic model. Then, the two evaluation indicators, i.e., flood overtopping duration and volume, are used to quantify flood risk.

2.1. Sensitivity Analysis Methods

2.1.1. General

Two global sensitivity analysis methods are employed in this paper for investigating parameter sensitivity in consideration with the interaction of variables and the sensitivity variation over the range of the variables. The model can be represented by a numerical function,

$$Y = f(X) = f(x_1, \ldots, x_n) \tag{1}$$

where Y is the model output (or objective function) and X is the variable set, $x_1, \ldots, x_n$.

2.1.2. Soboĺ Sensitivity Analysis

The Soboĺ sensitivity analysis [13,20] is a variance-based method, which uses variance decomposition to derive a variance ratio. It can provide a quantitative description of how individual variables and their interactions affect model performance [21]. An individual model parameter and its interaction with other parameters contribute to the total output variance, and the function is shown as follows:

$$V_S = \sum_{i=1}^{n} V_i + \sum_{i=1}^{n} \sum_{j<i}^{n} V_{ij} + \cdots + V_{1,2,\ldots,n} \tag{2}$$

where V_S is the total variance of the output variable Y; V_i is given by the variance of the conditional expectation $V_i = V[E(Y|x_i)]$ and $V_{ij}(V_{ij} = V[E(Y|x_i, y_i)] - V_i - V_j)$ to $V_{1\ldots k}$ the interactions among k parameters. To assess the role of each variable or interaction between variables, sensitivity measures are needed. The chosen measures are known as Soboĺ indices. The indices represent the bias in the variance of the output, which is attributed to a variable or a combination of variables. The first-order index (S_i) is

$$S_i = \frac{V_i}{V(Y)} \tag{3}$$

and the second-order index (S_{ij}) is

$$S_{ij} = \frac{V_{ij}}{V(Y)} \tag{4}$$

The total order sensitivity index of a single parameter and this parameter's interaction with other parameters, at least one index being $j \neq i$ from 1 to k, is as follows:

$$S_{Ti} = \sum S_i + \sum_{j \neq i} S_{ij} + \cdots + S_{1\ldots k} \tag{5}$$

The first-order sensitivity index only represents the individual contribution of variable x_i to the model output. The second-order index indicates the interaction effect of two variables $(x_i, x_j, i \neq j)$ on the model output. The total-order index (S_{Ti}) measures the main effect of parameter x_i and its interactions with all the other variables. The Soboĺ indices are obtained by a sampling process, e.g., Latin Hypercube.

2.1.3. Regionalized Sensitivity Analysis

Regionalized sensitivity analysis (RSA) is proposed by Spear and Hornberger [17] and further extended by Beven and Binley [18]. RSA method is broadly applied in hydrology and environmental system analysis [19,22,23]. The approach is based on the Monte Carlo simulation considering possible combination of uncertain parameters with the given possibility density function. The parameters sampling process can cover the whole distribution range, so RSA also belongs to the global sensitivity analysis category. The sampled parameter sets are divided into behavioral or non-behavioral. If the computational result of a parameter set (objective function evaluations) satisfies the prescribed condition (e.g., less than a threshold), the parameter set is behavioral, vice versa.

RSA results are expressed by the cumulative distribution. The difference between the behavioral and non-behavioral cumulative distributions is larger, and then the parameter is more sensitive.

Kolmogorov-Smirnoff (K-S) test is used to show the maximum vertical distance ($d_{m,n}$) between the behavioral and non-behavioral cumulative distributions. The K-S test function is given as

$$d_{m,n} = \sup_{x} |S_B(x) - S_{NB}(x)| \quad (6)$$

where $S_B(x), S_{NB}(x)$ are the behavioral and non-behavioral cumulative distributions, respectively.

2.2. *Definition of Evaluation Indicators*

If the water level at the inlet of the inverted siphon exceeds the embankment crest elevation of the canal (denoted by Z_s), then the flooding water will flow into the main trunk canal, i.e., flood overtopping happens. The event occurrence represents the inverted siphon hydraulic failure. This failure will bring the risks that are the embankment erosion and the water quality pollution for example. The longer duration of flood overtopping leads to more severe hazards and exaggerating the impacted extent on the canal. Therefore, we adopt the flood overtopping duration and flood overtopping volume as the indicators to evaluate the risk of flood overtopping for the inverted siphon.

It is assumed that the water level at the water inlet of the inverted siphon can rise, even if it exceeds the crest elevation of the canal Z_s (i.e., there is a virtual water pond with unlimited crest elevation). The time periods, when water level exceeds Z_s, are defined as the duration of flood overtopping. The flood overtopping volume can be calculated by the difference of maximum flood volume at the inlet and the volume that corresponds to the embankment crest elevation. The volumes are calculated by the water level–storage relationship at the inlet of the inverted siphon.

2.3. *System Modeling*

2.3.1. General

The input parameters that need to be tested in the sensitivity analysis are assigned by a set of random values ($p_1, p_2, p_3, \cdots, p_n$). The hydrologic model simulates the inflow of the inverted siphon in a given watershed. Sediment transport model is introduced to model the sand movement and deposition. Siphon hydraulic model is used to calculate the flow of the siphon. These three models are integrated and convey the parameter values, as shown in Figure 1. The outputs of the models are runoff, flow, sand content and water level. Two evaluation indexes are formulated by these outputs. The whole flow chart of the methodology is shown in Figure 1.

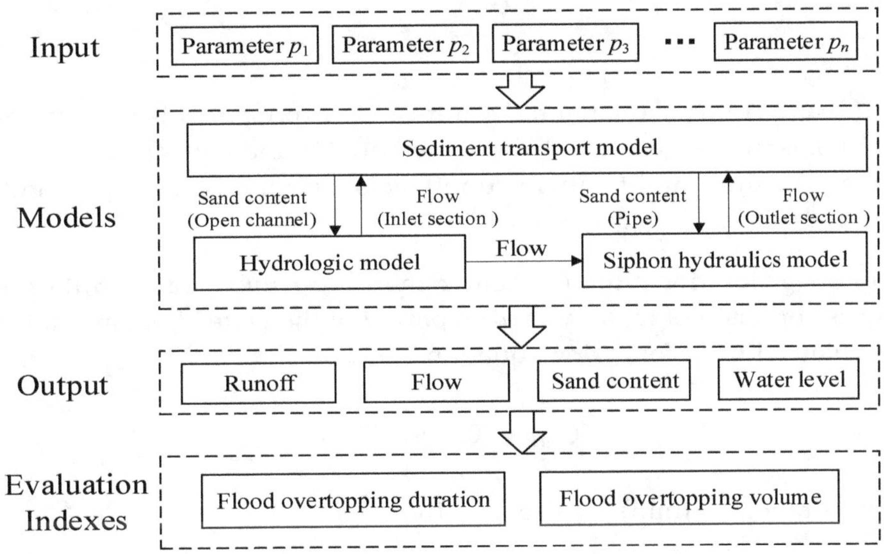

Figure 1. Flow chart of the models.

2.3.2. Hydrologic Model

Due to the lack of monitoring data in small watersheds, the Inferential Formula Method is often used to determine the design floods [24]. This study uses the Inferential Formula Method specified in China's design flood regulations and proposed by Chen [25]. The flood is derived from design rainfall in the assumption that rainfall frequency is the same with flood frequency. The flood calculation includes the following steps:

(1) The maximum rainfall

The maximum rainfall during t h (for small watersheds, t is set to be 24 h) can be calculated by

$$x_{t,P} = \overline{x_t}(1 + C_v \cdot \Phi_p) \tag{7}$$

where $x_{t,P}$ represents the amount of rainfall with the frequency P for t h (mm); $\overline{x_t}$ is the average of maximum rainfall for t h (mm); Φ_p is a frequency factor; and C_v is the coefficient of variation of maximum rainfall for t h.

(2) The peak discharge

Rational method [26] is one of the earliest methods to estimate the peak discharge according to the rainfall data. When the runoff generation time (denoted by t_c) is larger than the runoff concentration time (denoted by τ), i.e., $t_c \geq \tau$, the peak discharge is consisted of the runoff from the whole watershed. Assuming the runoff intensity is evenly distributed both in spatial and temporal, and the concentration is irrespective to the channels and slops, then the peak discharge is calculated by

$$Q_{mP} = 0.278\left(\frac{h_\tau}{\tau}\right)F \tag{8}$$

in which, F denotes the area of the watershed (km^2) and τ is the runoff concentration time (h), calculated based on the terrain characteristic for the given area,

$$\tau = 0.278\frac{L}{mJ^{1/3}Q_{mP}{}^{1/4}} \tag{9}$$

h_τ is the amount of runoff generation during τ hours (mm), given by the following equation under the assumption of excess infiltration, that is, there is runoff generation only when the rainfall intensity is larger than the infiltration rate.

$$h_\tau = x_{t,p}t^{\gamma-1}\tau^{1-\gamma} - \mu\tau \tag{10}$$

In Equations (9) and (10), γ is the rainfall diminishing exponent; μ is the average infiltration rate during τ h (mm/h); L is the longest length from the outlet of main river to watershed outline (km); J is the average slope for the longest path of runoff; m is the runoff parameter; and Q_{mP} is the peak discharge for the flood with the frequency P (m^3/s).

When the runoff generation time is smaller than the runoff concentration time, i.e., $t_c < \tau$, the peak discharge is consisted of the runoff from partial of the watershed, and the peak discharge is approximately calculated by the following equation [25]

$$Q_{mP} = 0.278\left(\frac{h_R}{\tau}\right)F \tag{11}$$

where the runoff generation h_R (mm) is calculated by

$$h_R = x_{t,p}t^{\gamma-1}t_c{}^{1-\gamma} - \mu t_c \tag{12}$$

and the runoff generation time can be calculated by the following equation:

$$t_c = \left[\frac{(1-\gamma)x_{t,p}t^{\gamma-1}}{\mu} \right]^{\frac{1}{\gamma}} \tag{13}$$

(3) The flood hydrograph

The generalized triangle hydrograph method [25], including the following steps, is applied to obtain the flood hydrograph.

Firstly, the total runoff during t hours is allocated to T periods in terms of the designed runoff hydrograph and the runoff concentration time (τ). The rainfall duration for each period (denoted by t_i and $I = 1, 2, \ldots T$) is larger than the runoff concentration time, that is $t_i \geq \tau$.

Secondly, assume the runoff generation time equals the rainfall time, and the concentration time in each period equals that of the peak discharge for the whole hydrograph (calculated by Equation (9)). Then, the peak discharge resulting from rainfall in ith period can be calculated by the following equation with the generalized triangle hydrograph.

$$Q_{ti} = 0.556\left(\frac{h_{ti}}{t_i + \tau}\right)F \tag{14}$$

where h_{ti} is the runoff of rainfall in ith period (mm); Q_{ti} is the peak flow at t_i (m^3/s). It is worth mentioning that, when $t_i = \tau$ (h), Equation (14) is the same with Equation (8).

Thirdly, the flood hydrograph resulting from rainfall in each period can be obtained with the peak flow, and the time for the flood hydrograph is the sum of rainfall duration and runoff concentration time, i.e., ($t_i + \tau$). When $t_i = \tau$, the hydrograph resulting from runoff ith periods is symmetric, i.e., equilateral triangle hydrograph. Otherwise, when $t_i > \tau$, the resulting hydrograph is asymmetric. Finally, the flood hydrograph for rainfall during t h can be convoluted.

2.3.3. Sediment Transport Model

The sensitivity analysis is conducted under the worst condition that the water with the given velocity contains the maximum amount of sands. If the velocity in the pipe shows down, the deposition can occur. It is therefore assumed that the flood can carry the maximum amount of sediments for the given flow velocity. Since the fluid regime is complicated and diversified inside the structure, the regression analysis to formulate an empirical formula is more practical. Here the capacity of sediment transport is fitted based on the field data. The Guojunke formula [27], which uses the logarithmic function to fit the field data and performs well for the Yellow River, is adopted here. The Guojunke formula for the open channel is formulated as,

$$S_* = \frac{\frac{1}{20}\left(\frac{V_p^3}{gR\omega}\right)^{1.5}}{1 + \left(\frac{1}{45}\frac{V_p^3}{gR\omega}\right)^{1.15}} \tag{15}$$

where S_* is the sand content, i.e., the mass of sands in a unit volume (kg/m^3); V_p is the average velocity of the section (m/s); g is the gravity acceleration (m/s^2); R is the hydraulic radius; and ω is the sand deposition rate which is calculated by Zhu-Cheng formula [28] (m/s),

$$\frac{\omega d}{\nu} = \frac{-24\cos^3\alpha + \sqrt{576\cos^6\alpha + \left(18\cos^3\alpha + 3.6\sin^2\alpha\right)d_*^3}}{9\cos^3\alpha + 1.8\sin^2\alpha} \tag{16}$$

$$d_* = \left(\frac{(\rho_s - \rho)g}{\nu^2 \rho}\right)^{\frac{1}{3}} d \tag{17}$$

$$\alpha = \begin{cases} 0 & , d_* \leq 1 \\ \pi/\left(2 + 2.5(\log d_*)^{-3}\right) & , d_* > 1 \end{cases} \tag{18}$$

where d is the diameter of sand, d = 0.033 mm for the silty-fine sand; ν is the fluid viscosity (m^2/s); ρ_s is sand density (kg/m^3), ρ_s = 2650 kg/m^3; and ρ is water density (kg/m^3), ρ = 1000 kg/m^3.

When the floods go into the inverted siphon, the capacity of sediment transport changes. The capacity that the pipe transports sand is determined by the non-deposition velocity. The critical non-deposition velocity (V_c) is given by Wasp equation [29],

$$V_c = 3.28 S_v{}^{0.243} \left[2gD\left(\frac{\rho_s - \rho}{\rho}\right)\right]^{1/2} \left(\frac{d}{D}\right)^{1/6} \tag{19}$$

$$S_v = \frac{S_*}{\rho_s} \tag{20}$$

where D is the diameter of the pipe (m) and S_v is the ratio of water to sand in the unit volume. When the velocity is larger than V_c, no sediment is deposited, otherwise sediment deposition happens. The diameter parameter of the circular pipe in the original Wasp equation is transformed into the equivalent hydraulic radius of the square pipe.

2.3.4. Siphon Hydraulics Model

The siphon is calculated in terms of the quasi steady flow (i.e., steady flow is calculated in each time interval, and the all simulation snapshots comprise the simulation process over time). The inverted siphon flow computation includes the downstream and upstream open channel flows and pressurized pipe flow. The calculation of the conveyance capacity of the inverted siphon includes flood regulation and hydraulic routing. The flood regulation is based on the water balance equation, given as,

$$Q_A - Q_B = \frac{\Delta W}{\Delta t} \tag{21}$$

where Q_A is the flow at the upstream channel of the inverted siphon (m^3/s), which is known according to the hydrologic model results; Q_B is the flow at downstream channel of the inverted siphon (m^3/s); and ΔW is the volume of retained water at upstream channel during the time of Δt (m^3). The water level at the upstream and downstream channels and flow in the siphon pipe are unknown, i.e., Q_B and ΔW are both unknown, but they meet the hydraulics conditions. Then, the iterative method is applied, and a water level at downstream is given before each iteration and thus the iterative process is implemented as follows,

Step 1. Given a water level Z_B at downstream channel, the flow rate at downstream channel can be calculated with the Chezy equations

$$Q_B = A_c C_c \sqrt{R_c J} \tag{22}$$

$$C_c = \frac{1}{n_c} R_c{}^{1/6} \tag{23}$$

where Q_B is the flow at downstream channel (m^3/s); C_c is Chezy coefficient of the channel (m$^{1/2}$/s); A_c is the section area of the flow (m^2); n_c is the roughness coefficient at downstream channel which is set to be 0.035 in this study (s/m$^{1/3}$); and R_c is the wetted perimeter of the channel; J is the hydraulic slope.

The iterative initial water level at downstream is assumed to be zero. With the flow rate Q_B, the retaining volume at the inlet ΔW can be calculated with Equation (12). Then, with the relationship between water level and water volume at upstream channel (denoted by F(Z)) and the initial water level at the inlet (denoted by $Z_{A,t-1}$, equaling zero at the first time period, otherwise, equaling the ending water level of the last period), the ending water level ($Z_{A,t}$) at upstream channel can be derived by $Z_{A,t} = F^{-1}[\Delta W + F(Z_{A,t-1})]$. This water level minus the presumed water level at downstream can obtain the water level difference ΔZ_1.

Step 2. The difference between downstream and upstream channel water levels (ΔZ_2) is given as,

$$\Delta Z_2 = h_f + h_j + \frac{{V_A}^2 - {V_B}^2}{2g} \tag{24}$$

where V_A and V_B are the velocity at upstream and downstream channel (m/s), and calculated by the ratio of flow to flow area. h_f and h_j are pipe friction headloss and local headloss, respectively (m), and given by

$$h_f = \frac{L_p {V_p}^2}{{C_p}^2 R_p} \tag{25}$$

$$h_j = (K_1 + K_2 + K_3)\frac{{V_p}^2}{2g} \tag{26}$$

where L_p is the length of the inverted siphon (m); C_p is Chezy coefficient of the inverted siphon ($m^{1/2}/s$); R_p is the wetted perimeter of the inverted siphon; $K_1, K_2,$ and K_3 are the local loss coefficients at inlet, outlet and inside of the inverted siphon, respectively; V_p is the velocity of inverted siphon (m/s), calculated by the ratio of flow (Q_p) to flow area (A_p) of the inverted siphon; and A_p can be calculated by

$$A_p = (H_p - z_p) \times W_p \tag{27}$$

where H_p and W_p are the height and width of inverted siphon, respectively (m); z_p is the deposition height (m).

Step 3. In comparison of ΔZ_1 (obtained from water balance) and ΔZ_2 (obtained from energy balance), if the difference between them is less than a threshold, then the calculation process terminates. Otherwise, a new water level at downstream is given following: if $\Delta Z_1 > \Delta Z_2$, the water level at downstream should be increased by a step; if $\Delta Z_1 < \Delta Z_2$, the water level should be decreased by a step. The step is determined according to the search method. Then, the Steps from 1 to 3 are repeated until the stop criteria.

3. Case Description

3.1. Overview of the Study Area

The Central Route of the South-to-North Water Diversion Project, shown in Figure 2, transfers water from Danjiangkou Reservoir on the Han River (a tributary of Yangtze River) to Beijing and Tianjin Cities. This project links up four major basins, including Yangtze River, Huai River, Yellow River and Hai River, and crosses Hebei, Henan, and Hubei Provinces. The main trunk canal is a total length of 1277 km, and crosses 205 rivers with cross-river buildings. The buildings are called river-canal crossing structures.

This study is targeted to the inverted siphon—A typical river-canal crossing structure. The inverted siphon, located on the intersection between main trunk canal of Central Route of the South-to-North Water Diversion Project and Meihe tributary, is taken as an illustrated case study. The drainage area of Meihe tributary is 10.80 km^2, the longest length from the outlet of the main river to watershed outline is 5 km, and the average slope for the longest path is 0.017. The inverted siphon consists of pipe section, upstream channel and downstream channel sections, as shown in Figure 3. The upstream channel and downstream channel sections are the trapezoidal open channels and the lengths are 55 m and 68 m, respectively. The pipe section includes four 3×3 m^2 square barrels, which have equal heights at the entrance. The horizontal projection length of each pipeline is 111.6 m, and the slopes of the rising and descending legs are 1:5 and 1:4, respectively. Since there is no gate control of the pipelines, the four pipelines operate simultaneously. The peak discharges for a 50-year return period of flood design criterion and 200-year flood check criterion are 209 m^3/s and 294 m^3/s, respectively.

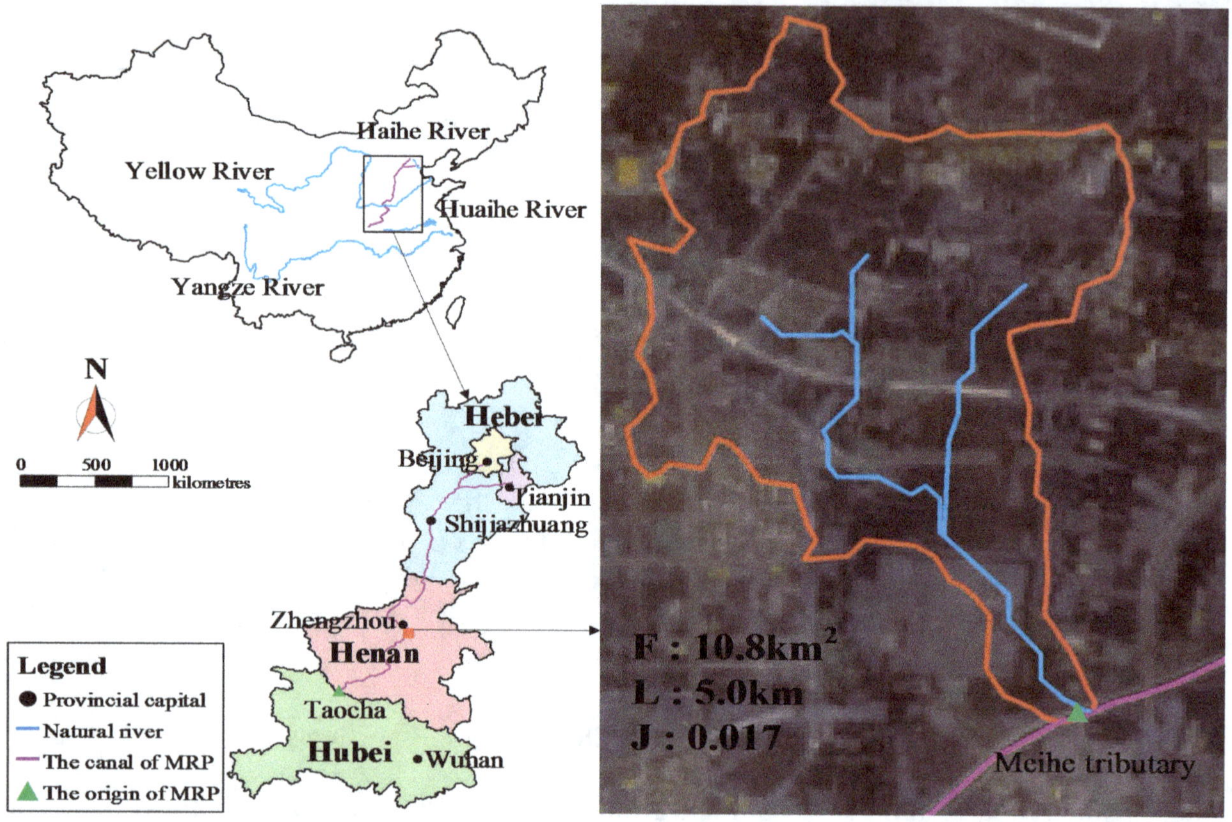

Figure 2. The sketch map of the study area.

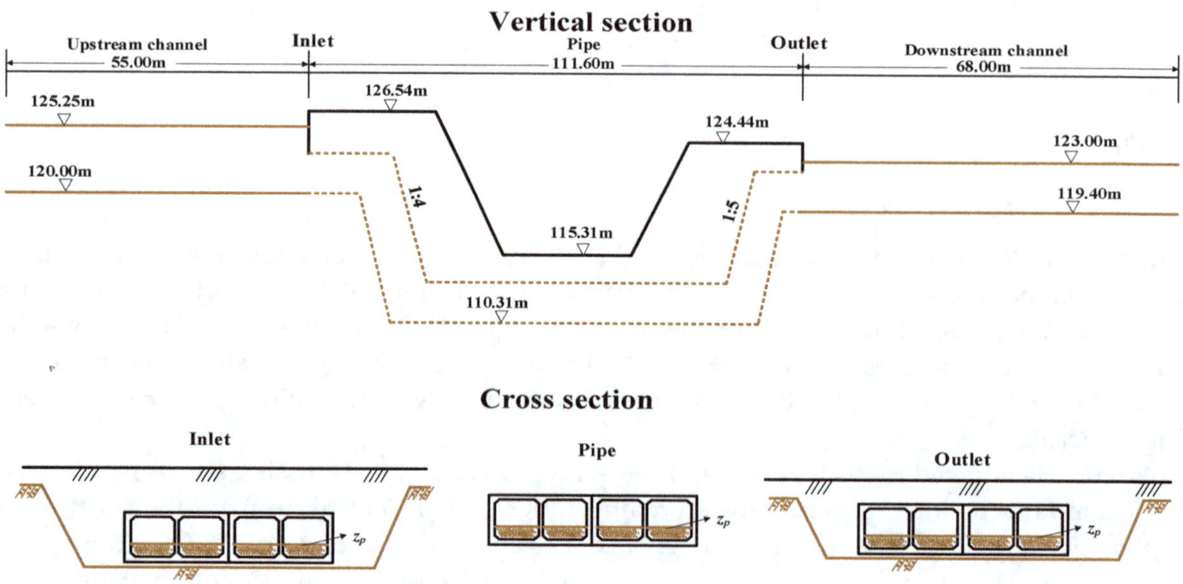

Figure 3. The schematic diagram of the inverted siphon.

3.2. Parameter Uncertainty Description

In the flood risk analysis of the inverted siphon, the parameter uncertainties in the integrated model, including rainfall module ($\overline{x}_{24}$, C_v, and γ), runoff generation module (m), runoff concentration module (u), and hydraulic module (n_c, z_p), are considered. The distribution and feasible range of the parameters are listed in Table 1.

Table 1. Parameters studied for the sensitivity analysis.

Model Response	Parameter	Description	Range	Distribution	Unit
Rainfall	$\overline{x_{24}}$	Mean annual maximum rainfall in 24 h	90–110	uniform	mm
	C_v	Variation coefficient of annual maximum rainfall in 24 h	0.5–0.6	uniform	–
	γ	Rainfall diminishing exponent	0.75–0.80	uniform	–
Runoff	u	Mean infiltration rates	2–5	uniform	mm/h
	m	Confluence coefficient	0.95–1.05	uniform	–
Conduit flow	n_c	Roughness coefficient of the pipe	0.014–0.020	uniform	–
	z_p	Initial deposition height	0–3	truncated normal	m

3.2.1. Mean Value ($\overline{x_{24}}$) and Variation Coefficient (C_v) for the Maximum 24-h Rainfall

According to the "Atlas of the Design Storm and Flood for the Medium and Small-Sized Basins in Henan Province" edited in 1984, the mean value and variation coefficient for the maximum 24-h are 100 and 0.55, respectively. However, they are both closely related to the length of the rainfall data. At the beginning of the design stage for South to North Water Diversion Project, the statistical rainfall parameters for 24 rain gauge stations, which are distributed on different rivers but along the main trunk canal, are validated with the rainfall data prolonged to 2000. The results show that the mean value $\overline{x_{24}}$ ranges within -10%–10% of the designed value and C_v ranges within -0.05–0.05. Regardless of the impact of human activities, it is assumed that the statistical parameters change within the above ranges. That is, the ranges of $\overline{x_{24}}$ and C_v are 90–110 and 0.5–0.6, respectively, and $\overline{x_{24}}$ and C_v are assumed to be uniformly distributed.

3.2.2. The Rainfall Diminishing Exponent (γ), Runoff Concentration Parameter (m) and the Mean Filtration Rate (u)

Statistic parameters of γ, m, and u are obtained from the corresponding contour map in "Atlas of the Design Storm and Flood for the Medium and Small-Sized Basins in Henan Province". The uncertainties in these parameters are caused by observation. The ranges for γ, m and u are 0.75–0.80, 2–5, and 0.95–1.05, respectively, by the upper and lower contour curve evaluation. γ, m and u are assumed to follow the uniform distribution.

3.2.3. The Roughness Coefficient (n_c)

The roughness coefficient of the pipe for the inverted siphon (n_c) changes with sediment deposition. The more sediments, the greater roughness coefficient. The roughness coefficient can reach as large as 0.020 according to the empirical data [24]. However, the inverted siphon was designed according to a fixed value, i.e., 0.014. Therefore, we consider the uncertainty of n_c within range 0.014–0.020 obeying a uniform distribution.

3.2.4. The Initial Deposition Height of Sediment (z_p)

The initial deposition height of sediment in the model is set at the beginning of the flood process. It can be as large as the pipe width, 3 m in this study. z_p is typically small with the larger probability, while small probability corresponds to a large z_p. Thus, the truncated normal distribution, which is widely used when there is little information about the distribution, is assumed for z_p within range 0–3.

4. Results and Discussion

4.1. RSA Results

The parameters in the integrated model, listed in Table 1, are assumed to be independent, and Latin Hypercube sampling method [30] is used. It is noted that the parameters of the hydraulic model (roughness coefficient and initial deposition height) for each conduit (the culvert consists of four squared conduits) are set to be identical. The sample size is 100,000, and one sample includes all parameters values that are randomly assigned. The distributions of runoff concentration time and peak discharge are shown in Figure 4. As can be seen, the peak discharge considering the uncertainties of parameters are all larger than the original designed value, 294 m^3/s.

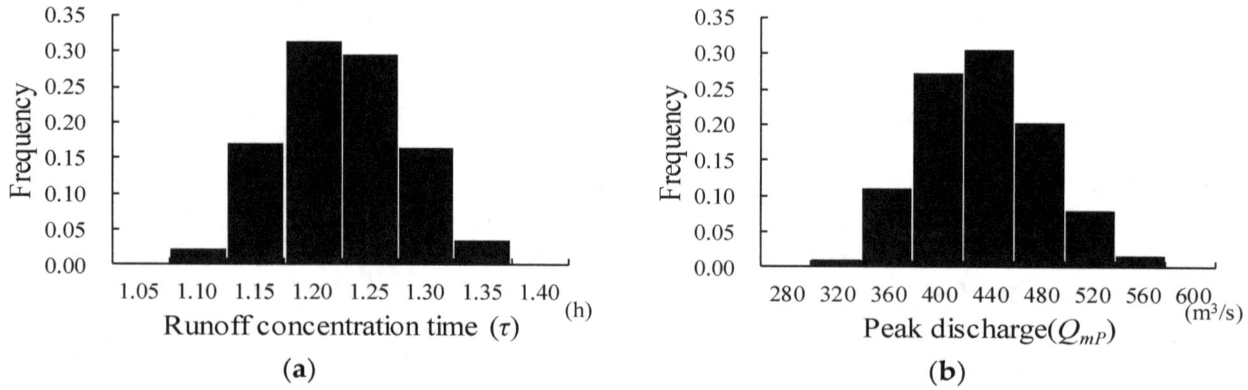

Figure 4. The distribution of runoff concentration time and peak discharge. (**a**) The distribution of runoff concentration time; (**b**) The distribution of peak discharge.

The parameters sets are divided into two subsets in terms of the inverted siphon failure (i.e., non-behavioral) or operation (behavioral), with cumulative distributions shown in Figure 3. The behavioral sample (S_B) is that no water flows into the canal, i.e., the water level at the water intake is smaller than the embankment crest elevation. In contrast, the non-behavioral sample (S_{NB}) is that water flows into the canal, i.e., the water level at the water intake is higher than the crest elevation. In Figure 3, the diagonal line (D-line) represents the parameter has a uniform distribution and the model is not sensitive to this parameter in terms of the chosen likelihood measure. Any deviation from the "D-line" shows a non-uniform distribution and the model is sensitive to this parameter. The larger distance between the S_B and S_{NB} indicates more sensitive range of this parameter.

As shown in Figure 5, all parameters exhibit an obvious shift. In addition, the S_{NB} curve for all the parameters except z_p is close to "D-line", indicating that the effects of parameters within the whole range on the failure of the inverted siphon are almost identical. The cumulative distributions for the mean rainfall ($\overline{x_{24}}$), variation coefficient (C_v), rainfall diminishing exponent (γ), runoff concentration parameter (m) and the roughness coefficient (n_c), show that the values at the lower end of the tested ranges contribute to the greatest number of behaviors, i.e., lower value of these parameters lead to lower flood risk of the inverted siphon. Conversely, the greatest number of behaviors occurs at the higher end of the range for mean filtration rate (u). For initial deposition height (z_p), the greatest number of behaviors and non-behaviors comes from values at the lower end of the range. Meanwhile, the initial deposition height (z_p) has the least impact on flood risk of the inverted siphon, with the smallest shift from the straight line in comparison with the others.

Each parameter sensitivity is tested by the two-sample Kolmogorov–Smirnov (K-S) test method with the confidential level of 95%. Results show that the mean rainfall ($\overline{x_{24}}$) is most sensitive, followed by the roughness coefficient (n_c). Furthermore, the other parameters related with the rainfall, i.e., variation coefficient (C_v), rainfall diminishing exponent (γ), are more sensitive, and thus, the rainfall is the most important factor for the flood risk of the inverted siphon. The roughness coefficient (n_c) is related with the conveyance capacity of the pipe, thus the conveyance capacity of the pipe also is the most important factor for the flood risk of the inverted siphon.

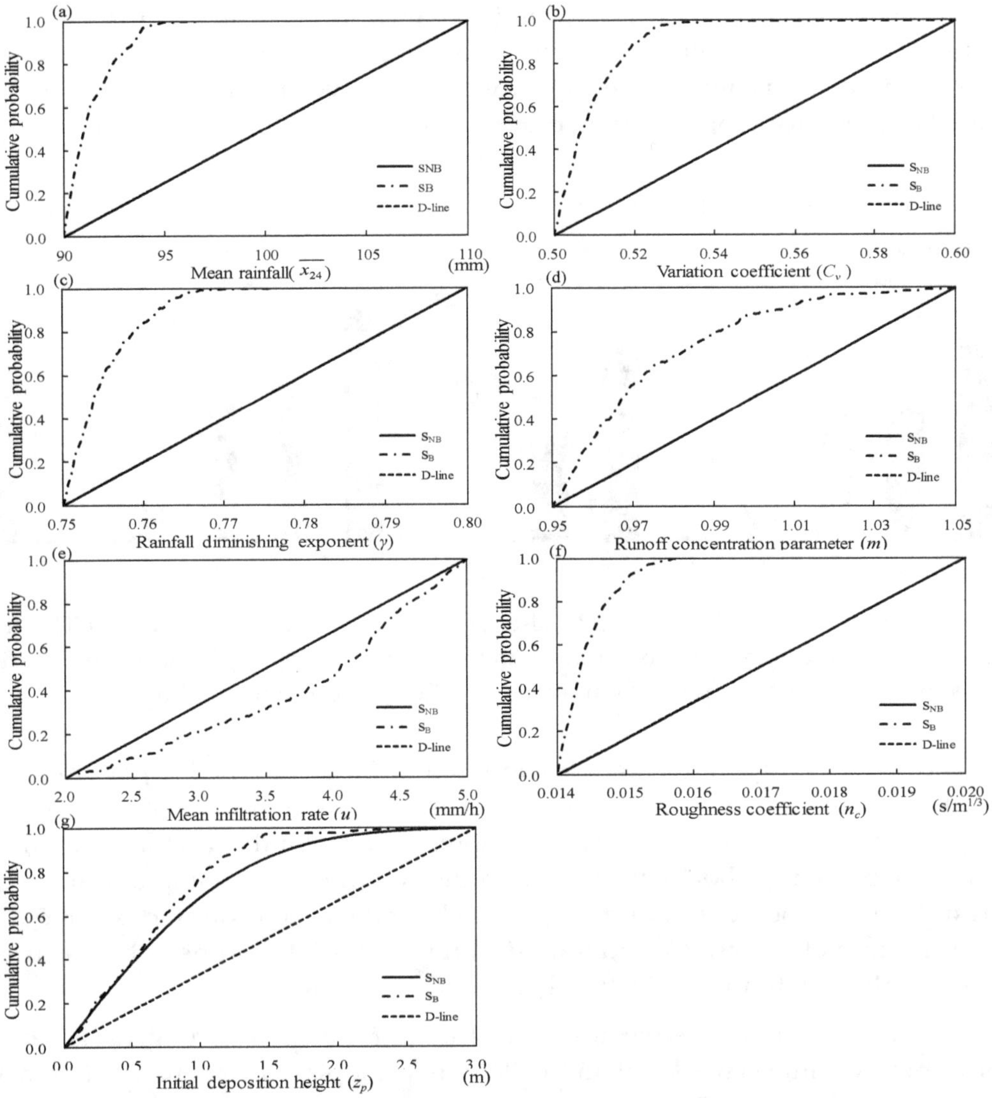

Figure 5. Cumulative distribution of the seven parameters with regard to failure of the inverted siphon. S_B and S_{NB} represent the behavioral and non-behavioral groups, respectively. (**a**) Cumulative distribution of mean rainfall; (**b**) Cumulative distribution of variation coefficient ; (**c**) Cumulative distribution of rainfall diminishing exponent; (**d**) Cumulative distribution of runoff concentration parameter; (**e**) Cumulative distribution of mean filtration rate; (**f**) Cumulative distribution of roughness coefficient; (**g**) Cumulative distribution of initial deposition height.

4.2. Soboĺ Sensitivity Analysis Results

The first-order and total-order sensitivity indices of seven parameters are shown in Figure 4. The black bars represent the first-order index values, which measure the individual parameter contributions to the duration and volume of flood overtopping. The white bars represent the interactive indices, which demonstrate the total interactive contribution of one parameter with all other parameters. The parameter is identified as sensitive when total-order sensitivity index is larger than 0.1.

As can be seen in Figure 6, the total-order index for roughness coefficient (n_c), mean rainfall ($\overline{x_{24}}$), variation coefficient (C_v) and rainfall diminishing exponent (γ), are all larger than 0.1, i.e., sensitive parameters for flood overtopping duration and volume. Parameters m and u, which are related with runoff generation and concentration, are not sensitive. The reason is that flood hydrograph, flood volume, and the conveyance capacity of the pipe are the most important factors for the flood overtopping. A larger flood volume or smaller conveyance capacity leads to more water retained at the inlet, and thus leads to a larger risk of flood overtopping. The magnitude of rainfall is the most important factor of flood volume and flood hydrograph. Thus, the parameters related with the magnitude of rainfall and conveyance capacity are sensitive. The initial deposition height (z_p) is sensitive for the flood overtopping duration, but it is not sensitive for the flood overtopping volume.

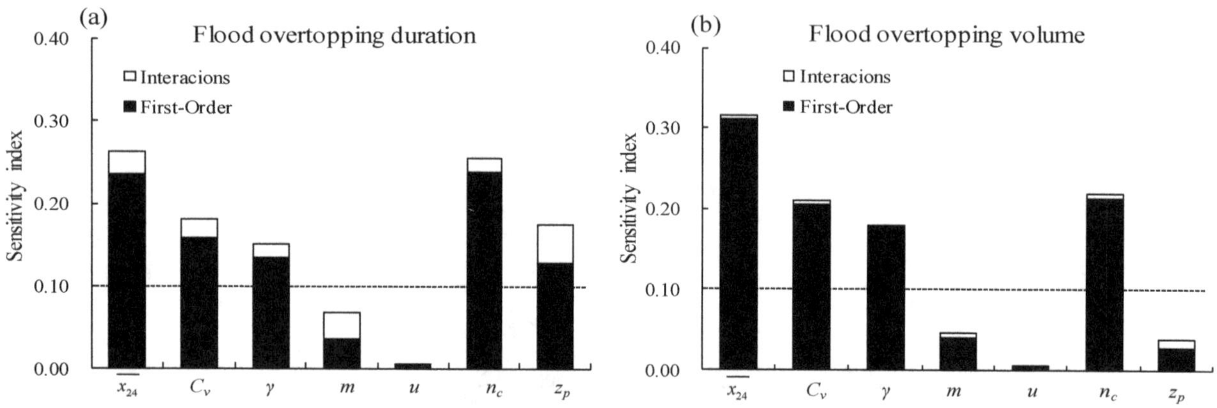

Figure 6. The first-order sensitivity and total-order sensitivity of parameters. (**a**) The first-order sensitivity and total-order sensitivity of parameters for flood overtopping duration; (**b**) The first-order sensitivity and total-order sensitivity of parameters for flood overtopping volume.

The results of the overtopping duration show that $\overline{x_{24}}$ obtains the largest value of total-order index, followed by n_c. These two parameters determine the magnitude of rainfall and the water conveyance capacity of the pipe, respectively. Therefore, the result indicates that rainfall and the water conveyance capacity of the pipe both have a great impact on the overtopping duration in the inverted siphon. The results of the flood overtopping volume show the total-order index for $\overline{x_{24}}$, C_v, γ as well as n_c are sensitive parameters. $\overline{x_{24}}$, C_v and γ determine the runoff volume, which is the input of the inverted siphon, while n_c determines the flood volume transported.

Figure 6 show the interaction between z_p and other parameters is strong, and the interaction for the flood overtopping volume is weaker than the flood overtopping duration. The interaction between any two parameters is shown in Figure 7. As can be seen in Figure 7a, the sum of the second-order index between z_p and other parameters is the largest, and the interaction between z_p and m is strongest for the flood overtopping duration. As for flood overtopping volume, the sum of second-order index between n_c and other parameters is largest. However, Figure 6b shows that the total interactions of z_p is larger than that of n_c, which indicates that higher order interactions (third-order, four-order, etc.) exist between z_p and other parameters. The second-order index value for n_c and γ is strongest, followed by the value for n_c and $\overline{x_{24}}$, indicating that the interaction between n_c and γ, n_c and $\overline{x_{24}}$ are strong. This is because $\overline{x_{24}}$ is the most important factor that determines the flood volume, and a larger $\overline{x_{24}}$ result in a larger flood. A larger γ results in a larger runoff generation under the same magnitude of rainfall, and a larger $\overline{x_{24}}$ results in a larger and thus a larger peak discharge and flood volume, while a larger n_c results in a smaller conveyance capacity.

In conclusion, results of RSA and Sobol sensitivity analysis both indicate that the mean rainfall and roughness coefficient of the pipe are two important parameters, which determine the rainfall and water conveyance capacity of the siphon pipe, respectively. These results demonstrate that lower value of the mean rainfall or roughness coefficient of the pipe lead to lower flood risk of the inverted siphon. Thus, the effective measures for reducing the flood risk of the inverted siphon to clean the inverted siphon periodically, i.e., reduce the roughness coefficient of pipe.

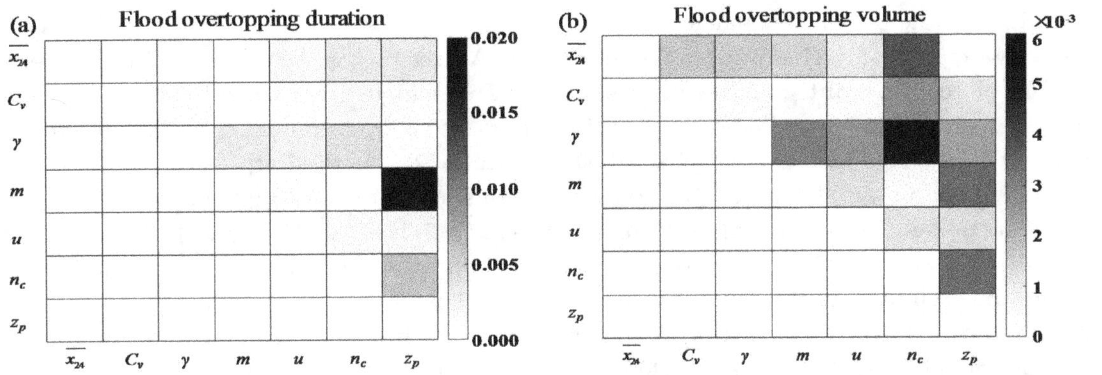

Figure 7. The interaction between each two parameters. (**a**) The interaction between each two parameters for flood overtopping duration; (**b**) The interaction between each two parameters for flood overtopping volume.

5. Conclusions

This paper investigated, by using an integrated system coupling the hydrology and hydraulic models, the impact of rainfall, sediments and pipe roughness coefficient on the failure of inverted siphon through RSA and Soboĺ sensitivity analysis. The flood risk of the inverted siphon is evaluated by the flood overtopping duration and volume. The conclusions are summarized as follows:

(1) RSA and Soboĺ sensitivity analyses both indicate the mean rainfall and the roughness coefficient of the pipe are most sensitive for the flood risk of the inverted siphon. These results imply that lower value of the mean rainfall or roughness coefficient of the pipe lead to lower flood risk of the inverted siphon. Thus, periodically cleaning the inverted siphon is an effective measure for reducing the flood risk of the inverted siphon, i.e., reduce the roughness coefficient of pipe.

(2) The RSA identifies the sensitivity range of safety and failure for the inverted siphon. Effects of all parameters except initial deposition height throughout the feasible range on the failure of the inverted siphon are almost identical. For the safety of the inverted siphon, the smaller values of variation coefficient, rainfall diminishing exponent, runoff concentration parameter, and roughness coefficient of pipe, the more safety of the inverted siphon. For the mean filtration, a higher value leads to the more security of the inverted siphon.

(3) Soboĺ sensitivity analysis reveals the individual and interactive effects of the parameters. The effects of all parameters on flow overtopping duration and volume all parameters are all dominated by the individual effects. For the flood overtopping duration, the interactions between the initial deposition height and other parameters are high, with the largest interaction between initial deposition height and confluence coefficient. For flood overtopping volume, the interaction between rainfall model parameters and hydraulic model parameters are both significant.

Acknowledgments: This research was supported by the National Key Research and Development Program of China (Grant No. 2016YFC0402203) and partly funded by the National Science and Technology Major Project (Grant No. 2014ZX03005001) and National Science Foundation of China (Grant No. 91547116).

Author Contributions: Wei Ding and Haixing Liu conceived and designed the experiments; Sifan Jin performed the experiments; SiFan Jin, Haixing Liu and Wei Ding analyzed the data; Hua Shang and Guoli Wang contributed reagents/materials/analysis tools; Sifan Jin wrote the paper.

References

1. Liu, C.; Zheng, H. South-to-north Water Transfer Schemes for China. *Int. J. Water Resour. Dev.* **2002**, *18*, 453–471. [CrossRef]
2. Koutsoyiannis, D. Scale of water resources development and sustainability: Small is beautiful, large is great. *Hydrol. Sci. J.* **2011**, *56*, 553–575. [CrossRef]

3. Tyralis, H.; Aristoteles, T.; Delichatsiou, A.; Mamassis, N.; Koutsoyiannis, D. A perpetually interrupted interbasin water transfer as a modern Greek drama: Assessing the Acheloos to Pinios interbasin water transfer in the context of integrated water resources management. *Open Water J.* **2017**, *4*, 11.
4. Zhang, Q. The South-to-North Water Transfer Project of China: Environmental Implications and Monitoring Strategy. *JAWRA J. Am. Water Resour. Assoc.* **2009**, *45*, 1238–1247. [CrossRef]
5. Tang, C.; Yi, Y.; Yang, Z.; Cheng, X. Water pollution risk simulation and prediction in the main canal of the South-to-North Water Transfer Project. *J. Hydrol.* **2014**, *519*, 2111–2120. [CrossRef]
6. Wei, G.; Zhang, C.; Li, Y.; Liu, H.; Zhou, H. Source identification of sudden contamination based on the parameter uncertainty analysis. *J. Hydroinform.* **2016**, *18*, 919–927.
7. Pappenberger, F.; Beven, K.J.; Ratto, M.; Matgen, P. Multi-method global sensitivity analysis of flood inundation models. *Adv. Water Resour.* **2008**, *31*, 1–14. [CrossRef]
8. Van Griensven, A.; Meixner, T.; Grunwald, S.; Bishop, T.; Diluzio, M.; Srinivasan, R. A global sensitivity analysis tool for the parameters of multi-variable catchment models. *J. Hydrol.* **2006**, *324*, 10–23. [CrossRef]
9. Chu, J.; Zhang, C.; Fu, G.; Li, Y.; Zhou, H. Improving multi-objective reservoir operation optimization with sensitivity-informed dimension reduction. *Hydrol. Earth Syst. Sci.* **2015**, *19*, 3557–3570. [CrossRef]
10. Song, X.; Zhang, J.; Zhan, C.; Xuan, Y.; Ye, M.; Xu, C. Global sensitivity analysis in hydrological modeling: Review of concepts, methods, theoretical framework, and applications. *J. Hydrol.* **2015**, *523*, 739–757. [CrossRef]
11. Ruppert, D.; Shoemaker, C.A.; Wang, Y.; Li, Y.; Bliznyuk, N. Uncertainty Analysis for Computationally Expensive Models with Multiple Outputs. *J. Agric. Biol. Environ. Stat.* **2012**, *17*, 623–640. [CrossRef]
12. Gan, Y.; Duan, Q.; Gong, W.; Tong, C.; Sun, Y.; Chu, W.; Ye, A.; Miao, C.; Di, Z. A comprehensive evaluation of various sensitivity analysis methods: A case study with a hydrological model. *Environ. Model. Softw.* **2014**, *51*, 269–285. [CrossRef]
13. Soboĺ, I. Quasi-Monte Carlo methods. *Prog. Nucl. Energy* **1990**, *24*, 55–61. [CrossRef]
14. Saltelli, A.; Tarantola, S. On the Relative Importance of Input Factors in Mathematical Models. *J. Am. Stat. Assoc.* **2002**, *97*, 702–709. [CrossRef]
15. Fu, G.; Kapelan, Z.; Reed, P. Reducing the complexity of multiobjective water distribution system optimization through global sensitivity analysis. *J. Water Resour. Plan. Manag.* **2012**, *138*, 196–207. [CrossRef]
16. Zhang, C.; Chu, J.; Fu, G. Soboĺ sensitivity analysis for a distributed hydrological model of Yichun River Basin, China. *J. Hydrol.* **2013**, *480*, 58–68. [CrossRef]
17. Spear, R.C.; Hornberger, G.M. Eutrophication in peel inlet—II. Identification of critical uncertainties via generalized sensitivity analysis. *Water Res.* **1980**, *14*, 43–49. [CrossRef]
18. Beven, K.; Binley, A. The future of distributed models: Model calibration and uncertainty prediction. *Hydrol. Process.* **1992**, *6*, 279–298. [CrossRef]
19. Pappenberger, F.; Iorgulescu, I.; Beven, K.J. Sensitivity analysis based on regional splits and regression trees (SARS-RT). *Environ. Model. Softw.* **2006**, *21*, 976–990. [CrossRef]
20. Saltelli, A.; Tarantola, S.; Chan, K.-S. A quantitative model-independent method for global sensitivity analysis of model output. *Technometrics* **1999**, *41*, 39–56. [CrossRef]
21. Tang, Y.; Reed, P.; Van Werkhoven, K.; Wagener, T. Advancing the identification and evaluation of distributed rainfall-runoff models using global sensitivity analysis. *Water Resour. Res.* **2007**, *43*. [CrossRef]
22. Yang, J. Convergence and uncertainty analyses in Monte-Carlo based sensitivity analysis. *Environ. Model. Softw.* **2011**, *26*, 444–457. [CrossRef]
23. Massmann, C.; Holzmann, H. Analysis of the behavior of a rainfall–runoff model using three global sensitivity analysis methods evaluated at different temporal scales. *J. Hydrol.* **2012**, *475*, 97–110. [CrossRef]
24. Wu, C. *Hydraulics*, 4th ed.; Higher Education Press: Beijing, China, 2008. (In Chinese)
25. Chen, J.; Zhang, G. *Storm-Flood Computation of Small Watershed*; China Conservancy and Hydropower Press: Beijing, China, 1985. (In Chinese)
26. Mulvaney, T.J. *On the Use of Self-Registering Rain and Flood Gauges in Making Observations of the Relations of Rain Falland Flood Discharges in a Given Catchment, Proceedings of the Institute of Civil Engineers of Ireland, Session 1850-1*; Transactions of the Institution of Civil Engineers of Ireland: Dublin, Ireland, 1851; Volume 4, pp. 18–33.
27. Guo, J. Logarithmic matching and its applications in computational hydraulics and sediment transport. *J. Hydraul. Res.* **2002**, *40*, 555–565. [CrossRef]

28. Zhu, L.; Cheng, N. *Settlement of Sediment Particles*; Nanjing Hydraulic Research Institute: Nanjing, China, 1993. (In Chinese)
29. Wasp, E.J.; Kenny, J.P.; Gandhi, R.L. *Solid-liquid Flow Slurry Pipeline Transportation*; Ser. Bulk Materials Handling; Trans Tech Publications: Zurich, Switzerland, 1977; Volume 1, p. 4.
30. McKay, M.D.; Beckman, R.J.; Conover, W.J. A comparison of three methods for selecting values of input variables in the analysis of output from a computer code. *Technometrics* **2000**, *42*, 55–61. [CrossRef]

13

Heating Impact of a Tropical Reservoir on Downstream Water Temperature

Bo Jiang, Fushan Wang * and Guangheng Ni

Department of Hydraulic Engineering, Tsinghua University, Beijing 100084, China; jiangbo117@gmail.com (B.J.); ghni@tsinghua.edu.cn (G.N.)
* Correspondence: wfs14@mails.tsinghua.edu.cn

Abstract: Reservoirs change downstream thermal regimes by releasing water of different temperatures to that under natural conditions, which may then alter downstream biodiversity and ecological processes. The hydropower exploitation in the mainstream Lancang-Mekong River has triggered concern for its potential effects on downstream countries, especially the impact of the released cold water on local fishery production. However, it was observed recently that the annual water temperature downstream of the Jinghong Reservoir (near the Chinese border) has increased by 3.0 °C compared to its historical average (1997–2004). In this study, a three-dimensional (3D) model of the Jinghong Reservoir was established to simulate its hydro- and thermodynamics. Results show that: (1) the impoundment of the Jinghong Reservoir contributed about 1.3 °C to the increment of the water temperature; (2) the solar radiation played a much more important role in comparison with atmosphere-water heat exchange in changing water temperatures; and (3) the outflow rate also imposed a significant influence on the water temperature by regulating the residence time. After impoundment, the residence time increased from 3 days to 11 days, which means that the duration that the water body can absorb solar radiation has been prolonged. The results explain the heating mechanism of the Jinghong Reservoir brought to downstream water temperatures.

Keywords: tropical reservoir; heating impact; Langcang-Mekong River

1. Introduction

The riverine environment can be influenced by reservoirs and dams, as well as their operations in many forms, including the changing of riverine thermal regimes and downstream water temperatures [1–3]. The health, distribution, and functions of aquatic creatures can be influenced by water temperatures [4–7], so the extensive construction of dams worldwide has long drawn attention to potential effects of damming on downstream thermal regimes [2,8].

Generally, the impact degree of a dam on downstream thermal regimes is decided by its mode of operation and specific mechanism of water release [9]. Many large dams release water from deep portals which are located under the thermocline, namely the hypolimnetic layer of a reservoir. As a result, the cold water is released to the downstream thermal regimes [2,10–12]. This case was rarely reported, but some small dams release water from above the thermocline, namely the epilimnetic layer, so temperatures of the downstream water increase [7].

In addition to annual water temperature, reservoirs also influence seasonal thermal patterns of downstream water. In general, in large reservoirs, the moderate temperatures of downstream water in spring and summer are lower than those in winter when the seasonal fluctuations decrease, while they also display a delay of maxima in comparison with natural rivers [2]. Such phenomena were observed at many dams and reservoirs across the world, such as the Dartmouth Dam on the Mitta Mitta River and Burrendong Dam on the Macquarie River, located in Australia [13,14], the regulated Lyon River located in Scotland [15], and the Hills Creek Dam on the Willamette River, located in America [16].

As the upstream part of the Lancang-Mekong River, the Lancang River is the largest international river in Southern Asia [17]. Since the 1950s, to fully exploit the river's resources, the Chinese government has come up with proposals concerning damming of mainstreams on the Lancang River [18]. At present, six large dams have been put into operation along the mainstream of the Lancang River. These dams include Nuozhadu, Manwan, Gongguoqiao, Dachaoshan, Xiaowan, and Jianghong dams (ranging from upstream to downstream), wherein the Nuozhadu Reservoir was recently put into operation in 2015 [19]. The cascade of dams is constructed with the primary purpose of hydropower generation. According to the hydropower development plan by Huaneng Lancang River Hydropower Company, the hydropower installed capacity will reach 30.0 GW in the Lancang mainstream by 2020 [19]. Nevertheless, the potential hydrological and environmental effects of reservoir impoundment and water release have drawn more attention to downstream countries [17,20–22]. Among these potential effects, altered downstream water temperature has become an important focus as it plays a significant role in influencing the combination of structure, growth, reproduction, distribution, and stream productivity of aquatic organisms [23], in addition to which fish in freshwater serve as the major protein source for local animals which live in downstream countries [19]. In view of those deep and large dams located along the Lancang River, the low-temperature outflow has drawn major concern. In addition, the Nuozhadu Dam is equipped with a multi-level stop log door so as to moderate the downstream water temperature to a pre-dam level [24]. Nevertheless, after the impoundment of six large reservoirs, the water temperatures downstream of the Jinghong Reservoir (the most downstream one of the cascade dams) increased, rather than decreased, in all seasons, which is against what has been observed in other large dams. In other words, the upstream multi-level intake structures could probably be unnecessary. As numerical modeling is a powerful and useful tool, we applied it in investigating the reasons why the Jinghong Reservoir had a unique heating impact on downstream water.

Some one-dimensional (1D) models are applicable to simulate the vertical distribution of the water temperature and chemical/biological materials in a lake or reservoir through time [25]. In general, these models are established under the one-dimensional assumption. Specifically speaking, variations in the lateral directions are relatively small in comparison with those in the vertical directions [26] and are widely used because they are fast enough to facilitate long-term simulation and have performed well in simulating the seasonal and inter-annual variations of lake water temperatures [27]. These models include the bulk model of Kraus and Turner [28], DYRESM [29], PROBE [30], the Hostetler Model [31,32], SEEMOD [33], LIMNMOD [34], MASAS and CHEMSEE [35], Minlake [36], GOTM [37], SIMSTRAT [38], LAKE [39,40], CLM4-LISSS [41], and WRF-Lake [42].

Although 1D lake models are more efficient in computation, under certain conditions (such as long, deep reservoirs) water mass exchange in both vertical and longitudinal directions and temperature gradients may be important [25]. Thus, many two-dimensional (2D) models have been developed through the integration of the reservoir, with the aim to reach the motion equations with lateral averaging [43]. Laterally averaged 2D models are applicable to modeling of long and relatively narrow reservoirs with no or negligible lateral inflow or outflows. These models include the model of Box Exchange Transport Temperature and Ecology of Reservoirs (BETTER) [44]; the Computation of Reservoir Stratification (COORS) model [45]; the Laterally Averaged Reservoir Model (LARM) [46];

the model of Generalized Longitudinal-Vertical Hydrodynamics and Transport (GLVHT) [47], which was developed from LARM; the CE-QUAL-W2 developed through GLVHT expansion to include water quality constituents [48]; the Laterally-Averaged Hydrodynamics Model (LAHM) [43]; and the MIKE21 Flow Model, developed by Danish Hydraulic Institute (DHI) Water and Environment [49–52], which is widely applied. In addition to the laterally-averaged models, vertically-averaged 2D models are also applied when vertical flow variations are not as important as the lateral case, e.g., when the water is shallow [53–55]. These models include the North Sea Model [56], the Tokyo Bay Model [57], the Haringvliet Model, and the depth-averaged mathematical model developed by McGuirk and Rodi [55].

Despite the extensive use of 1D and 2D models, the limit in one-dimensional or vertical and lateral averaging has long been realized and the needs for 3D modeling of lakes and reservoirs have raised more discussion [58–61]. In the past decades, the complex water quality models and 3D hydrodynamic models have been developed thanks to the high-performing super computers at a reasonable cost, as well as the efficient numerical algorithms [62]. The mentioned models include the Curvilinear-grid Hydrodynamics model in three dimensions (CH3D) [63], the Estuarine, Coastal, and Ocean Model (ECOM) [64]; the Environmental Fluid Dynamics Code (EFDC) [65]; the CE-QUAL-ICM model [66]; the model of Water Quality Analysis Simulation Program (WASP) [67]; the Delft3D-FLOW model [68]; the Row-Column AESOP (RCA) model [69]; and the MIKE 3 model by DHI Water and Environment [70].

In order to analyze the complex 3D hydrodynamic and thermodynamic processes more deeply, it is necessary to establish 3D models. Nevertheless, researchers have so far rarely established 3D models for simulation of the environmental and thermal results of damming along the Lancang River [19] in view of the high cost of computation and the lack of observational data. Nowadays, with computational resources and enough pre-dam and post-dam measurement data, we are capable of conducting 3D simulations at the Jinghong Reservoir. The Delft3D-FLOW model was selected for the simulations as it has been widely used around the world and evidence proves that it can simulate sediment transport, flows, water quality, waves, morphological developments, and ecological courses in coastal regions, rivers, and lakes [71–78]. After model calibration, the 3D model of the Jinghong Reservoir was used to simulate the thermodynamic and hydrodynamic situations in the reservoir, as well as the related outflow. The research also carried out some numerical experiments to analyze the heating impact brought by the Jinghong Reservoir to downstream water.

2. Study Area

The Jinghong Reservoir system is located in the lower reaches of the Lancang River, which is, in general, a south-flowing river in southwestern China, and called the Mekong River when entering Laos (Figure 1). This system is defined by the Nuozhadu Dam (22°38′ N, 100°26′ E), the Jinghong Dam (22°03′ N, 100°46′ E), and the 105-km long water body between the two (i.e., the Jinghong Reservoir). The Jinghong Dam went into service in 2009 with a height of 108 m, while the Nuozhadu Dam was put into operation in 2015 with a height of 262 m. The Jinghong Reservoir is a long and narrow channel-shaped reservoir with a length of 105 km and an average width of 312 m. It has a normal water level of 602 m above sea level, a maximum depth of 70 m and a total reservoir capacity of 1.14 billion m^3. Since the operation of the reservoir, the water surface area has increased from 12.7 km^2 to 32.8 km^2. The mean annual inflow is about 1674 $m^3\ s^{-1}$. A tropical reservoir, the Jinghong Reservoir, has a perennial mean air temperature reaching 22.2 °C at the Jinghong Dam.

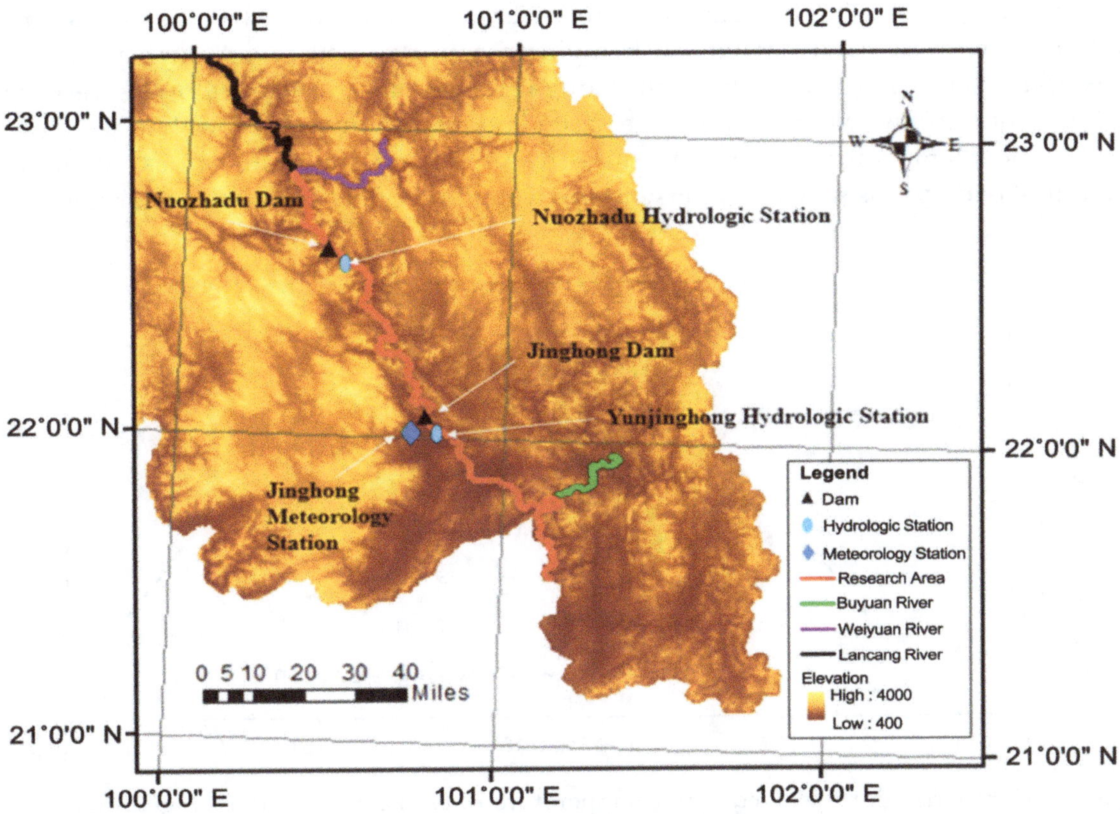

Figure 1. Location and topography of the Jinghong Reservoir system.

2.1. *Observation Data*

The two hydrological stations involved in the study area are the Nuozhadu Hydrological Station (NHS) and the Yunjinghong Hydrological Station (YHS), both located on the right bank of the river, with the former 5 km downstream of the Nuozhadu Reservoir, and the latter 3 km downstream of the Jinghong Reservoir. Both NHS and YHS are located at the riverine section of the water body, and the fully mixing assumption can be applied. Accordingly, point-measured water temperature can be used as an indicator of the cross-sectional water temperature. To measure the water temperature, water pressure, and air pressure, a Hobo Onset U20-001-02 water level data logger (U20 hereafter) is adopted at each hydrological station on an hourly basis throughout the study period. The water level at each station is computed by:

$$P_{water} = P_{total} - P_{air} = \rho_{water} g H_{U20}, \quad (1)$$

where P_{total} (Pa) is the U20-measured pressure underwater (water pressure and air pressure), P_{air} (Pa) is the U20-measured atmospheric pressure by the river bank, P_{water} (Pa) is the pressure of pure water, and H_{U20} is the depth of the U20, from which water levels are computed. The discharge at YHS is further computed using its rating curve obtained from the Water Resources Department of Yunnan Province, China [79].

2.2. *Observation Data Analysis*

2.2.1. Comparison with Historical Water Temperature

To illustrate the annual variations of water temperature at NHS and the Jinghong Reservoir, in situ observations were conducted (Figure 2). Water temperatures almost changed synchronously at the two sites throughout the year, with NHS always displaying a lower temperature (a mean difference of 2.2 °C) than the other. This systematic temperature difference reflected the latitude difference between the two sites as water flowed south, a phenomenon not observed at reservoirs flowing in other directions. According to a study on the Cougar Reservoir (an eastern-flowing reservoir in the U.S.), the peak of outflow water temperature lagged about 2 months behind that of inflow, but the inflow and outflow temperatures shared the same maximum and minimum values [80], indicating no large amount of energy was taken in, or lost, as the water body flowed into the reservoir.

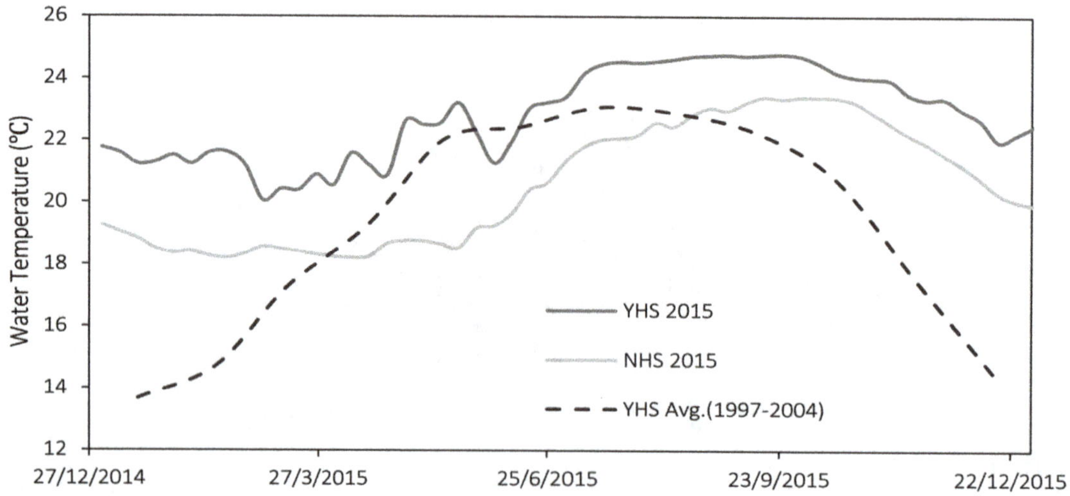

Figure 2. Comparison of observed water temperature data and historical averaged data at YHS. The temporal resolution of the observation data is daily.

Obtained and averaged from 1997 to 2004 at YHS, historical water temperature observations (Figure 2) were shown to range between 13.7 and 23.1 °C annually. However, after the impoundment of the Jinghong Reservoir, the water temperature range was changed to between 19.8 and 25.0 °C, with the annual variation reduced from 9.5 °C to 5.2 °C. This reduction in annual variation mainly resulted from a 6.1 °C increase in the minimum temperature. Overall, the annual average water temperature rose from 19.2 °C to 22.8 °C after impoundment.

2.2.2. Discharge and Water Temperature

Observed water temperatures at YHS were compared with outflow rates from the reservoir in Figure 3. As they were inversely correlated, the discharge rate of the Jinghong Dam could also be a potential factor behind the water temperature alteration.

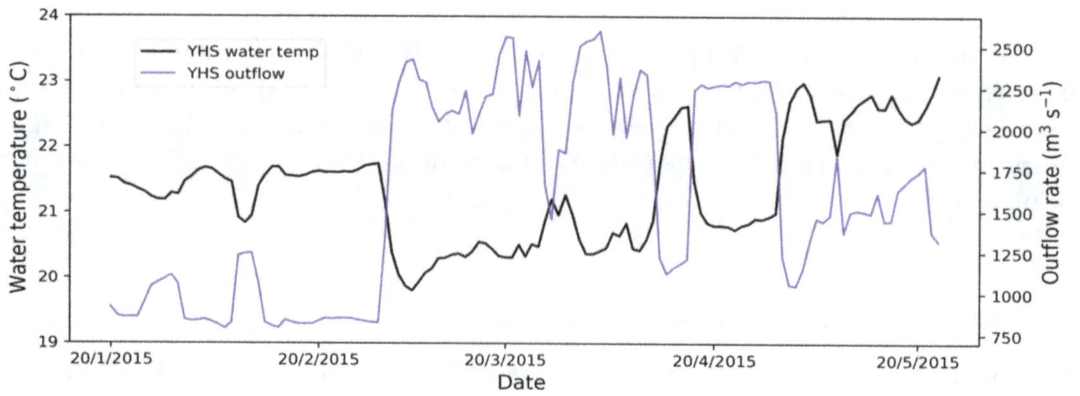

Figure 3. Comparison of the observed YHS water temperature and the YHS discharge rate.

3. Methods

3.1. The Delft3D-FLOW Model

The numerical hydrodynamic modeling system—the Delft3D-FLOW model—was developed by Deltares [68], a Dutch-based research institute. The Delft3D-FLOW software used for this study is based on Fortran and acquired from Delft3D website (https://oss.deltares.nl/web/delft3d/). With Delft3D-FLOW, the 2D (based on averaged depth) and 3D shallow water equations that were unsteady could be solved. The equation system comprises momentum equations in the horizontal plane, the continuity equation and the transport equations of conservative elements [68]. The basic equations are retained in their original form and are not changed in this study, which are introduced briefly as follows.

As for the hydrodynamic equations, Delft3D-FLOW is used for the solution of the Navier-Stokes equations with regard to a fluid which is incompressible under the assumption of Boussinesq and the shallow water environment [81]. In the equation of vertical momentum, the accelerations in the vertical direction are omitted, leading to the equation of hydrostatic pressure. Vertical speeds are computed according to the continuity equation in 3D models. The set of partial differential equations, together with a proper set of boundary and initial situations is solved via a mesh with finite differences. Delft3D-FLOW uses orthogonal curvilinear Cartesian coordinates (ξ, η) in the horizontal direction. Two different vertical grid systems are offered by Delft3D-FLOW, vertically: the Cartesian Z coordinate system (Z-model) and the σ coordinate system (σ-model). The σ coordinate system was first introduced by Phillips [82] and it is applied in this research.

To obtain the continuity equation with averaged depth, the continuity equation shall be integrated for incompressible fluids ($\nabla \bullet \vec{u} = 0$) based on the total depth, wherein the conditions of the kinematic boundary on the bed level and at the water surface are taken into account. It is expressed as follows:

$$\frac{\partial \zeta}{\partial t} + \frac{1}{\sqrt{G_{\xi\xi}}\sqrt{G_{\eta\eta}}}\frac{\partial((d+\zeta)U\sqrt{G_{\eta\eta}})}{\partial \xi} + \frac{1}{\sqrt{G_{\xi\xi}}\sqrt{G_{\eta\eta}}}\frac{\partial((d+\zeta)V\sqrt{G_{\xi\xi}})}{\partial \eta} = (d+\zeta)Q, \tag{2}$$

where ζ (m) is the water level on a number of horizontal planes of reference (datum), the coefficients $\sqrt{G_{\xi\xi}}$ and $\sqrt{G_{\eta\eta}}$ (m) are used for transformation from curvilinear coordinates to rectangular coordinates, t (s) is the time, d (m) is the depth under a number of horizontal planes of reference (datum), U (m s^{-1}) is the depth-averaged speed in ξ-direction, V (m s^{-1}) is the depth-averaged speed in η-direction, and Q (m s^{-1}) is the contributions made for each unit area because of the withdrawal or release of water, evaporation and precipitation.

The momentum equations in ξ-direction and η-direction are expressed as follows:

$$\begin{aligned}\frac{\partial u}{\partial t} &+ \frac{u}{\sqrt{G_{\xi\xi}}}\frac{\partial u}{\partial \xi} + \frac{v}{\sqrt{G_{\eta\eta}}}\frac{\partial u}{\partial \eta} + \frac{\omega}{d+\zeta}\frac{\partial u}{\partial \sigma} - \frac{v^2}{\sqrt{G_{\xi\xi}}\sqrt{G_{\eta\eta}}}\frac{\partial\sqrt{G_{\eta\eta}}}{\partial \xi} + \frac{uv}{\sqrt{G_{\xi\xi}}\sqrt{G_{\eta\eta}}} \\ &\frac{\partial\sqrt{G_{\xi\xi}}}{\partial \eta} - fv = -\frac{1}{\rho_0\sqrt{G_{\xi\xi}}}P_\xi + F_\xi + \frac{1}{(d+\zeta)^2}\frac{\partial}{\partial \sigma}\left(\nu_V\frac{\partial u}{\partial \sigma}\right) + M_\xi,\end{aligned} \tag{3}$$

$$\begin{aligned}\frac{\partial v}{\partial t} &+ \frac{u}{\sqrt{G_{\xi\xi}}}\frac{\partial v}{\partial \xi} + \frac{v}{\sqrt{G_{\eta\eta}}}\frac{\partial v}{\partial \eta} + \frac{\omega}{d+\zeta}\frac{\partial v}{\partial \sigma} - \frac{uv}{\sqrt{G_{\xi\xi}}\sqrt{G_{\eta\eta}}}\frac{\partial\sqrt{G_{\eta\eta}}}{\partial \xi} + \frac{u^2}{\sqrt{G_{\xi\xi}}\sqrt{G_{\eta\eta}}} \\ &\frac{\partial\sqrt{G_{\xi\xi}}}{\partial \eta} - fu = -\frac{1}{\rho_0\sqrt{G_{\eta\eta}}}P_\eta + F_\eta + \frac{1}{(d+\zeta)^2}\frac{\partial}{\partial \sigma}\left(\nu_V\frac{\partial v}{\partial \sigma}\right) + M_\eta,\end{aligned} \tag{4}$$

where u, v, and ω (m s^{-1}) are speeds in ξ, η, and σ-directions. f (s^{-1}) is the Coriolis parameter (inertial frequency), ρ_0 (kg m^{-3}) is the referential density of water, P_ξ and P_η (kg m^{-2} s^{-2}) are the gradient-based hydrostatic pressure in ξ-direction and η-direction, F_ξ and F_η (m s^{-2}) are the imbalances existing in horizontal Reynold's stresses in the ξ-direction and η-direction, ν_V (m^2 s^{-1}) is the vertical eddy viscosity, M_ξ and M_η (m s^{-2}) are source or sink of momentum in the ξ-direction and η-direction. Reynold's stresses F_ξ and F_η are simulated on the basis of the concept of eddy viscosity [83]. The eddy viscosities in vertical and horizontal directions are as follows:

$$v_H = v_{3D} + v_H^{back}; v_V = \nu_{mol} + \max\left(v_{3D}, v_V^{back}\right), \tag{5}$$

where v_{3D} is the viscosity computed with the 3D k-ε closure scheme, v_H^{back} is the user-defined background horizontal viscosity, v_V^{back} is the user-defined background vertical viscosity, ν_{mol} (m^2 s^{-1}) is the kinematic viscosity (molecular) coefficient of water.

Transport of heat and matter is simulated by an equation of advection diffusion in three coordinate directions, and is expressed as follows:

$$\begin{aligned} &\frac{\partial(d+\zeta)c}{\partial t} + \frac{1}{\sqrt{G_{\xi\xi}}\sqrt{G_{\eta\eta}}}\left\{\frac{\partial\left[\sqrt{G_{\eta\eta}}(d+\zeta)uc\right]}{\partial\xi} + \frac{\partial\left[\sqrt{G_{\xi\xi}}(d+\zeta)vc\right]}{\partial\eta}\right\} + \frac{\partial\omega c}{\partial\sigma} \\ &= \frac{d+\zeta}{\sqrt{G_{\xi\xi}}\sqrt{G_{\eta\eta}}}\left\{\frac{\partial}{\partial\xi}\left(D_H\frac{\sqrt{G_{\eta\eta}}}{\sqrt{G_{\xi\xi}}}\frac{\partial c}{\partial\xi}\right) + \frac{\partial}{\partial\eta}\left(D_H\frac{\sqrt{G_{\xi\xi}}}{\sqrt{G_{\eta\eta}}}\frac{\partial c}{\partial\eta}\right)\right\} + \frac{1}{d+\zeta}\frac{\partial}{\partial\sigma}\left(D_V\frac{\partial c}{\partial\sigma}\right) \\ &- \lambda_d(d+\zeta)c + S, \end{aligned} \tag{6}$$

where c (kg m^{-3}) is the concentration of mass, D_H (m^2 s^{-1}) is the total diffusion coefficient in the horizontal direction, D_V (m^2 s^{-1}) is the diffusion coefficient in the vertical direction, λ_d (s^{-1}) is the first order decay course and S (kg m^{-2} s^{-1}) is the source and sink terms in the unit area generated from the withdrawal or release of water and/or heat exchange on the free surface. The diffusion coefficient in the horizontal direction is user-specified, whereas the coefficient in the vertical direction is computed as follows:

$$D_H = D_{SGS} + D_V + D_H^{back}; D_V = \frac{\nu_{mol}}{\sigma_{mol}} + \max\left(D_{3D}, D_V^{back}\right), \tag{7}$$

where D_{SGS} is the 2D part of diffusion based on the turbulence model of sub-grid scale, D_H^{back} is the user-defined eddy diffusivity in the horizontal direction, D_V^{back} is the eddy diffusivity in the vertical direction, σ_{mol} is the Prandtl-Schmidt number for molecular mixing, and D_{3D} is the diffusion due to turbulent eddy viscosity [68].

3.2. Model Set-Up

3.2.1. Grid Coverage

Along the river flow direction, the upper boundary of the model is set to the Nuozhadu Hydrological Station, whereas the lower boundary of the model is set to the Jinghong Dam. With a longitudinal length of about 95 km, the model covers most parts of the Jinghong Reservoir. The computational grid of the Jinghong Reservoir hydrodynamic model is shown in Figure 4. Grid density varies with topography, i.e., in order to increase the computation accuracy, finer horizontal grids have been placed at the part of the reservoir close to the Jinghong Dam, resulting in a total of about 11,000 horizontal grid cells per layer. Meanwhile, for the vertical direction, the σ coordinate system (σ-model) is applied, with the reservoir discretized into 40 vertical layers, with each layer less than 2 m in thickness.

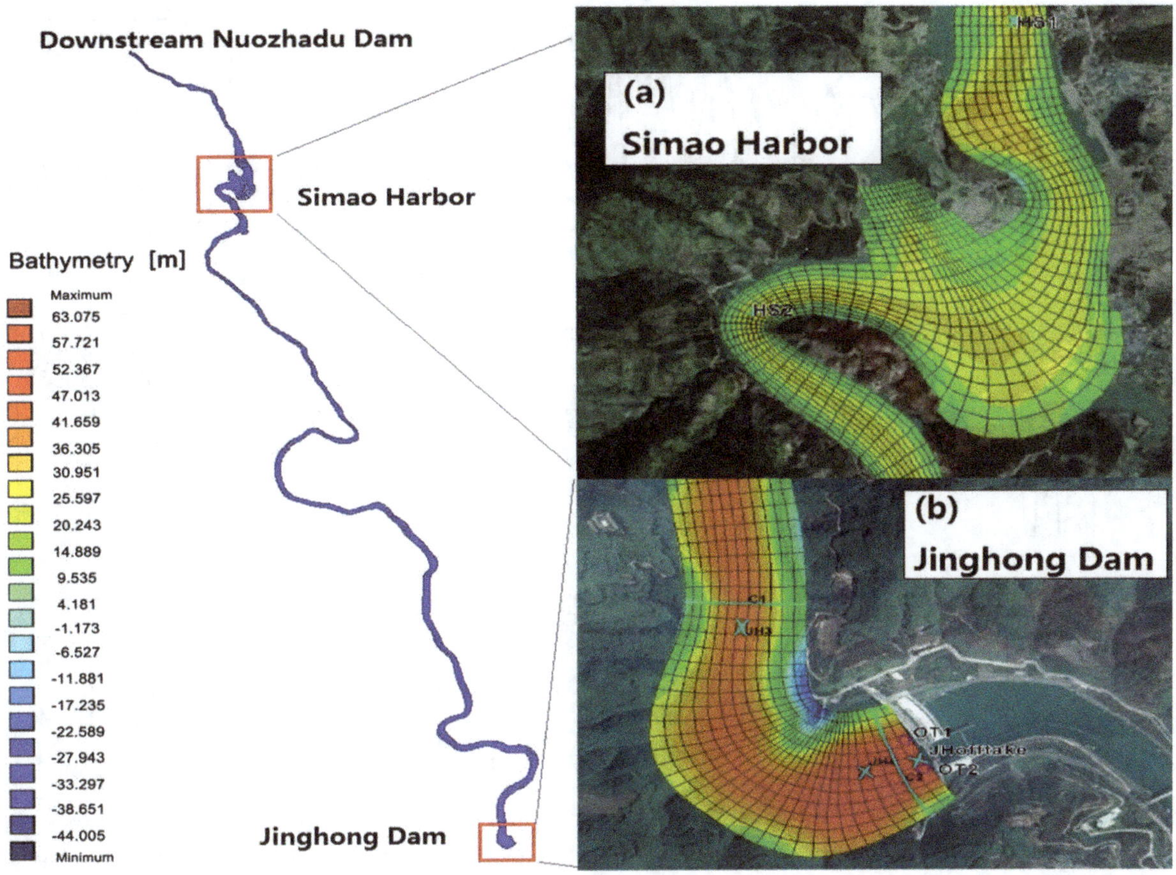

Figure 4. Horizontal discretization (about 11,000 horizontal grid points) and bathymetry of the Jinghong Reservoir with local enlargements at the site of (**a**) the Simao Harbor, and (**b**) the Jinghong Dam. We set 590 m above sea level as the datum (0 m) in the model so as to keep the water level a positive value through simulation period. Grid points with the riverbed higher than 590 m have a negative bathymetry value.

3.2.2. Boundary Conditions

The upper boundary is set at the water level at NHS and the lower boundary is set at the discharge rate downstream of the Jinghong Dam. The terrain data is extracted from a digital elevation model (DEM) with a spatial resolution of 25 m × 25 m.

3.2.3. Meteorological Data and Model Flow Chart

Restrained by the poor data availability, the simulation only covers a period from 20 December 2014 to 22 May 2015. The computational time step is set to 1 minute to meet the demand of numerical stability, as well as computational efficiency.

Meteorological data for this study, which is obtained from the Jinghong National Meteorological Station (Figure 5), is not readily available. As a result, such data is only gathered for a relatively short period, from 20 December 2014 to 20 May 2015, which also defines the study period for this research. These data include local air temperature data, relative humidity data, and solar radiation data with a temporal resolution of 1 day. We acknowledge that it would be better if a full year simulation could be conducted. However, due to solar radiation data availability, 5-month simulation is the best we can practice at the current stage. Similar simulations with temporal coverage of less than 1 year can also be found in the literature [84–86].

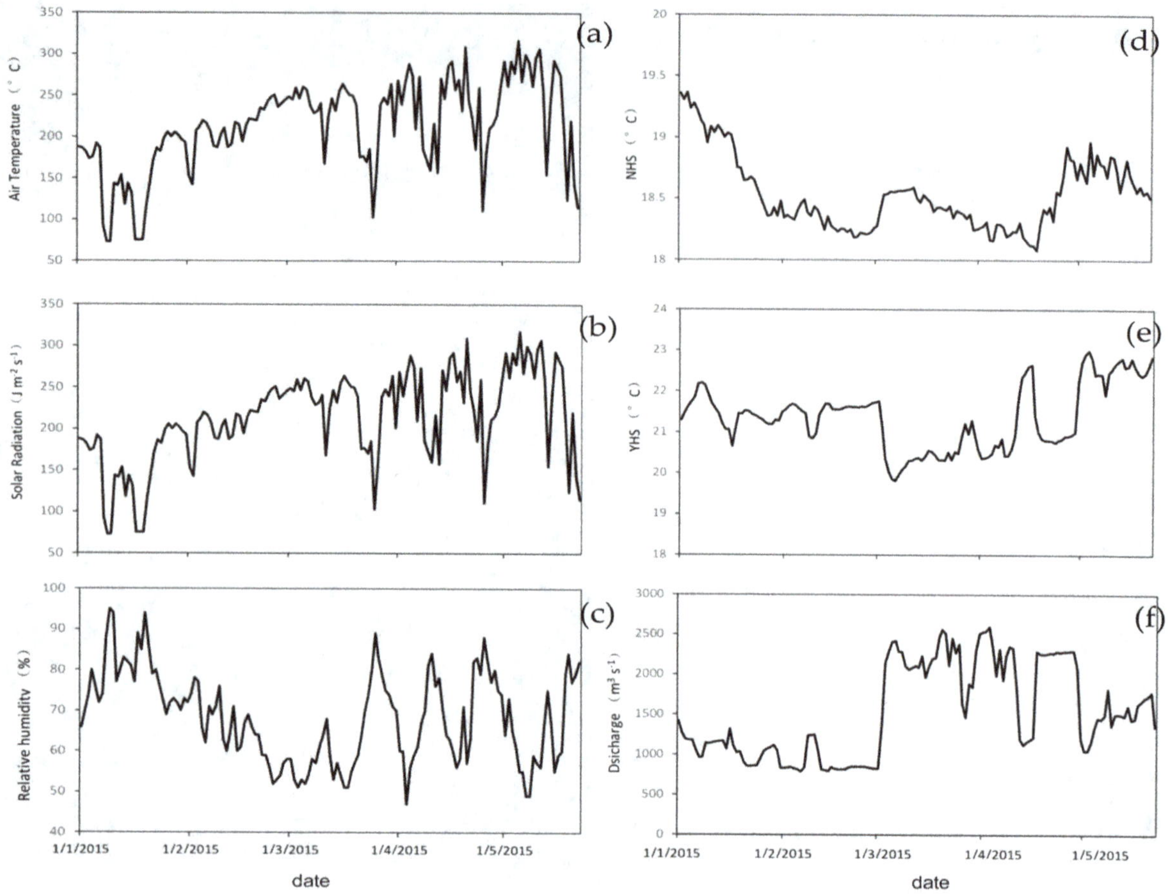

Figure 5. Daily meteorological data (1 January 2015 to 20 May 2015) measured at the Jinghong National Meteorological Station: (**a**) air temperature, (**b**) solar radiation, and (**c**) relative humidity. Water temperature data is measured at the NHS and YHS station: (**d**) water temperature at NHS, (**e**) water temperature at YHS, and (**f**) discharge at YHS.

A flowchart illustrating the Jinghong Reservoir hydrodynamic model is shown in Figure 6.

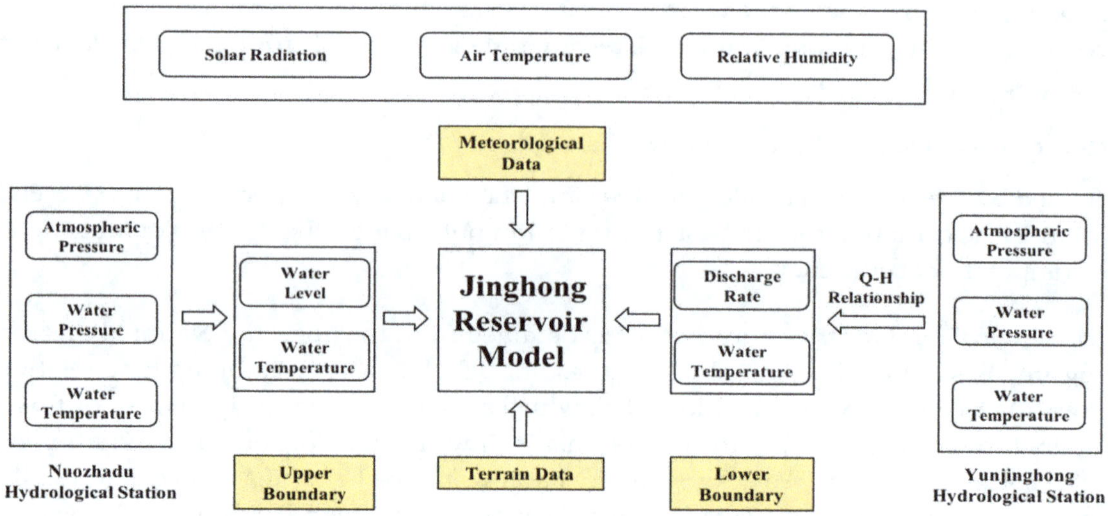

Figure 6. Flowchart of the Jinghong Reservoir model set-up.

3.3. Model Calibration and Validation

The warm-up period of this study occurs from 22 December 2014 to 31 January 2015.

For the purpose of verifying the hydro-dynamic condition of this simulation, the simulated water levels were firstly calibrated against measurements. As the model output temporal resolution was 10 minutes, the hourly average for simulated water levels was calculated for comparison. Data on 2 days' on-site observed water levels at Simao Harbor were provided by field experiment. The best match of the model output and measured data (Figure 7) were achieved by applying critical parameters, as shown in Table 1.

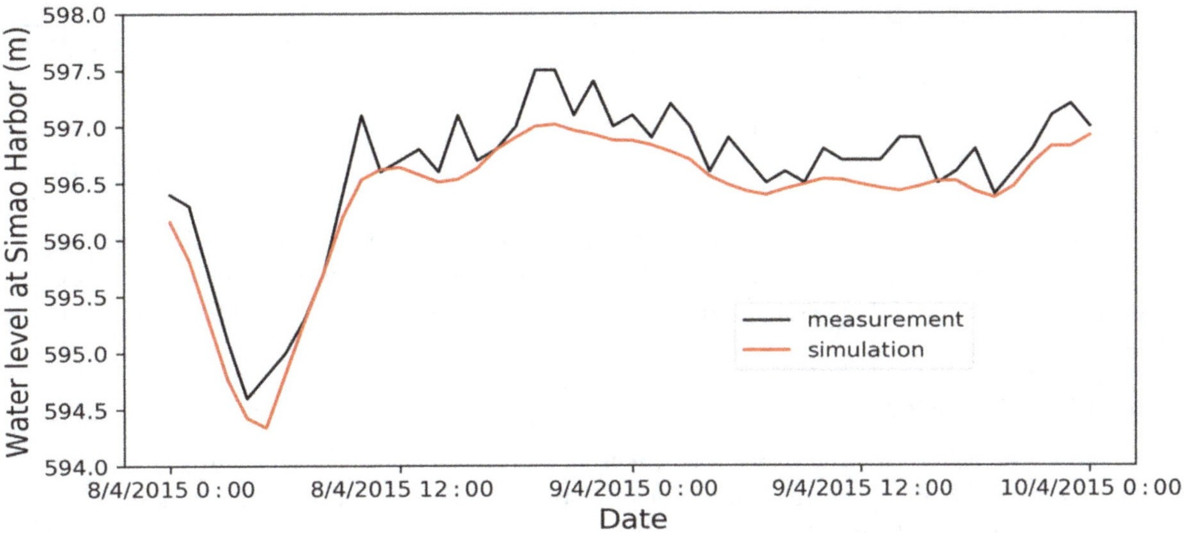

Figure 7. Comparison between simulated and measured water levels at Simao Harbor. Simulated water levels are plotted on an hourly average.

Five main parameters were calibrated by Chanudet et al. (2012): the Chezy's coefficient, which represents the roughness at the water bottom, background horizontal and vertical eddy viscosity, and background horizontal and vertical eddy diffusivity. The studied range of Chezy's coefficient can be referred to Chow (1959), while the range of the background eddy viscosity and diffusivity can also be referred to in previous studies [87–89].

Table 1. Parameters studied for the calibration of the Jinghong Reservoir model and their chosen value.

Parameter	Studied Range	Chosen Value
Chezy's coefficient	10–100 $m^{\frac{1}{2}}$ s^{-1}	65 $m^{\frac{1}{2}}$ s^{-1}
Background horizontal eddy viscosity	1×10^{-6}–1 m^2 s^{-1}	5×10^{-4} m^2 s^{-1}
Background horizontal eddy diffusivity	1×10^{-6}–1 m^2 s^{-1}	1×10^{-4} m^2 s^{-1}
Background vertical eddy viscosity	1×10^{-10}–1 m^2 s^{-1}	1×10^{-6} m^2 s^{-1}
Background vertical eddy diffusivity	1×10^{-10}–1 m^2 s^{-1}	1×10^{-6} m^2 s^{-1}

The focus of the calibration then shifted to outlet water temperatures downstream the Jinghong Dam. As YHS was within 3 km downstream of the Jinghong Reservoir in the riverine section of the water body, the simulated outlet water temperature was compared with the measured water temperature at YHS for calibration, with the difference between the two shown in Figure 8. The simulated results were in good agreement with measurements, with the absolute error limited to 1.5 °C.

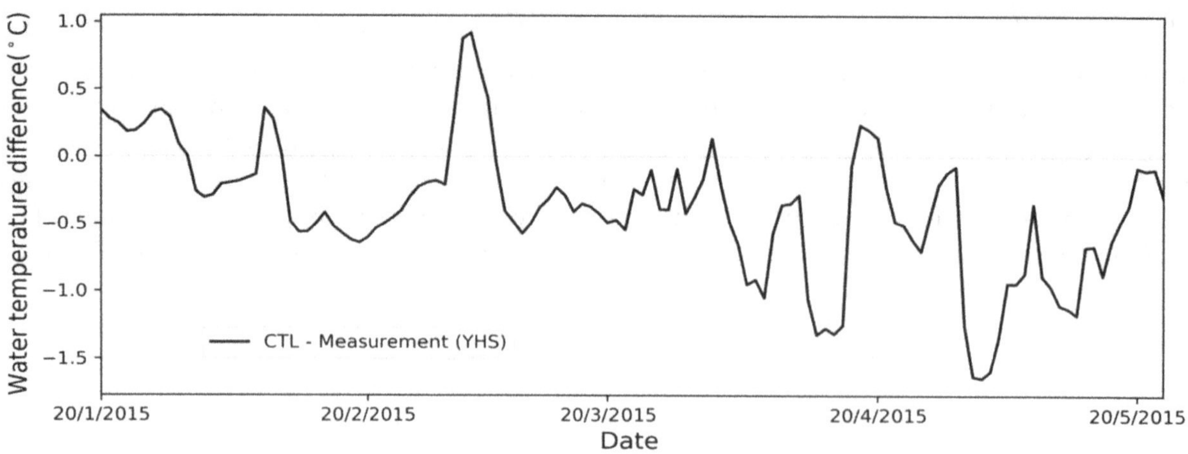

Figure 8. Outlet water temperature difference between the simulation results (CTL) and measurements at YHS.

Compared with measurements, the Jinghong Reservoir model produced good simulations of water temperatures. Nevertheless, since insufficient observations of the water velocity fields were available, no comparison concerning water velocity was carried out either for calibration or for validation.

3.4. Numerical Scenario Setups

As presented in Section 2.2, three main issues are ought to be found out through numerical modeling: the impact of impoundment of Jinghong Reservoir; the contributing factor to water temperature increment; and the relation between outflow rate and outflow temperature from Jinghong Dam. The wind speed and relative humidity were precluded as the driving forces and several numerical experiments were performed (Table 2).

- **Impact of Impoundment of the Jinghong Reservoir**

A control run (CTL) was set up as described in Sections 3.2 and 3.3.

S1 (i.e., scenario 1) was set up to examine the impact of the impoundment of the Jinghong Reservoir, in which the Jinghong Dam was removed and the water body returned to the natural river conditions. A depth-averaged two-dimensional flow model with the same horizontal grid discretization was set up, with the upper boundary condition as described in Section 3.2.2, the Q-H rating-curve adapted as the downstream boundary condition and the atmospheric data as for CTL.

- **Contributing Factors**

To examine the contribution of solar energy to the increase of water temperatures, S2 was imposed with no solar radiation. To further evaluate the impact of heat exchange with the overlying atmosphere on water temperature, atmosphere-water heat conduction, together with solar radiation was turned down in S3.

- **Outflow Rate and Outflow Temperature**

Different scenarios were further established so as to investigate the relationship between the outflow rate and the outflow water temperature at the Jinghong Dam. The water temperature measured at YHS was constantly higher than the one measured at NHS, suggesting that the water parcel experienced an energy absorption process travelling from NHS to YHS. In addition to the rate of energy transfer, the amount of energy absorbed by the water parcel was also determined by the duration of the absorption process. The outflow rate directly affected the time during which the water parcel absorbed energy. In order to measure the duration of the energy absorption process, the tracer function of the model was enabled. The duration of energy absorption is represented by the resident time of the tracer. Different scenarios were established to assess the relationship among the outflow rate, resident time, and inflow-outflow temperature difference. In addition, for the sheer size difference between the Nuozhadu Reservoir (upstream) and the Jinghong Reservoir (downstream), the Jinghong Reservoir's outflow rate was determined by the outflow rate of the Nuozhadu Reservoir. To further simplify the simulation, an assumption was made that the Jinghong Reservoir's inflow rate is equal its outflow rate. The inflow water temperature at the upper boundary was set to the same as for CTL. The simulation duration remained the same as CTL.

The real outflow rate for the Jinghong Reservoir ranged from 536 to 3398 $m^3 s^{-1}$ and with an annual average of 1674 $m^3 s^{-1}$ in 2015. Eight scenarios were established to examine the effect of outflow rates of the Jinghong Dam, with outflow rates prescribed from 500 to 4000 $m^3 s^{-1}$, with 500 $m^3 s^{-1}$ as intervals. The eight scenarios had been denoted as F1–F8 in the study.

Table 2. An overview of numerical experiments.

Experiments	The Jinghong Reservoir	Solar Radiation	Atmosphere-Water Heat Exchange	Outflow Rate ($m^3 s^{-1}$)	Tracer
CTL	ON	ON	ON	Real	OFF
S1	OFF	ON	ON	Real	OFF
S2	ON	OFF	ON	Real	OFF
S3	ON	OFF	OFF	Real	OFF
F1	ON	ON	ON	500	ON
F2	ON	ON	ON	1000	ON
F3	ON	ON	ON	1500	ON
F4	ON	ON	ON	2000	ON
F5	ON	ON	ON	2500	ON
F6	ON	ON	ON	3000	ON
F7	ON	ON	ON	3500	ON
F8	ON	ON	ON	4000	ON

4. Results and Discussion

4.1. Thermal Structure of the Jinghong Reservoir

Covering a distance of 105 km, the Jinghong Reservoir has a water depth roughly increasing along the direction of the river flow. The thermal structure along the reservoir from 10 km down the Nuozhadu Dam to the Jinghong Dam (about 90 km) is shown in Figure 9 (3 days corresponding to different months were selected for demonstration). Generally, the water temperature was almost homogeneous during the cool season from December 2014 to February 2015. As the air temperature increased from February on, a warmer water layer appeared at the surface, which eventually developed into a weakly stratified thermocline at the depth of approximately 5 m by the end of May 2015. Unlike many natural lakes, which experienced overturn in early spring, no overturn was observed during the study period at the Jinghong Reservoir, as it was a tropical one and the surface water temperature never went below 4 °C.

During the study period, only weak stratification was observed in the Jinghong Reservoir in May with the surface-bottom temperature difference reaching only 3 °C. In comparison, much stronger stratifications were reported at other deep reservoirs along the Lancang River after impoundment. For example, the Nuozhadu Reservoir (upstream reservoir to the Jinghong Reservoir) experienced a surface-bottom temperature difference of 10 °C in summer and the largest temperature difference in the Xiaowan Reservoir even reached 14 °C [90]. One main reason is the depth difference between these reservoirs. The Jinghong Reservoir is 70 m deep, while the Nuozhadu and Xiaowan Reservoirs are much deeper, each more than 200 m. By preventing solar radiation and wind-driven eddy penetration, large water bodies in these two deep reservoirs retard heat transfer and are difficult to reach a homogeneous temperature. Another reason is the water exchange index, or the α index, which is a widely used approach [91] in identifying reservoir thermal structures, defined by:

$$\alpha = \frac{w}{v}, \tag{8}$$

where w (m^3) is average annual inflow and v (m^3) is the total capacity of the reservoir. When $\alpha \leq 10$, the reservoir is stably stratified; when $10 < \alpha < 20$, the reservoir is unstably stratified; when $\alpha \geq 20$, the reservoir is mixed. The α index of the Jinghong Reservoir is much higher compared with that of the other two (Table 3). The large quantity of annual inflow (~25 times that of reservoir capacity) serves as a major disturbance, leaving the water body well mixed.

Table 3. The α index of three major reservoirs along the middle and lower reaches of the Lancang River and their thermal structure type according to the index.

Reservoir Name	Average Annual Inflow (billion m^3)	Total Capacity (billion m^3)	α Index	Type
Xiaowan	38.2	15.1	2.5	Stably stratified
Nuozhadu	55.5	23.7	2.3	Stably stratified
Jinghong	58.0	1.1	25.4	Mixed

Vertically, the water columns along the reservoir showed almost no stratification regardless of their location and depth. The water temperature difference between surface and bottom was less than 1 °C in the first 40 km along the reservoir in all simulated time, while its maximum of 3 °C was reached at the Jinghong Dam in May. Longitudinally, water surface temperature rose as the river flowed southward, with its value down the Nuozhadu Dam increasing from around 18.5 °C all the way down the river, to approximately 22 °C before the Jinghong Dam. Temporally, water temperature changed in accordance with air temperature. When the air temperature was lower than the water temperature in January and February, the water temperature experienced a drop, while when the air temperature was higher than the water temperature from March to May, the water temperature rose. However, the water temperature variation also increased longitudinally, with the variation down the Nuozhadu Dam being only 0.5 °C and reaching 3 °C at the Jinghong Dam.

(a)

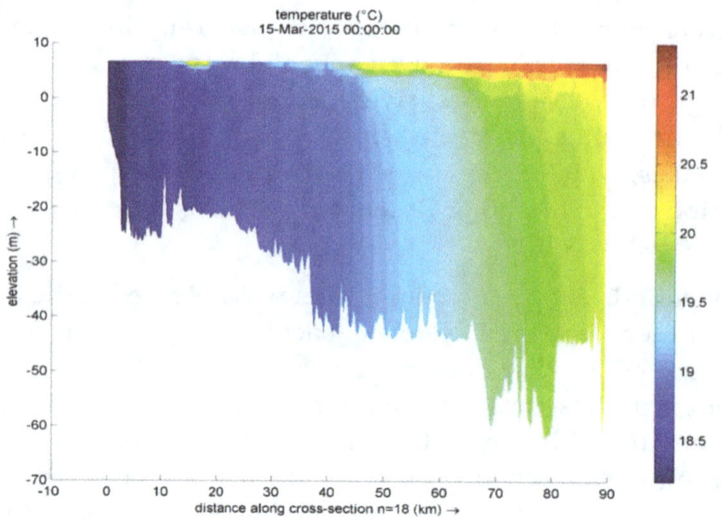

Figure 9. *Cont.*

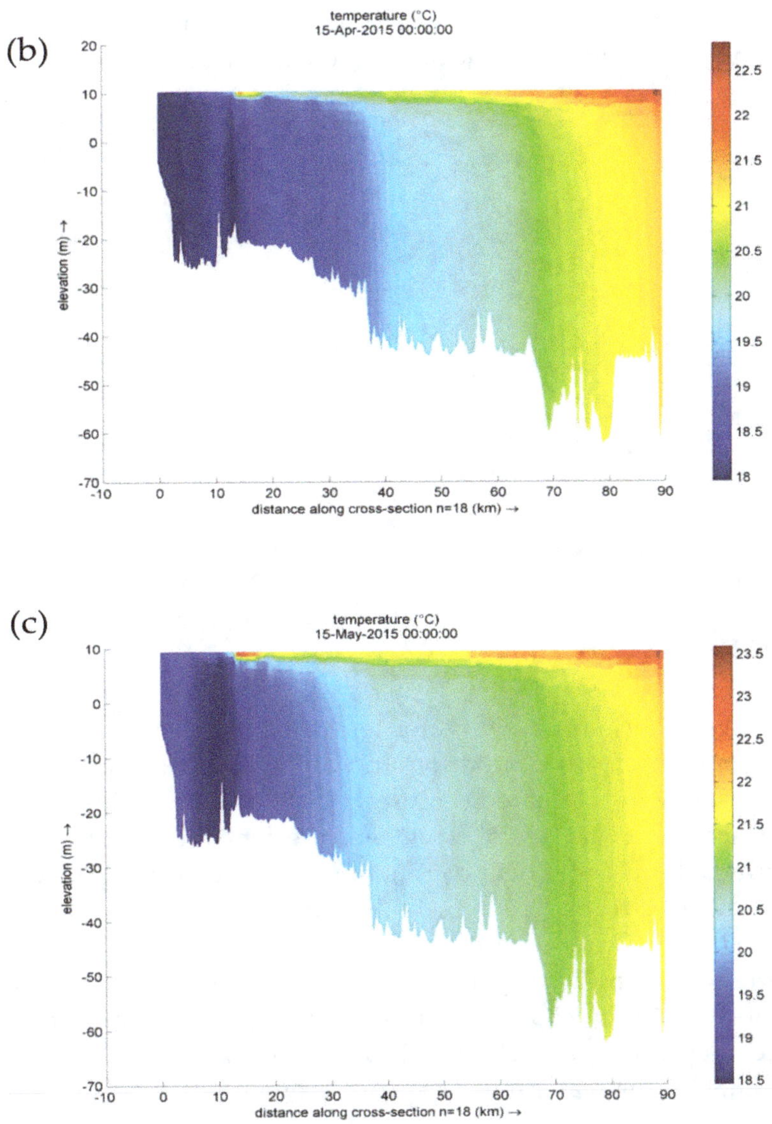

Figure 9. Simulated water temperature profile along the reservoir on (**a**) 15th March, (**b**) 15th April, and (**c**) 15th May of 2015.

4.2. Impact of Solar Radiation and Air Temperature

The absorption of solar radiation and atmosphere-water heat conduction are the two main heat exchange mechanisms governing a water body and the environment. Therefore, the contribution to water temperature alteration provided by solar radiation and atmosphere-water heat conduction can be revealed by comparing the model results of S2 and S3. By removing solar radiation, scenario S2, a notable water temperature drop with a mean value of 3 °C compared with CTL (Figure 10) has been observed. In fact, the water temperature simulated by S2 is rather close to the one measured at the NHS with a mean difference of only 0.4 °C. This indicates that the water parcel does not take in or emit substantial energy while traveling from the NHS to YHS. By further neglecting atmosphere-water heat conduction, S3 produces similar water temperature profiles to the ones of S2. The mean simulated temperature in these two scenarios differs about 0.3 °C, which suggests that the solar radiation, rather than the atmosphere-water heat conduction, plays the dominant role in altering water temperature. Moreover, the difference between S2 and S3 (S2 minus S3) varies with season. The difference reached its maximum (about 0.8 °C) in January and February when air temperature was significantly lower than the water temperature, signifying the water energy loss to the atmosphere. However, in May, when air temperature was substantially higher than the water temperature, the difference is positive (about 0.2 °C), indicating the overlying air heats up the water body.

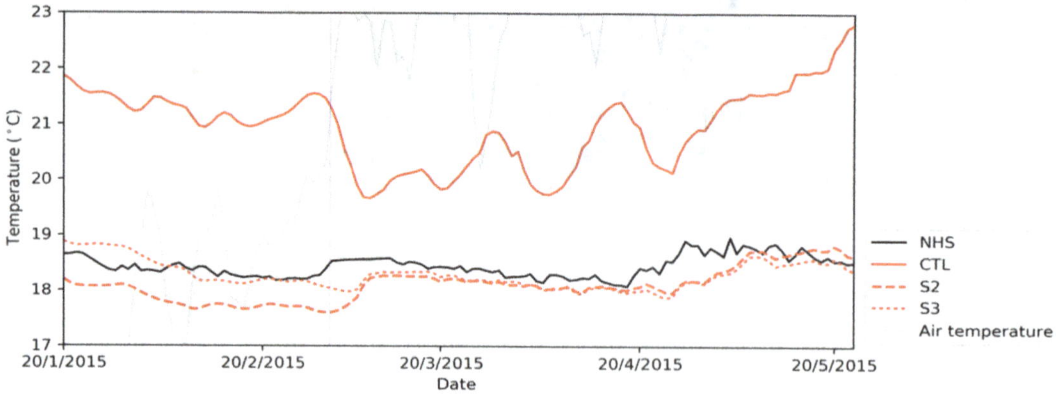

Figure 10. Simulated water temperature by CTL, S2, and S3 at the YHS. The measured water temperature at the NHS and air temperature at the Nuozhadu Dam are shown for reference. The vertical blue line indicates the boundary of the period when air temperature was significantly lower than water temperature at the YHS.

The difference between the simulated water temperature in scenarios CTL and S3 at the YHS was taken as the total impact of solar radiation and air temperature and the weekly contribution (percentage) of both factors were evaluated (Figure 11). Overall, the solar radiation's contribution to the water temperature increment was 106.7%. However, the contribution of air temperature was −6.7%, which suggests that atmosphere-water heat conduction has a decreasing effect on water temperature during the study period. In addition, it is worth observing that the contribution of air temperature is largely dependent on the air-water temperature difference and may vary with simulation periods. When the air temperature was significantly lower than the reservoir water temperature in January and February 2015 (left side of the vertical blue line in Figure 10), its contribution could reach as much as −24%. When the air temperature approached and eventually exceeded the water temperature, the air temperature's contribution to the water temperature increment, became more significant.

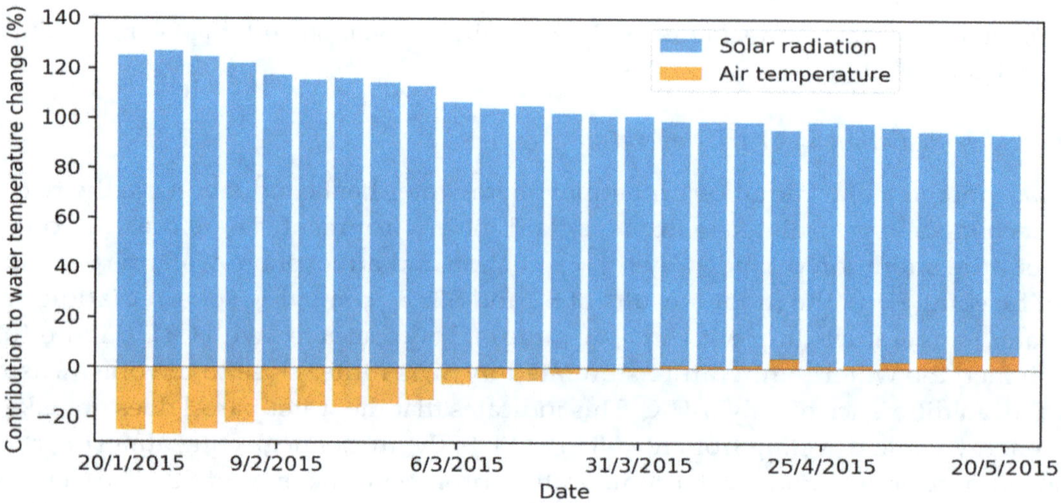

Figure 11. Respective contributions of solar radiation and atmosphere-water heat exchange to the water temperature at the YHS on a weekly scale. Negative contribution values mean that the factor had a decreasing effect on water temperature. Contribution values are computed based on the extent of water temperature changes, taking the water temperature difference between CTL and S3 as 100%.

4.3. Impact of Outflow Rates

A negative correlation between outflow rates and outflow temperature has been observed in Section 2.2.2. For reservoirs with small capacities, a negative relationship between outflow rates and outflow temperature has been reported [92,93]. To further analyze the impact of outflow rates, we carried out eight scenario experiments with different outflow rates ranging from 500 to 4000 $m^3 s^{-1}$.

One and a half period of the tracer concentration has been designed so as to easily differentiate the respective peak concentration value from outflow, which would increase the readability of the residence time. However, the temporal length of the period is not much restrained, as long as it is longer than the longest residence time studied and shorter than the simulation period, it would be sufficient in explaining the problem. Using seven different concentrations instead of a single concentration would also increase the readability. Additionally, under different scenarios, the diffusion effect is also associated with the residence time. By using seven different concentrations, the tracer effect could be observed from both temporal and amplitude points of view.

Outflow rates regulate the residence time of the reservoir. By definition, the residence time refers to how long a parcel, starting from a specific location within a water body, will remain in the water body before exiting [94]. Generally, the residence time decreases as the outflow increases, which results in a shorter duration available for the water parcel to interact with heat sources/sinks (mainly solar radiation and ambient). Thus, its temperature would result closer to the one of the upstream water body. Residence time has been widely used to describe the variability of the lake thermal structure, isotopic composition, alkalinity, dissolved organic carbon concentration, elemental ratios of heavy metals and nutrients, mineralization rates of organic matter, and primary production [95–100].

The tracer methodology mentioned in Section 3.4 was applied to measure the residence time in different scenarios. The tracer was released at the Nuozhadu Reservoir with cyclical concentration peaks ranging from 1 to 7 mg m^{-3} at 1 mg m^{-3} intervals (Figure 12a). Tracer concentrations at the YHS were recorded. We computed the residence time by:

$$T_{re} = \frac{1}{2}[(t_{o1} - t_{i1}) + (t_{o7} - t_{i7})], \quad (9)$$

where T_{re} is the residence time,t_{i1} is the time of inflow tracer peak of 1 mg m^{-3}, t_{i7} is that of 7 mg m^{-3}, t_{o1} is the time of outflow tracer peak corresponding to t_{i1} and t_{o7} is that corresponding to t_{i7}. When the outflow rate was extremely high at 4000 $m^3 s^{-1}$, the residence time was only about 4 days. On the other hand, when the outflow rate was relatively low at 1000 $m^3 s^{-1}$, it took about 11 days for a water parcel to exit the water body. Generally, the residence time decreased exponentially with the outflow rate and the two variables were highly correlated with a correlation coefficient (R^2) of 0.9999 (Figure 12b).

Afterwards, the simulated outflow water temperature at the YHS was averaged during the study period. We used the temperature difference, the averaged outflow temperature minus the averaged inflow temperature, to indicate the temperature changes under different scenarios (F1–F4). Results based on the eight scenario experiments suggested that water temperature difference and outflow rate were also exponentially correlated with an R^2 of 0.9997 (Figure 13a). Thus, water temperature difference increased linearly with residence time with an R^2 of over 0.9977 (Figure 13b).

To sum up, in the case of the Jinghong Reservoir, outflow rates regulate the residence time. The larger residence time means that the water body is able to absorb more solar energy and, thus, gain a larger temperature increment. Nevertheless, this conclusion is only valid under certain conditions: On one hand, the solar radiation and the atmosphere play a positive role in regulating water temperature; on the other hand, the reservoir capacity is relatively small (for example, the discharge/capacity ratio is larger than 20) so that the outflow rate is able to strongly influence the outflow temperature. The Jinghong Reservoir matches both requirements mentioned above.

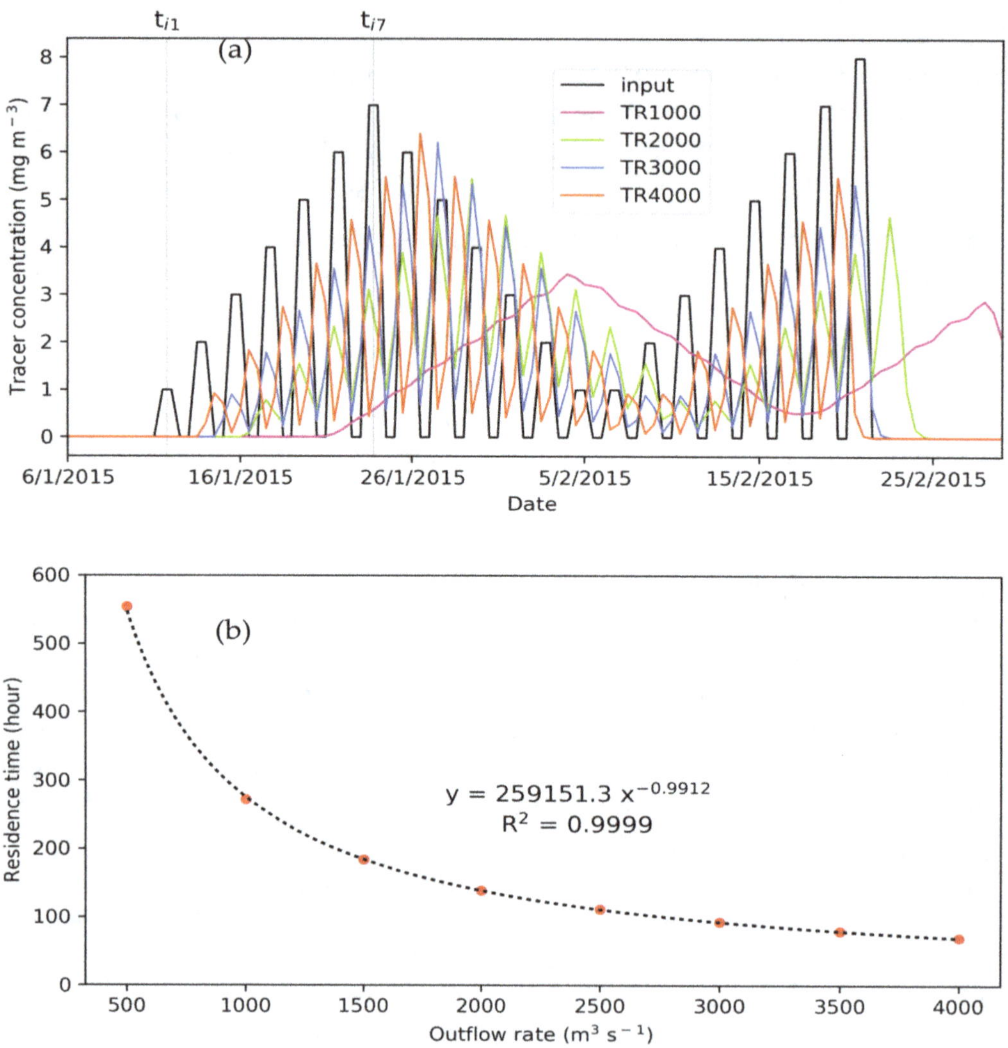

Figure 12. (**a**) Tracer concentration variations at the YHS corresponding to various outflow rates (F2, F4, F6, and F8 are demonstrated in the figure); and (**b**) the exponential relationship between residence time and outflow rate of the Jinghong Reservoir.

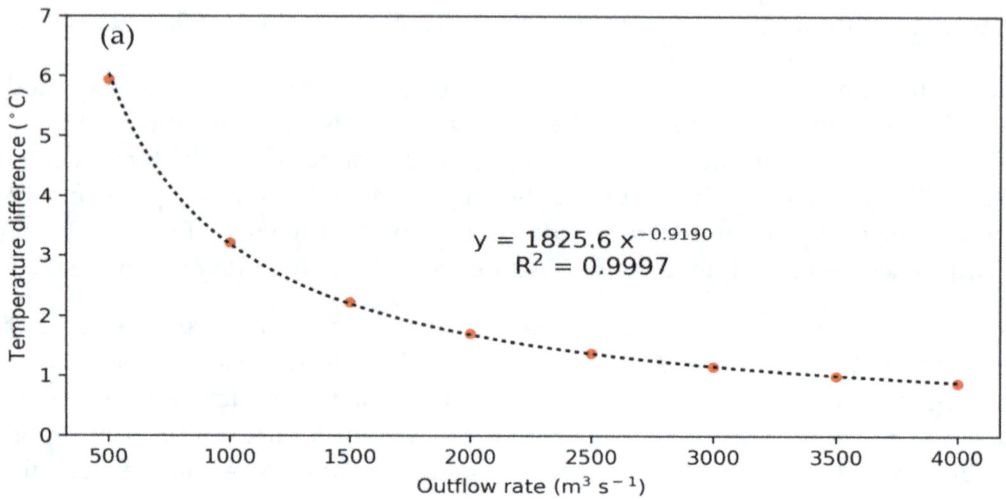

Figure 13. *Cont*

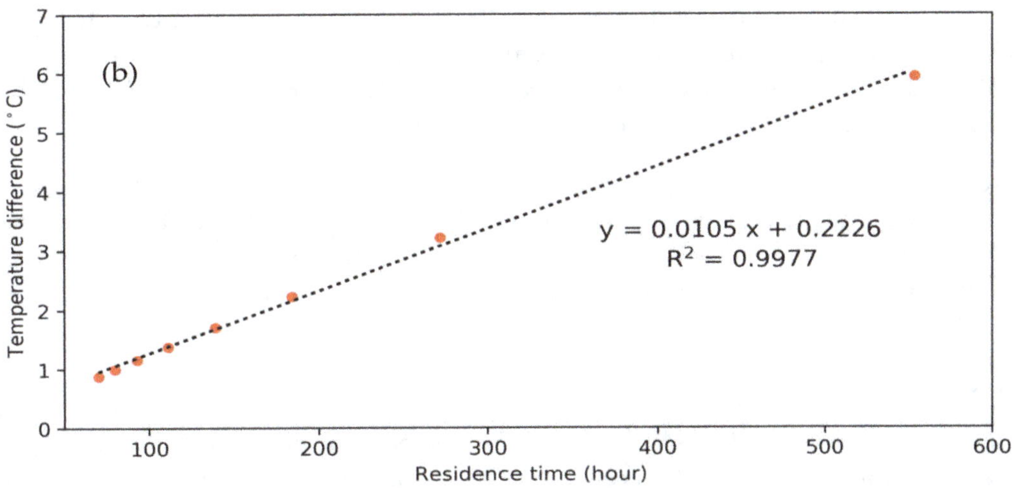

Figure 13. (**a**) Relationship between outflow rate and water temperature difference (outflow minus inflow); and (**b**) relationship between residence time and water temperature difference at the YHS.

4.4. Impact of the Impoundment

In order to examine the overall impact of the impoundment of the Jinghong Reservoir, we established scenario S1, where the Jinghong Dam was removed and the reservoir recovers to the pre-damming state. During the study period, the water temperature for CTL warmed up by about 3 °C on average. However, when the reservoir returned to the natural river state, the temperature increment was not as significant as for the post-impoundment, with an increment of only 1.7 °C, which was about half the case during the reservoir's operation (Figure 14). That is to say, the impoundment of the Jinghong Reservoir alone can explain nearly half (1.3 °C) of the water temperature increment from the NHS to the YHS. Based on the findings of Section 4.3, after the impoundment, the residence time extended from 3 days to 11 days. The period available to absorb solar radiation has been prolonged. Thus, a higher water temperature increment is reached.

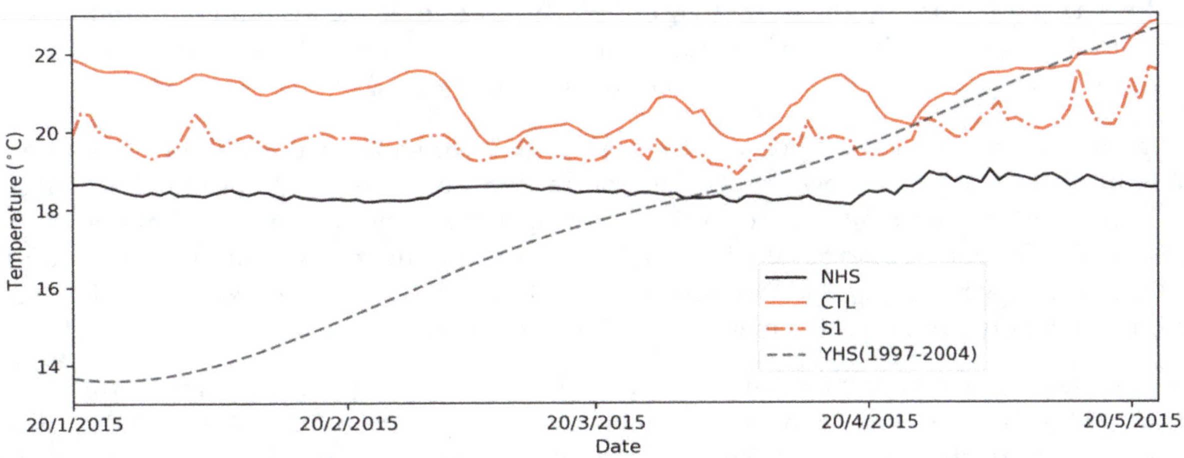

Figure 14. Simulated water temperature during the study period by CTL and S1 at the YHS, and measured water temperature at the NHS and historical water temperature (1997–2004) at the YHS.

5. Research Limitations

Though thorough investigation has been conducted, research limitations due to data availability are not ignorable. Firstly, no continuously measured water temperature data on the Lancang River can be found in the previous studies. Therefore, we conducted a field measurement so as to obtain the water temperature information from November 2014 to February 2016. On the other hand, we only obtained daily solar radiation values from January 2014 to May 2015. Therefore, the study period has been greatly shortened. However, the focus of this study is the explanation of the phenomenon, and a study duration shorter than a full year would not undermine its credibility.

In addition, the influence of the operation rules of the dam have not been directly assessed in this research. In this study, the operation rules have been lumped inside the outflow rate, which is only juxtaposed with outflow water temperature so as to indicate the relation among the two. In which way the operation rules would be most favorable to the downstream environment will be another direction of study in our future research.

6. Conclusions

The temperature characteristics of a river are important due to their strong influence on environmental conditions and aquatic creatures. The impact of reservoirs on downstream water temperature has triggered researchers' concerns worldwide as reservoirs usually release water from deep outlets, in most cases, resulting in a cold water thermal regime downstream. However, the water temperature downstream of the Jinghong Reservoir has increased by 3.0 °C annually compared with historical conditions, according to the observational data. In this study, a Jinghong Reservoir 3D model was established using the Delft3D-FLOW model to explore the unique heating processes of the Jinghong Reservoir on water temperature.

The simulated water temperatures at the outlet showed good agreement with measurements at the Jinghong dam. Comparing the actual operation scenario (CTL) with the pre-damming scenario (S1), it was indicated that the impoundment of the Jinghong Reservoir could explain 1.3 °C (about a half) of the water temperature increase compared to historical values, while the rest might be an accumulated effect by the impoundment of the upstream cascade reservoirs.

Further, this study examined the major factors (solar radiation, atmosphere-water heat conduction) which can potentially influence the water temperature. Numerical experiment results show the contribution of solar radiation to the water temperature increment is up to 106.7%, which consists in the dominating factor. On the other hand, the contribution of the air temperature to the water temperature increment is about −6.7%, which implies that the atmosphere-water heat conduction would result in a decrease of the water temperature during the study period.

Experiment results with different outflow rates (F1–F8) showed that the outflow rate could substantially influence the outflow water temperature because of the variations in the residence time. The residence time is determined by the outflow rate; higher residence time means that the water body absorbs more solar energy and results in a larger temperature increment. After the impoundment of the Jinghong Reservoir, the residence time extended from 3 days to 11 days. The prolonged solar radiation absorption duration is the main reason for its temperature raise.

In conclusion, the Jinghong Reservoir alone could not fully explain the temperature increases compared to historical data since this is also dependent on the accumulated effect by the cascade of dams along the mainstream Lancang River. In the future, a model of the entire cascade of reservoirs should be implemented, in order to improve the assessments of water temperature in pre-dam and post-dam eras.

Author Contributions: Conceptualization: B.J. and G.N.; methodology, validation, and software: B.J.; investigation: F.W. and B.J.; writing—original draft preparation: F.W.; writing—review and editing: B.J., G.N., and F.W.; visualization: F.W.; supervision: G.N.; and funding acquisition: G.N.

Acknowledgments: We would like to thank Daming He and Ying Lu of Yunnan University for their contribution in collecting water temperature data. Also, we would like to express sincere thanks to the editor and the three anonymous reviewers whose comments led to great improvement of this paper.

References

1. Carron, J.C.; Rajaram, H. Impact of variable reservoir releases on management of downstream water temperatures. *Water Resour. Res.* **2001**, *37*, 1733–1744. [CrossRef]
2. Olden, J.D.; Naiman, R.J. Incorporating thermal regimes into environmental flows assessments: Modifying dam operations to restore freshwater ecosystem integrity. *Freshw. Biol.* **2010**, *55*, 86–107. [CrossRef]
3. Rheinheimer, D.E.; Null, S.E.; Lund, J.R. Optimizing selective withdrawal from reservoirs to manage downstream temperatures with climate warming. *J. Water Resour. Plan. Manag.* **2015**, *141*, 04014063. [CrossRef]
4. Hynes, H.B.N. Lotic biology. (Book reviews: The ecology of running waters). *Science* **1971**, *172*, 251.
5. Sullivan, K.; Martin, D.J.; Cardwell, R.D.; Inc, P.; Toll, J.E.; Inc, P.; Duke, S. An analysis of the effects of temperature on salmonids of the pacific northwest with implications for selecting temperature criteria. Sustainable ecosystems institute. *Histochemistry* **2000**, *90*, 85–97.
6. Poole, G.C.; Berman, C.H. An ecological perspective on in-stream temperature: Natural heat dynamics and mechanisms of human-caused thermal degradation. *Environ. Manag.* **2001**, *27*, 787–802. [CrossRef]
7. Lessard, J.A.L.; Hayes, D.B. Effects of elevated water temperature on fish and macroinvertebrate communities below small dams. *River Res. Appl.* **2010**, *19*, 721–732. [CrossRef]
8. Murchie, K.J.; Hair, K.P.E.; Pullen, C.E.; Redpath, T.D.; Stephens, H.R.; Cooke, S.J. Fish response to modified flow regimes in regulated rivers: Research methods, effects and opportunities. *River Res. Appl.* **2008**, *24*, 197–217. [CrossRef]
9. Ward, J.V. Thermal characteristics of running waters. *Hydrobiologia* **1985**, *125*, 31–46. [CrossRef]
10. Brooker, M.P. The impact of impoundments on the downstream fisheries and general ecology of rivers. *Adv. Appl. Boil.* **1981**, *6*, 91–152.
11. Horne, B.D.; Rutherford, E.S.; Wehrly, K.E. Simulating effects of hydro-dam alteration on thermal regime and wild steelhead recruitment in a stable-flow lake michigan tributary. *River Res. Appl.* **2004**, *20*, 185–203. [CrossRef]
12. Krause, C.W.; Newcomb, T.J.; Orth, D.J. Thermal habitat assessment of alternative flow scenarios in a tailwater fishery. *River Res. Appl.* **2005**, *21*, 581–593. [CrossRef]
13. Ryan, T.F.; Webb, J.A.; Lennie, R.; Lyon, J. *Satus of Cold Water Releases from Victorian Dam*; Dept. of Natural Resources and Environment: Heidelberg, Australia, 2001.
14. Preece, R. *Cold Water Pollution below Dams in New South Wales: A Desktop Assessment*; Water management Division, Dept. of Infrastructure, Planning and Natural Resources: Sydney, Australia, 2004.
15. Jackson, H.M.; Gibbins, C.N.; Soulsby, C. Role of discharge and temperature variation in determining invertebrate community structure in a regulated river. *River Res. Appl.* **2007**, *23*, 651–669. [CrossRef]
16. Angilletta, M.J.; Steel, E.A.; Bartz, K.K.; Kingsolver, J.G.; Scheuerell, M.D.; Beckman, B.R.; Crozier, L.G. Big dams and salmon evolution: Changes in thermal regimes and their potential evolutionary consequences. *Evol. Appl.* **2008**, *1*, 286–299. [CrossRef] [PubMed]
17. Campbell, I.C. *The Mekong: Biophysical Environment of an International River Basin*; Elsevier Academic Press: Amsterdam, The Netherlands, 2009.
18. Zhao, A.G. Planning and development of lower and middle Lancang River's water power resources. *Pearl River* **2000**, *21*, 5–8.
19. Fan, H.; He, D.; Wang, H. Environmental consequences of damming the mainstream Lancang-Mekong River: A review. *Earth-Sci. Rev.* **2015**, *146*, 77–91. [CrossRef]
20. Lu, X.X.; Siew, R.Y. Water discharge and sediment flux changes over the past decades in the lower Mekong River: Possible impacts of the Chinese dams. *Hydrol. Earth Syst. Sci.* **2006**, *10*, 181–195. [CrossRef]
21. Campbell, I.C. Perceptions, data, and river management: Lessons from the Mekong River. *Water Resour. Res.* **2007**, *43*, 329–335. [CrossRef]
22. Lu, X.X.; Li, S.; Kummu, M.; Padawangi, R.; Wang, J.J. Observed changes in the water flow at Chiang Saen in the lower Mekong: Impacts of Chinese dams? *Quat. Int.* **2014**, *336*, 145–157. [CrossRef]
23. Haxton, T.J.; Findlay, C.S. Meta-analysis of the impacts of water management on aquatic communities. *J. Can. Sci. Halieutiques Aquat.* **2008**, *65*, 437–447. [CrossRef]
24. Gao, X.; Zhang, S.; Zhang, C. 3-D numerical simulation of water temperature released from the multi-level intake of Nuozhadu hydropower station. *J. Hydroelectr. Eng.* **2012**, *31*, 195–201+207.

25. Wurbs. *Computer Models for Water Resources Planning and Management*; S Army Corps of Engineers, Institute for Water Resources: Alexandria, VA, USA, 1994.
26. Imteaz, M.A.; Asaeda, T.; Lockington, D.A. Modelling the effects of inflow parameters on lake water quality. *Environ. Model. Assess.* **2003**, *8*, 63–70. [CrossRef]
27. Peeters, F.; Livingstone, D.M.; Goudsmit, G.; Kipfer, R.; Forster, R. Modeling 50 years of historical temperature profiles in a large central European lake. *Limnol. Ocean.* **2002**, *47*, 186–197. [CrossRef]
28. Kraus, E.B.; Turner, J.S. A one-dimensional model of the seasonal thermocline II. The general theory and its consequences. *Tellus* **1967**, *19*, 98–106. [CrossRef]
29. Imberger, J.; Patterson, J.; Hebbert, B.; Loh, I. Dynamics of reservoir of medium size. *J. Hydraul. Div.* **1978**, *104*, 725–743.
30. Svensson, U. A Mathematical Model of the Seasonal Thermocline. Ph.D. Thesis, University of Lund, Lund, Sweden, 1978.
31. Hostetler, S.W.; Bartlein, P.J. Simulation of lake evaporation with application to modeling lake level variations of Harney-Malheur Lake, oregon. *Water Resour. Res.* **1990**, *26*, 2603–2612.
32. Hostetler, S.W.; Giorgi, F.; Bates, G.T.; Bartlein, P.J. Lake-atmosphere feedbacks associated with paleolakes Bonneville and Lahontan. *Science* **1994**, *263*, 665. [CrossRef] [PubMed]
33. Zamboni, F.; Barbieri, A.; Polli, B.; Salvadè, G.; Simona, M. The dynamic model seemod applied to the Southern Basin of Lake Lugano. *Aquat. Sci.* **1992**, *54*, 367–380. [CrossRef]
34. Karagounis, I.; Trösch, J.; Zamboni, F. A coupled physical-biochemical lake model for forecasting water quality. *Aquat. Sci.* **1993**, *55*, 87–102. [CrossRef]
35. Ulrich, M. Modeling of Chemicals in Lakes-Development and Application of User-Friendly Simulation Software (MASAS & CHEMSEE) on Personal Computers. Ph.D. Thesis, Swiss Federal Institute of Technology, Zürich, Switzerland, 1991.
36. Fang, X.; Stefan, H.G. Long-term lake water temperature and ice cover simulations/measurements. *Cold Reg. Sci. Technol.* **1996**, *24*, 289–304. [CrossRef]
37. Burchard, H.; Bolding, K.; Villarreal, M.R. *GOTM, a General Ocean Turbulence Model: Theory, Implementation and Test Cases*; Space Applications Institute: Ispra, Italy, 1999.
38. Goudsmit, G.H.; Burchard, H.; Peeters, F.; Wüest, A. Application of k-ϵ turbulence models to enclosed basins: The role of internal seiches. *J. Geophys. Res. Ocean.* **2002**, *107*, C12. [CrossRef]
39. Stepanenko, V.M.; Lykossov, V.N. Numerical modeling of heat and moisture transfer processes in a system lake—Soil. *Russ. J. Meteorol. Hydrol.* **2005**, *3*, 95–104.
40. Stepanenko, V.M.; Machul'Skaya, E.E.; Glagolev, M.V.; Lykossov, V.N. Numerical modeling of methane emissions from lakes in the permafrost zone. *Izv. Atmos. Ocean. Phys.* **2011**, *47*, 252–264. [CrossRef]
41. Subin, Z.M.; Riley, W.J.; Mironov, D. An improved lake model for climate simulations: Model structure, evaluation, and sensitivity analyses in CESM1. *J. Adv. Model. Earth Syst.* **2012**, *4*, 1. [CrossRef]
42. Gu, H.; Jin, J.; Wu, Y.; Ek, M.B.; Subin, Z.M. Calibration and validation of lake surface temperature simulations with the coupled WRF-lake model. *Clim. Chang.* **2015**, *129*, 471–483. [CrossRef]
43. Karpik, S.R.; Raithby, G.D. Laterally averaged hydrodynamics model for reservoir predictions. *J. Hydraul. Eng.* **1990**, *116*, 783–798. [CrossRef]
44. Brown, R.T. *BETTER, A Two-Dimensional Reservoir Water Quality Model: Model User's Guide*; Tennessee Technological University: Cookeville, TN, USA, 1985.
45. Waldrop, W.R.; Ungate, C.D.; Harper, W.L. *Computer Simulation of Hydrodynamics and Temperatures of Tellico Reservoir*; TVA Water Systems Development Branch: Norris, TN, USA, 1980.
46. Edinger, J.E.; Buchak, E.M. *Developments in LARM2: A Longitudinal-Vertical, Time-Varying Hydrodynamic Reservoir Model*; US Army Engineer Waterways Experimental Station: Visburg, MS, USA, 1983.
47. Buchak, E.M.; Edinger, J.E. *Generalized Longitudinal-Vertical Hydrodynamics and Transport: Development, Programming and Application*; US Army Engineer Waterways Experimental Station: Visburg, MS, USA, 1984.
48. Cole, T.M.; Buchak, E.M. *CE-QUAL-W2: A Two-Dimensional, Laterally Averaged, Hydrodynamic and Water Quality Model, Version 2.0. User Manual*; US Army Engineer Waterways Experimental Station: Visburg, MS, USA, 1995.
49. Chen, X.F.; Wang, G.X. Mike 21 software and its application on the offshore reconstruction engineering of Changxing Islands. *J. Dalian Univ.* **2007**, *28*, 93–98. (In Chinese)
50. Wang, Z. Application of Mike 21 in ecological design of artificial lake. *Water Res. Power* **2008**, *26*, 124–127. (In Chinese)

51. Cheng, Y.U.; Ren, X.; Ban, X.A.; Yun, D.U. Application of two-dimensional water quality model in the project of the water diversion in East Lake, Wuhan. *J. Lake Sci.* **2012**, *24*, 43–50. [CrossRef]
52. Liang, Y.; Yin, J.; Zhu, X.; Huang, X. Application of Mike 21 hydrodynamic model in water level simulation of Hongze Lake. *Water Res. Power* **2013**, *31*, 135–137.
53. Leendertse, J.J. *Aspects of a Computational Model for Long-Period Water-Wave Propagation*; Rand Corporation: Santa Monica, LA, USA, 1967.
54. Kuipers, J.; Vreugdeuhil, C.B. *Calculations of Two-Dimensional Horizontal Flow*; Delft Hydraulics Laboratory Report: Delft, The Netherlands, 1973.
55. McGuirk, J.J.; Rodi, W. A depth-averaged mathematical model for the near field of side discharges into open-channel flow. *J. Fluid Mech.* **1978**, *86*, 761–781. [CrossRef]
56. Lauwerier, H.A. Some recent work of the Amsterdam Mathematical Centre on the hydrodynamics of the North Sea. *Sticht. Math. Cent. Toegep. Wiskund.* **1961**, *1*, 1–12.
57. Isozaki, I.; Unoki, S. The numerical computation of the Tsunami in Tokyo Bay caused by the Chilean Earthquake in May 1960. In *Studies on Oceanography—A Collection of Papers Dedicated to Koji Hidaka*; Tokyo University Press: Tokyo, Japan, 1964.
58. Ziegler, C.K.; Nisbet, B. Fine-Grained Sediment Transport in Pawtuxet River, Rhode Island. *J. Hydraul. Eng.* **1994**, *120*, 561–576. [CrossRef]
59. Ziegler, C.K.; Nisbet, B.S. Long-Term Simulation of Fine-Grained Sediment Transport in Large Reservoir. *J. Hydraul. Eng.* **1995**, *121*, 773–781. [CrossRef]
60. Chung, S.; Gu, R. Two-Dimensional Simulations of Contaminant Currents in Stratified Reservoir. *J. Hydraul. Eng.* **1998**, *124*, 704–711. [CrossRef]
61. Tufford, D.L.; Mckellar, H.N. Spatial and temporal hydrodynamic and water quality modeling analysis of a large reservoir on the South Carolina (U.S.) coastal plain. *Ecol. Model.* **1999**, *114*, 137–173. [CrossRef]
62. Ji, Z.G. *Hydrodynamics and Water Quality: Modeling Rivers, Lakes, and Estuaries*; John Wiley & Sons: Hoboken, NJ, USA, 2008.
63. Sheng, Y.P. *A Three-Dimensional Mathematical Model of Coastal, Estuarine and Lake Currents Using Bounary Fitted Grid*; Report No. 585; Aeronautical Research Associates of Princeton: Princeton, NJ, USA, 1986.
64. Blumberg, A.F.; Mellor, G.L. A description of a three-dimensional coastal ocean circulation model. In *Three-Dimensional Coastal Ocean Models Coastal and Estuarine Sciences*; American Geophysical Union: Washington, DC, USA, 1987.
65. Hamrick, J.M. *User's Manual for the Environmental Fluid Dynamics Computer Code*; College of William and Mary: Gloucester Point, VA, USA, 1996.
66. Cerco, C.F.; Cole, T.M. Three-dimensional eutrophication model of Chesapeake Bay. Volume 1: Main report. *Am. Soc. Civ. Eng.* **1994**, *119*, 1006–1025. [CrossRef]
67. Wool, T.A.; Ambrose, R.B.; Martin, J.L.; Cormer, E.A. *Water Quality Analysis Simulation Program (WASP), Version 6.0*; United States Environmental Protection Agency: Washington, DC, USA, 1995.
68. Hydraulics. *Delft3D-FLOW: Simulation of Multi-Dimensional Hydrodynamic Flows and Transport Phenomena, Including Sediments—User Manual*; WL | Delft Hydraulics: Delft, The Netherlands, 2003.
69. HydroQual. *User's Guide for RCA*; Technical Report; HydroQual, Inc.: Mahwah, NJ, USA, 2004.
70. Emma, J.; Lars-Göran, G.; Sten, B.; Tobias, L.; Jan-Olof, S.; Lillemor, C.L.; Destouni, G. Data evaluation and numerical modeling of hydrological interactions between active layer, lake and talik in a permafrost catchment, Western Greenland. *J. Hydrol.* **2015**, *527*, 688–703.
71. Harcourt-Baldwin, J.L.; Diedericks, G.P.J. Numerical modelling and analysis of temperature controlled density currents in Tomales Bay, California. *Estuar. Coast. Shelf Sci.* **2006**, *66*, 417–428. [CrossRef]
72. Maren, D.S.V. Grain size and sediment concentration effects on channel patterns of silt-laden rivers. *Sediment. Geol.* **2007**, *202*, 297–316. [CrossRef]
73. Bouma, T.J.; Duren, L.A.V.; Temmerman, S.; Claverie, T.; Blanco-Garcia, A.; Ysebaert, T.; Herman, P.M.J. Spatial flow and sedimentation patterns within patches of epibenthic structures: Combining field, flume and modelling experiments. *Contin. Shelf Res.* **2007**, *27*, 1020–1045. [CrossRef]
74. Tonnon, P.K.; Rijn, L.C.V.; Walstra, D.J.R. The morphodynamic modelling of tidal sand waves on the shoreface. *Coast. Eng.* **2007**, *54*, 279–296. [CrossRef]
75. Allard, R.; Dykes, J.; Hsu, Y.L.; Kaihatu, J.; Conley, D. A real-time nearshore wave and current prediction system. *J. Mar. Syst.* **2008**, *69*, 37–58. [CrossRef]

76. Lam, N.T. Hydrodynamics and morphodynamics of a seasonally forced tidal inlet system. *J. Water Res. Environ. Eng.* **2008**, *1*, 114–124.
77. Leeuwen, B.V.; Augustijn, D.C.M.; Wesenbeeck, B.K.V.; Hulscher, S.J.M.H.; Vries, M.B.D. Modeling the influence of a young mussel bed on fine sediment dynamics on an intertidal flat in the Wadden Sea. *Ecol. Eng.* **2010**, *36*, 145–153. [CrossRef]
78. Souliotis, D.; Prinos, P. Effect of a vegetation patch on turbulent channel flow. *J. Hydraul. Res.* **2011**, *49*, 157–167. [CrossRef]
79. Department of Water Resource of Yunnan Province, China. Available online: http://www.wcb.yn.gov.cn/ (accessed on 1 July 2015).
80. Khan, F.; Johnson, G.E.; Royer, I.M.; Phillips, N.R.; Hughes, J.S.; Fischer, E.S.; Ham, K.D.; Ploskey, G.R. *Imaging Evaluation of Juvenile Salmonid Behavior in the Immediate Forebay of the Water Temperature Control Tower at Cougar Dam, 2010*; Pacific Northwest National Laboratory: Richland, WA, USA, 2012.
81. Chanudet, V.; Fabre, V.; Kaaij, T.V.D. Application of a three-dimensional hydrodynamic model to the Nam Theun 2 Reservoir (Lao PDR). *J. Great Lakes Res.* **2012**, *38*, 260–269. [CrossRef]
82. Phillips, N.A. A coordinate system having some special advantages for numerical forecasting. *J. Atmos. Sci.* **1957**, *14*, 184–185. [CrossRef]
83. Rodi, W. *Turbulence Models and Their Application in Hydraulics—A State of the Art Review*; International Association for Hydraulic Research: Rotterdam, The Netherlands, 1993.
84. Hu, K.; Ding, P.; Wang, Z.; Yang, S. A 2D/3D hydrodynamic and sediment transport model for the Yangtze Estuary, China. *J. Mar. Syst.* **2009**, *77*, 114–136. [CrossRef]
85. Carrivick, J.L.; Brown, L.E.; Hannah, D.M.; Turner, A.G. Numerical modelling of spatio-temporal thermal heterogeneity in a complex river system. *J. Hydrol.* **2012**, *414–415*, 491–502. [CrossRef]
86. Majerova, M.; Neilson, B.T.; Schmadel, N.M.; Wheaton, J.M.; Snow, C.J. Impacts of beaver dams on hydrologic and temperature regimes in a mountain stream. *Hydrol. Earth Syst. Sci.* **2015**, *19*, 3541–3556. [CrossRef]
87. Chow, V.T. *Open-Channel Hydraulics*; McGraw Hill Book Co.: New York, NY, USA, 1959.
88. Elzawahry, A.E. Advection, Diffusion and Settling in the Coastal Boundary Layer of Lake Erie. Ph.D. Thesis, McMaster University, Hamilton, ON, Canada, 1985.
89. Tsanis, I.K.; Wu, J. Application and verification of a three-dimensional hydrodynamic model to Hamilton Harbour, Canada. *Glob. Nest Int. J.* **2000**, *2*, 77–89.
90. Chen, S. *Accumulation Impact on Water Temperature in a South-North Dammed River: A Case Study in the Middle and Lower Reaches of Lancang River*; Tsinghua University: Beijing, China, 2017. (In Chinese)
91. Huang, Z.; Wu, B. *Three Gorges Dam: Environmental Monitoring Network and Practice*; Science Press: Beijing, China, 2018.
92. Huang, T.; Li, X.; Rijnaarts, H.; Grotenhuis, T.; Ma, W.; Sun, X.; Xu, J. Effects of storm runoff on the thermal regime and water quality of a deep, stratified reservoir in a temperate monsoon zone, in Northwest China. *Sci. Total Environ.* **2014**, *485–486*, 820–827. [CrossRef] [PubMed]
93. Monsen, N.E.; Cloern, J.E.; Lucas, L.V.; Monismith, S.G. A comment on the use of flushing time, residence time, and age as transport time scales. *Limnol. Ocean.* **2002**, *47*, 1545–1553. [CrossRef]
94. Casamitjana, X.; Serra, T.; Colomer, J.; Baserba, C.; Pérez-Losada, J. Effects of the water withdrawal in the stratification patterns of a reservoir. *Hydrobiologia* **2003**, *504*, 21–28. [CrossRef]
95. Hamilton, S.K.; Lewis, W.M. Causes of seasonality in the chemistry of a lake on the Orinoco River floodplain, Venezuela1. *Limnol. Ocean.* **1987**, *32*, 1277–1290. [CrossRef]
96. Herczeg, A.L.; Imboden, D.M. Tritium hydrologic studies in four closed-basin lakes in the Great Basin, U.S.A. *Limnol. Ocean.* **1988**, *33*, 157–173. [CrossRef]
97. Eshleman, K.N.; Hemond, H.F. Alkalinity and major ion budgets for a Massachusetts reservoir and watershed1. *Limnol. Ocean.* **1988**, *33*, 174–185. [CrossRef]
98. Christensen, D.L.; Carpenter, S.R.; Cottingham, K.L.; Knight, S.E.; Lebouton, J.P.; Schindler, D.E.; Pace, M.L. Pelagic responses to changes in dissolved organic carbon following division of a seepage lake. *Limnol. Ocean.* **1996**, *41*, 553–559. [CrossRef]
99. Hecky, R.E.; Campbell, P.; Hendzel, L.L. The stoichiometry of carbon, nitrogen, and phosphorus in particulate matter of lakes and oceans. *Limnol. Ocean.* **1993**, *38*, 709–724. [CrossRef]

Permissions

All chapters in this book were first published in MDPI; hereby published with permission under the Creative Commons Attribution License or equivalent. Every chapter published in this book has been scrutinized by our experts. Their significance has been extensively debated. The topics covered herein carry significant findings which will fuel the growth of the discipline. They may even be implemented as practical applications or may be referred to as a beginning point for another development.

The contributors of this book come from diverse backgrounds, making this book a truly international effort. This book will bring forth new frontiers with its revolutionizing research information and detailed analysis of the nascent developments around the world.

We would like to thank all the contributing authors for lending their expertise to make the book truly unique. They have played a crucial role in the development of this book. Without their invaluable contributions this book wouldn't have been possible. They have made vital efforts to compile up to date information on the varied aspects of this subject to make this book a valuable addition to the collection of many professionals and students.

This book was conceptualized with the vision of imparting up-to-date information and advanced data in this field. To ensure the same, a matchless editorial board was set up. Every individual on the board went through rigorous rounds of assessment to prove their worth. After which they invested a large part of their time researching and compiling the most relevant data for our readers.

The editorial board has been involved in producing this book since its inception. They have spent rigorous hours researching and exploring the diverse topics which have resulted in the successful publishing of this book. They have passed on their knowledge of decades through this book. To expedite this challenging task, the publisher supported the team at every step. A small team of assistant editors was also appointed to further simplify the editing procedure and attain best results for the readers.

Apart from the editorial board, the designing team has also invested a significant amount of their time in understanding the subject and creating the most relevant covers. They scrutinized every image to scout for the most suitable representation of the subject and create an appropriate cover for the book.

The publishing team has been an ardent support to the editorial, designing and production team. Their endless efforts to recruit the best for this project, has resulted in the accomplishment of this book. They are a veteran in the field of academics and their pool of knowledge is as vast as their experience in printing. Their expertise and guidance has proved useful at every step. Their uncompromising quality standards have made this book an exceptional effort. Their encouragement from time to time has been an inspiration for everyone.

The publisher and the editorial board hope that this book will prove to be a valuable piece of knowledge for researchers, students, practitioners and scholars across the globe.

List of Contributors

Juliet Katusiime and Brigitta Schütt
Department of Earth Sciences, Freie Universität Berlin, Physical Geography, Malteserstr. 74-100, Haus H, 12249 Berlin, Germany

Mário David Sequeira and Alexandre Oliveira Tavares
Centre for Social Studies, Department of Earth Sciences, Faculty of Sciences and Technology, University of Coimbra, 3030-790 Coimbra, Portugal

Ana Maria Castilho
Geosciences Centre, Department of Earth Sciences, Faculty of Sciences and Technology, University of Coimbra, 3030-790 Coimbra, Portugal

Pedro Alexandre Dinis
MARE-Marine and Environmental Sciences Centre, Department of Earth Sciences, Faculty of Sciences and Technology, University of Coimbra, 3030-790 Coimbra, Portugal

Lei Ye, Wei Ding, Zhuohang Xin, Jian Wu and Chi Zhang
School of Hydraulic Engineering, Dalian University of Technology, Dalian 116023, China

Xiaofan Zeng
School of Hydropower and Information Engineering, Huazhong University of Science and Technology, Wuhan 430074, China

Fikru Fentaw Abera
School of Civil and Environmental Engineering, Addis Ababa Institute of Technology (AAiT), Addis Ababa 1000, Ethiopia
Department of Earth and Environment, Florida International University, Miami, FL 33199, USA

Dereje Hailu Asfaw and Agizew Nigussie Engida
School of Civil and Environmental Engineering, Addis Ababa Institute of Technology (AAiT), Addis Ababa 1000, Ethiopia

Assefa M. Melesse
School of Civil and Environmental Engineering, Addis Ababa Institute of Technology (AAiT), Addis Ababa 1000, Ethiopia

Taymaz Esmaeili
Department of Civil Engineering, Gorgan Branch, Islamic Azad University, Gorgan 49147-39975, Iran

Tetsuya Sumi and Sameh A. Kantoush
Disaster Prevention Research Institute (DPRI), Kyoto University, Uji, Kyoto 611-0011, Japan

Yoji Kubota
Hydro-Soft Technology Institute Co., Ltd., Nakanoshima, Osaka 530-6126, Japan

Stefan Haun
Institute for Modelling Hydraulic and Environmental Systems, University of Stuttgart, 70569 Stuttgart, Germany

Nils Rüther
Department of Civil and Environmental Engineering, Norwegian University of Science and Technology, S.P. Andersens vei 5, 7491 Trondheim, Norway

Xiaohao Wen
School of Hydropower and Information Engineering, Huazhong University of Science and Technology, Wuhan 430074, China
China Three Gorges Corporation, Beijing 100038, China

Jianzhong Zhou and Zhongzheng He
School of Hydropower and Information Engineering, Huazhong University of Science and Technology, Wuhan 430074, China
Hubei Key Laboratory of Digital Valley Science and Technology, Wuhan 430074, China

Chao Wang
China Institute of Water Resources and Hydropower Research, Beijing 100038, China

Mohammad Ezz-Aldeen, Rebwar Hassan, Nadhir Al-Ansari and Sven Knutsson
Department of Civil, Environmental and Natural Resources Engineering, Lulea University of Technology, 971 87 Lulea, Sweden

Ammar Ali
Department of Water Resources Engineering, Baghdad University, Baghdad 10071, Iraq

Gökçen Uysal
Institute of Hydraulic Engineering and Water Resources Management, University of Duisburg-Essen, 45141 Essen, Germany
Department of Civil Engineering, Anadolu University, 26555 Eskişehir, Turkey

Rodolfo Alvarado-Montero
Institute of Hydraulic Engineering and Water Resources Management, University of Duisburg-Essen, 45141 Essen, Germany
Deltares, Operational Water Management, Deltares, Rotterdamseweg 185, 26 MH Delft, The Netherlands

Dirk Schwanenberg
Institute of Hydraulic Engineering and Water Resources Management, University of Duisburg-Essen, 45141 Essen, Germany
KISTERS AG, Business Unit, Pascalstraße, 52076 Aachen, Germany

Aynur Şensoy
Department of Civil Engineering, Anadolu University, 26555 Eskişehir, Turkey

Xueping Zhu, Wenjun Cai, Xuehua Zhao and Xueni Wang
College of Water Resources Science and Engineering, Taiyuan University of Technology, Taiyuan 030024, China

Chi Zhang
School of Hydraulic Engineering, Dalian University of Technology, Dalian 116024, China

Wei Qi
School of Environmental Science and Engineering, South University of Science and Technology of China, Shenzhen 518055, China
State Key Laboratory of Water Resource & Hydropower Engineering Science, Wuhan University, Wuhan 430072, China

Li He
Key Laboratory of Water Cycle and Related Land Surface Processes, Institute of Geographic Sciences and
Natural Resources Research, Chinese Academy of Sciences, Beijing 100101, China

Stanislav Paseka and Daniel Marton
Institute of Landscape Water Management, Faculty of Civil Engineering, Brno University of Technology, 602 00 Brno, Czech Republic

Zoran Kapelan
Centre for Water Systems, College of Engineering, Mathematics and Physical Sciences, University of Exeter, Exeter EX4 4QF, UK

Sifan Jin, Haixing Liu and Guoli Wang
School of Hydraulic Engineering, Dalian University of Technology, Dalian 116024, China

Hua Shang
Faculty of Management and Economics, Dalian University of Technology, Dalian 116024, China

Bo Jiang, Fushan Wang and Guangheng Ni
Department of Hydraulic Engineering, Tsinghua University, Beijing 100084, China

Index

www.ingramcontent.com/pod-product-compliance
Lightning Source LLC
LaVergne TN
LVHW060057220926
848484LV00008BA/2075
* 9 7 9 8 8 8 8 3 6 0 7 9 8 *